I0034303

Lithium-Ion
Batteries and Solar Cells

Lithium-Ion
Batteries and Solar Cells
Physical, Chemical, and
Materials Properties

Edited by
Ming-Fa Lin, Wen-Dung Hsu,
and Jow-Lay Huang

CRC Press
Taylor & Francis Group
Boca Raton London New York

CRC Press is an imprint of the
Taylor & Francis Group, an **informa** business

MATLAB® is a trademark of The MathWorks, Inc. and is used with permission. The MathWorks does not warrant the accuracy of the text or exercises in this book. This book's use or discussion of MATLAB® software or related products does not constitute endorsement or sponsorship by The MathWorks of a particular pedagogical approach or particular use of the MATLAB® software

First edition published 2021
by CRC Press
6000 Broken Sound Parkway NW, Suite 300, Boca Raton, FL 33487-2742

and by CRC Press
2 Park Square, Milton Park, Abingdon, Oxon, OX14 4RN

© 2021 Taylor & Francis Group, LLC

CRC Press is an imprint of Taylor & Francis Group, LLC

Reasonable efforts have been made to publish reliable data and information, but the author and publisher cannot assume responsibility for the validity of all materials or the consequences of their use. The authors and publishers have attempted to trace the copyright holders of all material reproduced in this publication and apologize to copyright holders if permission to publish in this form has not been obtained. If any copyright material has not been acknowledged please write and let us know so we may rectify in any future reprint.

Except as permitted under U.S. Copyright Law, no part of this book may be reprinted, reproduced, transmitted, or utilized in any form by any electronic, mechanical, or other means, now known or hereafter invented, including photocopying, microfilming, and recording, or in any information storage or retrieval system, without written permission from the publishers.

For permission to photocopy or use material electronically from this work, access www.copyright.com or contact the Copyright Clearance Center, Inc. (CCC), 222 Rosewood Drive, Danvers, MA 01923, 978-750-8400. For works that are not available on CCC please contact mpkbookspermissions@tandf.co.uk.

Trademark notice: Product or corporate names may be trademarks or registered trademarks and are used only for identification and explanation without intent to infringe.

Library of Congress Cataloging-in-Publication Data
Names: Lin, Ming-Fa, editor. | Hsu, Wen-Dung, editor. | Huang, Jow-Lay, editor.
Title: Lithium-ion batteries and solar cells : physical, chemical, and
materials properties / edited by Ming-Fa Lin, Wen-Dung Hsu, Jow-Lay Huang.
Description: First edition. | Boca Raton, FL : CRC Press/ Taylor & Francis
Group, LLC, 2021. | Includes bibliographical references and index. |
Summary: "This book presents a thorough investigation of diverse
physical, chemical, and material properties and special functionalities
of lithium-ion batteries and solar cells. It covers theoretical
simulations and high-resolution experimental measurements that promote a
full understanding of the basic science to develop excellent device
performance. This book will be of interest to materials scientists and
engineers developing enhanced batteries and solar cells for peak
performance"—Provided by publisher.
Identifiers: LCCN 2020043473 (print) | LCCN 2020043474 (ebook) |
ISBN 9780367686239 (hardback) | ISBN 9781003138327 (ebook)
Subjects: LCSH: Lithium ion batteries—Materials. | Solar cells—Materials.
Classification: LCC TK2945.L58 L5554 2021 (print) | LCC TK2945.L58 (ebook) |
DDC 621.31/242—dc23
LC record available at https://lccn.loc.gov/2020043473
LC ebook record available at https://lccn.loc.gov/2020043474

ISBN: 9780367686239 (hbk)
ISBN: 9781003138327 (ebk)

Typeset in Times
by codeMantra

Contents

Preface

This book presents thorough investigations on the diversified physical/chemical/ materials properties and special functionalities of certain batteries, covering the development of theoretical simulations, high-resolution experimental measurements, and excellent device performance. It is robust in detail and thoroughly explores the basic sciences, application engineering, and commercial products. Concerning lithium-ion-based batteries, the first-principles method and the machine learning method are utilized to fully explore the rich and unique phenomena of cathodes, anodes, and electrolytes (solid and liquid states). Specifically, the theoretical frameworks are built from them. Parts of theoretical predictions are consistent with the measured results. Distinct experimental methods/techniques are developed to greatly enhance the performance of lithium-ion-based batteries and solar cells, such as the efficient syntheses of ion-battery core materials. Moreover, the emerging open issues are currently under systematic investigations. Very interestingly, the whole content is characterized by concise pictures, so that this work is suitable for senior graduate students and scientists.

We have successfully developed the first-principles theoretical framework and the machine learning method, as clearly illustrated for LiCo/NiO/LiSiO/LiTiO/graphite alkali-intercalation compounds and various liquid electrolytes, respectively. Such methods are able to thoroughly explore the geometric structures, electronic properties, and magnetic configurations. Accurate VASP calculations and analyses are achieved from the optimal geometric symmetries, atom-dominated electronic energy spectra, spatial charge density distributions, atom- and orbital-projected density of states, spin distribution configurations, spin-split or spin-degenerate energy bands, and the atom- and orbital-dependent net magnetic moments. Consequently, the concise physical/chemical properties, the single- and/or multi-orbital hybridizations in various chemical bonds and the spin configurations due to distinct atoms and orbitals are clearly identified in this work.

The successful syntheses, fabrications, and measurements are outlined in this book. Enhancing the Li^+-transport performance is achieved by examining the chemisorption of ions on the undoped/doped (B&N on vacancies)/porous graphene systems for anode materials. Different nanocarbon structures are efficiently generated using submerged liquid plasma. This provides a simple, low capital cost, high yield, and an environmentally friendly approach. The advantages and near-future challenges, which are similar to the use of ionic liquids as electrolytes for distinct chemical batteries, are discussed in detail. The gel polymer electrolytes of lithium-ion batteries are successfully synthesized and measured. The silicon-nanowire-based hybrid solar cells are modified to greatly enhance the photovoltaic performance under the various chemical treatments, such as the doping mechanisms, conformational changes of the

PEDOT segment, arrangement alteration of PEDOT:PS (poly(3,4-ethylenedioxythio phene):poly(styrenesulfonate)), and the removal of excess insulating PSS chains from the PEDOT:PSS film. Moreover, back-contact perovskite solar cells are realized to present a high photon-to-electron conversion efficiency.

MATLAB® is a registered trademark of The MathWorks, Inc. For product information, please contact:
The MathWorks, Inc.
3 Apple Hill Drive
Natick, MA 01760-2098 USA
Tel: 508-647-7000
Fax: 508-647-7001
E-mail: info@mathworks.com
Web: www.mathworks.com

Acknowledgments

This work was financially supported by the Hierarchical Green-Energy Materials (Hi-GEM) Research Center, from the Featured Areas Research Center Program within the framework of the Higher Education Sprout Project by the Ministry of Education (MOE) and the Ministry of Science and Technology (MOST 109-2634-F-006-020) in Taiwan.

Editors

Ming-Fa Lin, PhD, is a distinguished professor in the Department of Physics, National Cheng Kung University (NCKU), Taiwan. He earned his PhD in physics in 1993 from the National Tsing-Hua University, Taiwan. His main scientific interests focus on the essential properties of carbon-related materials and low-dimensional systems. He is a member of the American Physical Society, American Chemical Society, and the Physical Society of Republic of China (Taiwan).

Wen-Dung Hsu, PhD, is an associate professor in the Department of Materials Science and Engineering, National Cheng Kung University (NCKU). His expertise is utilizing computational materials science methods including first-principles calculations, molecular dynamics simulations, Monte Carlo methods, and finite-element methods to study materials issues. His research interests are mechanical properties of materials from atomic to macro scale, lithium-ion battery, solid-oxide fuel cell, ferroelectrics, solid catalyst design for biodiesel, and processing design for single-crystal growth. He earned his PhD from the Department of Materials Science and Engineering, University of Florida, in 2007. He then served as a post-doctoral researcher in the Department of Mechanical Engineering at the University of Michigan. He joined National Cheng Kung University in 2008.

Jow-Lay Huang, PhD, is a chair professor in the Department of Materials Science and Engineering, National Cheng Kung University (NCKU), Taiwan. He is the director in Hierarchical Green-Energy Materials (Hi-GEM) Research Center, NCKU, Taiwan. He earned his PhD in Materials Science and Engineering in 1983 from the University of Utah, Salt Lake City, Utah, USA. His research interests include the fabrication, development, and application of ceramic nanocomposites, piezo-phototronic thin films for photodetector devices, piezoelectric thin films for high-frequency devices, metal oxide/graphene and SiC_x nanocomposites utilized as anode materials for lithium-ion battery, and 2D nanocrystal materials for photoelectrochemical application.

Contributors

Sanjaya Brahma
Hierarchical Green-Energy Materials
(Hi-GEM) Research Center
National Cheng Kung University
Tainan, Taiwan

Chia-Yun Chen
Department of Materials Science and
Engineering
Hierarchical Green-Energy Materials
(Hi-GEM) Research Center
National Cheng Kung University
Tainan, Taiwan

Peter Chen
Department of Photonics
Hierarchical Green-Energy Materials
(Hi-GEM) Research Center
National Cheng Kung University
Tainan, Taiwan

Vo Khuong Dien
Department of Physics
National Cheng Kung University
Tainan, Taiwan

Nguyen Thi Han
Department of Physics
National Cheng Kung University
Tainan, Taiwan
and
Department of Chemistry
Thai Nguyen University of Education
Thai Nguyen City, Vietnam

Po-Hsuan Hsiao
Department of Materials Science and
Engineering
National Cheng Kung University
Tainan, Taiwan

Jeng-Shiung Jan
Department of Chemical Engineering
National Cheng Kung University
Tainan, Taiwan

Chin-Lung Kuo
Department of Materials Science and
Engineering
National Taiwan University
Taipei, Taiwan

Linh T. M. Le
Applied Physical Chemistry Key
Laboratory
University of Science, Vietnam National
University
Ho Chi Minh City (VNU HCM),
Vietnam

Phung My Loan Le
Applied Physical Chemistry Key
Laboratory
University of Science, Vietnam National
University
Ho Chi Minh City (VNU HCM),
Vietnam

Alex Chinghuan Lee
Hierarchical Green-Energy Materials
(Hi-GEM) Research Center
National Cheng Kung University
Tainan, Taiwan

Ming-Hsien Li
Department of Photonics
National Cheng Kung University
Tainan, Taiwan

Wei-bang Li
Department of Physics
National Cheng Kung University
Tainan, Taiwan

Chih-Ao Liao
Department of Materials Science and
    Engineering
National Cheng Kung University
Tainan, Taiwan

Joey Lin
Department of Photonics
National Cheng Kung University
Tainan, Taiwan

Pei-Ying Lin
Department of Photonics
National Cheng Kung University
Tainan, Taiwan

Shih-Yang Lin
Department of Physics
National Chung Cheng University
Chiayi, Taiwan

Tai-Fu Lin
Department of Photonics
National Cheng Kung University
Tainan, Taiwan

Hsin-Yi Liu
Department of Physics
National Cheng Kung University
Tainan, Taiwan

Duy Khanh Nguyen
Advance Institute of Materials Science
Ton Duc Thang University
Ho Chi Minh City, Vietnam

Hoang V. Nguyen
Applied Physical Chemistry Key
    Laboratory
University of Science, Vietnam National
    University
Ho Chi Minh City (VNU HCM),
    Vietnam

Thi Dieu Hien Nguyen
Department of Physics
National Cheng Kung University
Tainan, Taiwan

Trung T. Nguyen
Applied Physical Chemistry Key
    Laboratory
University of Science, Viet Nam
    National University
Ho Chi Minh City (VNU HCM),
    Vietnam

Hai Duong Pham
Department of Physics
National Cheng Kung University
Tainan, Taiwan

Ilham Ramadhan Putra
Department of Materials Science and
    Engineering
National Cheng Kung University
Tainan, Taiwan

Itaru Raifuku
Department of Photonics
National Cheng Kung University
Tainan, Taiwan

Jaganathan Senthilnathan
Environmental and Water Resources
    Engineering Division
Department of Civil Engineering
Indian Institute of Technology Madras
Chennai, India

Anupama Surenjan
Department of Civil Engineering
National Institute of Technology
  Karnataka
Surathkal, India

Man V. Tran
Applied Physical Chemistry Key
  Laboratory
University of Science, Vietnam National
  University
Ho Chi Minh City (VNU HCM),
  Vietnam

Ngoc Thanh Thuy Tran
Hierarchical Green-Energy Materials
  (Hi-GEM) Research Center
National Cheng Kung University
Tainan, Taiwan

Sing-Jyun Tsai
Department of Physics
National Cheng Kung University
Tainan, Taiwan

Yu-Jen Tsai
Department of Materials Science and
  Engineering
National Taiwan University
Taipei, Taiwan

Yu-Chao Tseng
Department of Chemical Engineering
National Cheng Kung University
Tainan, Taiwan

Thanh D. Vo
Applied Physical Chemistry Key
  Laboratory
University of Science, Vietnam National
  University
Ho Chi Minh City (VNU HCM),
  Vietnam

Ming-Hsiu Wu
Department of Materials Science and
  Engineering
National Cheng Kung University
Tainan, Taiwan

Masahiro Yoshimura
Promotion Centre for Global Materials
  Research (PCGMR)
Department of Material Science and
  Engineering
National Cheng Kung University
Tainan, Taiwan

1 Introduction

Sanjaya Brahma, Ngoc Thanh Thuy Tran, and Wen-Dung Hsu
National Cheng Kung University

Chin-Lung Kuo
National Taiwan University

Shih-Yang Lin, Jow-Lay Huang, and Masahiro Yoshimura
National Cheng Kung University

Phung My Loan Le
University of Science, Vietnam National University

Jeng-Shiung Jan, Chia-Yun Chen, Peter Chen, and Ming-Fa Lin
National Cheng Kung University

CONTENTS

1.1 INTRODUCTION

How to get and use energies very efficiently is the mainstream research topic in terms of the basic sciences/advanced engineering and potential applications. The various theoretical models [1–5] and experimental syntheses [6–11] have been proposed to fully present the essential properties, outstanding functionalities, and commercialized products of green energy materials. The LIBs principally consist of cathode, electrolyte, and anode materials, in which the second systems might be either in solid [12] or in liquid states [13,14]. The numerical simulations, the first-principles calculations [15], neural network, and molecular dynamics [16] are frequently utilized to investigate their rich and unique properties, such as the growth processes, optimal geometric structures within large unit cells, electronic energy spectra, van Hove singularities in density of states, orbital hybridizations of chemical bonds, magnetic configurations, and special optical absorption spectra. The close combinations of core components in batteries are required to display the high-performance characteristics: low cost, lightweight, high safety, short charging time, long operation time, controllable temperature, and wide voltage range.

Up to now, there exist a lot of well-established products that cover the battery-driven cell phones and electric vehicles (EVs), the solar cell companies, the hydrogen-based cars, the water-induced electric power, the wind turbines, and so on. By the delicate numerical calculations, successful syntheses, detailed analysis, and well-behaved designs, this book thoroughly explores the diversified physical, chemical, and material phenomena of fundamental properties and the unusual functionalities in LIBs [1–11], Si nanowire-based solar cells [17], and perovskite solar cells [18]. Furthermore, the relations between the theoretical predictions and the high-resolution measurements are fully discussed. It provides very useful information about science bases, integrated engineering, and real applications.

Table 1.1 provides the diversified materials of anode, cathode, and electrolyte. The anode materials require the large capability of lithium intercalation/adsorption, high efficiency of charge/discharge, excellent cyclability, low reactivity against electrolyte, fast reaction rate, low cost, environmental-friendly, and nontoxic [19–23]. Graphite, which is one of the primary carbon materials, can serve as anode of Li^+-ion-based batteries and is predominantly used in commercial products [23–25]. Lithium ions are electrochemically intercalated into the space between the graphitic sheets during the charging process and de-intercalated in the discharging process. A practical reversible capacity is greater than 360 mAh g^{-1} (theoretically at 372 mAh g^{-1}) with the high discharge/charge efficiency [26–28]. However, graphite has a huge backward in volume expansion. There are new carbon materials, such as carbon nanofibers (CNF) [33–35] and carbon nanotubes (CNT) [31,32], where the single-walled CNTs are expected to exhibit reversible capacities about 300–600 mAh g^{-1} [32]. Besides the above materials, $Li_4Ti_5O_{12}$ is known as a potential anode material for the next-generation LIBs [39–41]. In the current work, $Li_4Ti_5O_{12}$ has been focused on the rich and unique essential properties with highly nonuniform environments, clearly revealing the thermodynamic stability, high cycle life performance, and safety, compared to other anode material candidates. The other materials are also available in anode electrode, e.g., TiO_2 [19–22], patterned Si [29], Si film [30], Si nanowires [36,37], Si nanotubes [38], MoO_3 [42], SnO_2 [43], ZnO [44], Fe_3O_4/carbon foam [45], MnO [46], Co_3O_4 [47], GaS_x [48,49], and MoS_2 [50].

The most common cathode materials are $LiCoO_2$ [55,56], Li-Mn-O [58], $LiFePO_4$ [68], and lithium-layered metal oxides, mainly owing to the very high capacity. In general, such complicated lithium oxide compounds contain the first-row transition metals (including Mn, Fe, Co, and Ni), for example, $LiCoO_2$ is 140 mAh g^{-1}, $LiMn_2O_4$ and its variants (LMO) are 100–120 mAh g^{-1}, and $LiFePO_4$ (LFP) is 150–170 mAh g^{-1} [51,52]. Cathode materials should meet other following requirements: high discharge voltage, high energy capacity, long cycle life, high power density, lightweight, low self-discharge, and absence of environmentally hazardous elements. V_2O_5 is also one of the potential materials used for cathode electrode, which is successfully synthesized by several methods, such as green route, green method, sol-gel, chemical precipitation, ultrasound-assisted method, and thermal decomposition. For the lithium-ion battery, V_2O_5 from the microwave-assisted green route clearly displays the specific capacity (225 mAh g^{-1}) at 1.5–4.0 V [53,54].

As potential electrolyte candidates, these materials exhibit the large ionic conductivity of above 10^{-4} S cm^{-1} at room temperature, the negligible electronic conductivity with a high ionic transference number [69], and the wide electrochemical stability windows [70]. Many types of Li-ion solid electrolytes are verified to possess the above-mentioned properties through the long-time experimental tests, including Na super-ionic conductor (NASICON) [73], garnet [71], perovskite [72], LISICON [74], LiPON [78], Li$_3$N [82], sulfide [84], argyrodite [85], anti-perovskite [86], and so on. Among these systems, LISICON and LiPON types obviously show the low ionic conductivities (~10^{-6} to 10^{-5} S cm^{-1}), but good at the electrochemical stability. Li$_3$N type exhibits the reasonable ionic conductivities (~10^{-4} S cm^{-1}), but the narrow electrochemical stability [82]. The sulfide-, argyrodite-, and anti-perovskite-type solid electrolytes might be unstable in the ambient atmosphere, although they present the high ionic conductivities (~10^{-5} to 10^{-2}, ~10^{-7} to 10^{-2}, and ~10^{-4}–10^{-2} S cm^{-1}, respectively). In general, NASICON-, garnet-, and perovskite-type solid electrolytes are promising for applications in solid-state batteries because of their performances in electrochemical stability and acceptable ionic conductivities (~10^{-4} to 10^{-2}, ~10^{-4} to 10^{-3}, and ~10^{-5} to 10^{-4} S cm^{-1}, respectively). For the battery safety, the suitable selection of solid electrolytes has to be done carefully, according to the great reduction in interface resistance. However, Li$^+$-ion-based batteries are required to be encapsulated in avoiding contamination from the ambient environment, which could lead to a decrease in ionic conductivity and change of structure. Solid electrolytes must improve their properties and make compatibly between solid electrolytes and electrodes.

Basic science research will be very useful in greatly enhancing the performance of various batteries and their high potential applications. The critical physical, chemical, and material mechanisms, which are closely related to the efficiencies of cathodes, electrolytes, and anodes, cannot be clearly identified from the previous published works up to now, such as the significant roles of the nonuniform environment (the Moire superlattice; [89–91]), the multi-/single-orbital hybridizations of chemical bonds, and the intercalation/substitution of transport ions. The theoretical frameworks, being suitable for the systematic investigations, could be developed in the near-future studies. From the physical points of view, the essential properties have been successfully explored by using the first-principles simulations [92,93] and the phenomenological models [94]. However, such methods need to be strongly modified in most of the emergent materials. The former is available for studying the optimal geometries, electronic properties, magnetic configurations, mechanical stains, phonon spectra, optical absorptions, and Coulomb excitations. It evaluates a lot of reliable data, in which the very close integration of all the useful data would become the most important strategy in the near-future studies. For example, Chapters 2 and 5–7 develop a theoretical framework and can account for the diverse phenomena through the identified orbital hybridizations and spin distributions. On the contrary, the tight-binding model [94,95], the static/dynamic Kubo formula [96], and the random-phase approximations [97], respectively, deal with the electronic, transport/optical, and Coulomb excitation properties. Each model might be very difficult for most of the complicated compounds with many atoms in a unit cell. The direct combination of the first and second methods would play the critical roles for the full understanding of ion transports among the different materials. Both physical

TABLE 1.1

Various Anodes, Electrolytes, and Cathodes in Li-Ion-Based Batteries

	Materials	References
Anode	TiO_2	[1922]
	Graphite	[2328]
	Patterned Si	[29]
	Si film	[30]
	Carbon nanotubes	[31,32]
	Carbon nanofibers	[3335]
	Si nanowires	[36,37]
	Si nanotubes	[38]
	$Li_4Ti_5O_{12}$	[3941]
	MoO_3	[42]
	SnO_2	[43]
	ZnO	[44]
	Fe_3O_4/carbon foam	[45]
	MnO	[46]
	Co_3O_4	[47]
	GaS_x	[48,49]
	MoS_2	[50]
Cathode	V_2O_5	[53,54]
	$LiCoO_2$	[55,56]
	Nano-$LiCoO_2$	[57]
	$LiMn_2O_4$	[58]
	$Li[Li_{0.20}Mn_{0.54}Ni_{0.13}Co_{0.13}]O_2$	[59]
	$LiNi_{1/3}Mn_{1/3}Co_{1/3}O_2$	[60,61]
	$LiMn_{1.5}Ni_{0.5}O_4$	[62,63]
	$0.5Li_2MnO_3 \cdot 0.5LiNi_{0.375}Mn_{0.375}Co_{0.25}O_2$	[64]
	$Li_{1.2}Ni_{0.2}Mn_{0.6}O_2$	[65]
	$LiNi_{0.5}Mn_{1.5}O_4$	[66]
	$FePO_4$	[67]
	LiFe(Co/Ni)PO_4	[68]
Solid-state electrolyte	Garnet ($Li_7La_3Zr_2O_{12}$)	[71]
	Perovskite ($Li_{3x}La_{2/3-x}TiO_3$)	[72]
	Na super-ionic conductor (NASICON)	[73]
	LISICON	[74]
	$(LiM^{IV}{}_2(PO_4)_3$ (M^{IV} = Ti, Zr, Ge, and Hf)	[75]
	$LiAlO_x$	[76,77]
	Li_3PO_4	[78]
	Lithium silicate	[79]
	Li (Ta/Nb)O_3	[80,81]
	Li_3N	[82]
	$LiSiAlO_2$	[83]
	Sulfide (Li_4GeS_4, $Li_{10}GeP_2S_{12}$, Li_2S-P_2S_5	[84]
	Argyrodite (Li_6PS_5X (X = Cl, Br, I))	[85]
	Anti-perovskite (Li_3OX (X = Cl, Br, I))	[86]
	LiSi/Ge/SnO	[87,88]

interactions and chemical actions frequently come to exist at any position and time during the charging and discharging processes. Their simultaneous considerations in the generalized model must be an extreme challenge.

The organization of this work is as follows. Chapter 1 includes the total motivations, especially for the whole developments in the theoretical and experimental researches. As for the former, the machine learning, which is built from the close association of the SMILES format and the neural network, is utilized to explore the best liquid-state electrolytes with the high-voltage lithium-ion batteries. Electron affinity and ionization energy are two critical parameters to choose the most outstanding one from a lot of potential materials. The first-principles calculations are covered in Chapters 3–7. Specifically, Chapter 3 thoroughly explores the nitrogen doping on the vacancies of monolayer graphene and its effects on the storage capacity of Li/Na. The various defects and N-concentrations are taken into consideration. Chapters 4–7, being, respectively, done for LiCo/NiO, LiSiO, LiTiO, and graphite alkali-intercalation compounds, fully investigate the geometric structures, electronic properties, and magnetic configurations of the solid-state cathode/electrolyte/anode materials. The accurate VASP calculations and analyses are achieved from the optimal geometric symmetries, the atom-dominated electronic energy spectra, the spatial charge density distributions, the atom and orbital-projected density of states, the spin distribution configurations, the spin-split or spin-degenerate energy bands, and the atom- and orbital-dependent net magnetic moments. Consequently, the concise physical and chemical pictures, the single- and/or multi-orbital hybridizations in various chemical bonds and the spin configurations due to distinct atoms and orbitals, are clearly identified under the developed framework.

The experimental syntheses and measurements are covered in Chapters 8–13. Chapter 8 is focused on how to enhance the Li^+-transport performance by examining the chemisorption of ions on the un-doped/doped (B&N on vacancies)/porous graphene systems (anode materials). The different nanocarbon structures are efficiently generated through the submerged liquid plasma, as clearly illustrated in Chapter 9. This provides a simple, low capital cost, high-yield, and an environmental-friendly approach. Chapter 10 reveals an experimental overview of the usage of ILs as electrolytes for distinct chemical batteries, including the whole advantages and near-future challenges. The polyelectrolytes of lithium-ion batteries, which consist of the imidazolium-based ionogels with various amounts of IL additives, are successfully synthesized and measured in Chapter 11. Concerning solar cells, the Si-nanowire-based hybrids in Chapter 12 are modified for the great enhancement of the photovoltaic performance under the various chemical treatments, such as the doping mechanisms, the conformational changes of poly(3,4-ethylenedioxythiophene) (PEDOT) segment, the arrangement alteration of poly(3,4-ethylenedioxythiophene): poly(styrenesulfonate) (PEDOT:PSS), and the removal of excess insulating PSS chains from the PEDOT:PSS film. Chapter 13 presents the enhanced photon-to-electron conversion efficiency for the back-contact perovskite solar cells. Chapters 14 and 15, respectively, cover concluding remarks and open issues, especially for the latter in the near-future researches. Finally, the battery-related practice problems are stated in Chapter 16.

In recent years, there has been a dramatic proliferation of research concerned with machine learning and application of that technology to material science. It is considered as a powerful method, which allows the screening of huge amount of materials to find out the candidates – a process that would take several years or even decades by conventional calculations or experimental techniques. Nowadays, machine learning technology has almost been applied to all aspects of human living to get a smarter and better life. Batteries is one of its potential applications, especially LIBs that are in high demand to solve the environmental issue. Machine learning has been utilized for battery designs based on the properties such as C-rate, electrode thickness, and porosity, which affect the specific power and specific energy of LIBs [98]. Moreover, it has been used to search for high ionic conductivity solid electrolytes [99] and predict the suitable inorganic solid electrolytes to prevent the dendrite formation with Li metal anode in LIBs [100]. It is worth knowing that high-performance electrolyte molecules that can enable high-voltage battery are also an attracted issue that machine learning can involve in. For a more detailed discussion on the metrics for molecular selection, different training and predicting models can thus help to accelerate the research progress.

Nitrogen doping into graphene-based anode materials has been found to effectively enhance the specific capacity of Li/Na-ion batteries, but the mechanisms remain unclear so far. Our results based on the first-principles calculations indicate that introducing N-dopants in graphene can lower down the formation energy of defects and in turn increase the amount of active sites in graphene for Li/Na-adsorption. However, our calculations also found that adding too many N-dopants in graphene may lead to N-aggregation near defect sites and further reduce the reversible Li/Na-storage capacity. Despite the disadvantage coming from the excess amount of N-dopants, an appropriate amount of N-dopants can effectively lift the migration energy barrier of vacancy defects in graphene, which can avoid vacancy aggregation and suppress the loss of reversible Li/Na-capacity.

Batteries [101,102], which are driven by the electric supply and chemical energy, have become one of the mainstream energy resources. Up to date, the whole experimental progress clearly shows that the Li^+- [101–103], Al^+- [104,105], and Fe^+-based systems [106] present the rich merits in the separate functionalities. Specifically, there exist a lot of frequently commercialized products for the first kind of ion transports, e.g., radios [107,108], clocks, cell telephones, notebooks, iPods, modulated planes [107,108], and buses/vehicles [109,110]. All the batteries consist of the critical anode, cathode, and electrolyte materials. Very interesting, the Li^+-ion-based batteries might be made up of LiTiO/LiScO/graphite [111–113], LiFe/Co/NiO [114–117], and LiSi/Ge/LiSnO [118–120], respectively. The different combinations of three critical components would have many merits and drawbacks, according to the high-performance criteria. The previous studies show that most of the subsystems have been investigated by the first-principles calculations for their essential properties. For example, the pristine and Li-atom-intercalation graphite systems exhibit the semimetallic and metallic behaviors [121], respectively, in which the flexible interlayer spacings provide the ion/atom intercalation during the charging and discharging processes [122]. Another anode material, LiTiO, is thoroughly explored in Chapter 4 for the geometric, electronic, and magnetic properties. Maybe, this system provides the excellent heterojunction using the electrolyte of lithium oxide compounds [123]. The important differences with graphite are worthy of detailed discussions [124].

According to the viewpoints of experimental developments and potential applications, batteries [125] have become one of the mainstream energy resources [126]. Their charging and discharging processes mainly arise from the rapid ion transports in the internal circuit, such as the quick flowing of lithium/aluminum/iron ions [127–129], depending on their special functionalities. Up to now, a lot of high-resolution experimental measurements on the Li^+-ion-based batteries can provide the necessary information for their fundamental properties; furthermore, they are frequently used in commercialized products. Each battery, at least, consists of three independent components, namely, electrolyte [130], cathode [131], and anode compounds. Specifically, the first ones could be in the solid or liquid state, while the latter are very popular in everyday living. All the solid-state Li^+-ion-based batteries are thoroughly illustrated in Chapter 4, especially for the electrolyte of 3D ternary Li_2SiO_3 compound. The previous few theoretical studies are conducted on its geometric and electronic properties through the first-principles calculations. The delicate results and analyses are thoroughly absent up to now. That is to say, the calculated results are insufficient and there are no critical mechanisms (concise physical pictures) in comprehending the diversified phenomena. These drawbacks will be fully solved by developing a generalized theoretical framework (details in Chapter 4). The predicted properties could be further verified by the experimental examinations, e.g., lattice symmetries and valence band energy spectra, respectively, through the X-ray diffraction [132] and angle-resolve photoemission spectroscopy. Most important, the close and complex relations among electrolyte, cathode, and anode of LiXO-related systems are also discussed in detail. They are worthy of the systematic investigations in the near-future studies.

Recently, there are plenty of experimental [133,134] and theoretical researches [135] on the emergent green energy materials for the great reduction of the room temperature effects [136]. Obviously, the current results show that the successful applications have covered the battery-driven electrochemical devices [137], the solar energy companies [138], the hydrogen-operated cars [139], the methane-gas-created biological energies [140], the water-induced electric power [141], the wind turbines [142], and so on. Specifically, the chemical batteries mainly originate from the ion transports, in which both Li^+- and Al^+-related [143–145] ones are frequently utilized the different commercialized products, strongly depending on their functionalities. Their critical components consist of the cathode, electrolyte, and anode materials [146–148]. Most important, the charging process starts from the release of ions due to the first subsystem, transports through the second one, and then intercalates into the third one. The opposite is true for the discharging process. The high performance will be evaluated according to the large current density of ions, the negligible electronic current, the wide voltage range, the outstanding matches of different subsystems, the rapid charging/discharging processes, the long lifetime, and the high safety. Three kinds of subsystems, which belong to the lithium oxide compounds, are suitable and reliable in the practical applications, for example, the 3D ternary compounds of LiFe/Co/NiO, LiSi/Ge/SnO, and LiTi/ScO/graphite, respectively, serve as cathode, electrolyte, and anode materials. The first systems are chosen for a model study in Chapter 6 for their essential properties through the VASP calculations. The previous few studies are insufficient in fully comprehending the diversified physical, chemical, and material phenomena. However, the important multi-orbital hybridizations and the atom- and orbital-dependent spin configurations will be proposed under the developed theoretical framework. The created viewpoints could be generalized to the emergent materials even in the presence of external fields and chemical modifications.

A bulk graphite is frequently utilized in the commercialized anode material of Li$^+$-ion-based batteries [149], mainly owing to its layered structure. This condensed matter system has been extensively studied both experimentally and theoretically. In the natural environment, most of the pure graphitic materials exhibit the AB-stacked configuration; that is, they belong to the Bernal graphite [150]. Only a few of them have the ABC-stacking configuration [151]. Recently, the simple hexagonal graphite, with the AA stacking, could be synthesized in the experimental laboratories. Three kinds of graphite systems are semimetals, since the van der Waals interlayer interactions can create a significant overlap of valence and conduction band near the Fermi level. The free conduction electrons and valence holes are responsible for the diversified physical, chemical, and material phenomena. The AA [152] and ABC stackings, respectively, corresponding to the highest and lowest symmetries, possess the highest and lowest free carrier densities. By the delicate theoretical calculations and analyses, each graphitic sheet presents a very strong σ bonding of C-($2s$, $2px$, $2py$) orbitals, being accompanied with the relatively weak π bondings of C-$2pz$ orbitals along the perpendicular direction. Apparently, the interlayer spacings are easily modulated by the atom, molecule, and ion intercalations. Such metastable geometric structures might be very suitable for the Li$^+$-ion (or Li-atom) intercalation/de-intercalation during the charging/discharging processes. Graphite intercalation compounds have been fully explored more than forty, in which they are classified into the donor- and acceptor-type behaviors under the n- and p-type dopings, respectively. For example, the alkali-atom intercalation leads to high electrical conductivities, as best as copper. Up to now, there are some numerical simulations and phenomenological models on the essential properties of stage-n graphite alkali-intercalation compounds. However, the unusual relations among the stacking configurations, the intercalant concentrations, the atom-dominated band structures, the atom- and orbital-decomposed density of states, and the spatial charge density distributions are worthy of the further investigations through the first-principles calculations (details in Chapter 7).

Recently, graphene [153–155] has attracted a lot of interest among the LIB community because of their outstanding electrochemical performance. Graphene is attributed to high surface area, excellent electrical conductivity, flexibility, and Li-ions can intercalate on both the sides and edges of the graphene to yield high electrochemical performance. Graphene can also be doped with other elements such as B/N [156] or the overall microstructure can be modified [157,158] to enhance the overall capacity and rate capability of the battery. Here, we discuss the variety of synthesis procedure used to prepare graphene and investigate the structure/microstructure/bond vibration and electrochemical properties.

Fossil fuels are the most widely used energy resource worldwide. Risks related to resource depletion, environmental pollution, and political unrest with regard to fossil fuel production have led to the rapid emergence of a variety of intermittent renewable and cleaner energy sources such as wind, solar, and wave. In order to integrate these renewable energies into the electrical grid, a large-scale energy storage system (ESS) is vital to peak shift operation [159]. Among various energy storage technologies, using an electrochemical secondary battery is a promising method for large-scale storage of electricity due to its flexibility, high energy conversion efficiency, and simple maintenance [160]. Lithium-ion batteries (LIBs), which have become common power sources in the portable electronic market since their first commercialization by Sony in the early 1990s, are the primary candidates for ESSs [161]. The introduction of LIBs into the automotive market as the battery of choice for powering hybrid electric vehicles (HEVs), plug-in hybrid electric vehicles (PHEVs), and EVs could reduce dependence on fossil fuels.

Although LIBs are the most commonly used for consumer portable electronic devices and are promising for the next-generation large-scale charge storage due to their outstanding energy density and power capacity [162–167], there are still some drawbacks regarding lithium battery safety, lifetime, poor low-temperature performance, and cost. Moreover, the popularity, as well as increasing demand for LIBs with geographically, constrained that Li-mineral resources will drive up the price. Therefore, findings to replace LIBs in the future have attracted a lot of attention of scientists.

With the large natural abundance and low cost of sodium, sodium-ion batteries (SIBs) have become a promising candidate for large-scale usage related to energy storage belong to renewable energy sources. In addition, Na and Li have similar chemical properties, and thus, comprehensive knowledge about LIBs can be applied, then making SIBs become competitive to LIBs in cell markets [168–170]. However, the performance of SIBs is still a challenge with respect to scientists, and research conducted on Na-based chemistry is remarkably less compared to that on Li-based chemistry. Further study on electrode materials, electrolytes, and electrode–electrolyte interphase as well as in-depth mechanism, especially safety is required to improve the behavior of SIBs [171,172].

A serious issue in lithium-ion cell technology is safety. Electrodes and electrolytes are both hazard factors in Li-ion cells. In particular, the use of graphite-based electrodes easily leads to the release upon cycling of gaseous products or even more dangerously to lithium plating on the electrode surface at high current regimes. However, electrolytes are the most critical components for the control of the safety of LIBs because of the high vapor pressure and the flammability of the organic solvents in electrolytes. One of the best ways to improve the safety and reliability of the Li-ion battery electrolytes is the use of an emerging class of electrolytes based on ionic liquids (ILs), namely low-temperature molten salts. The ILs are nonvolatile, nonflammable, highly conductive, environmentally compatible, and can safely operate in a wide temperature range. This unique combination of favorable properties makes ILs very appealing materials for stable and safe electrolytes in lithium batteries to reduce the risk of thermal runaways and, eventually, fire accidents, thus drastically improving the overall safety of the cell.

Batteries have been developed for more than 100 years and have a great impact on our daily life. Recently, due to the use of mobile devices, a battery with high capacity and high safety is especially in demand. Apart from mobile devices, EVs are now ready to market and also in need of batteries with high capacity as well as high energy density. The LIBs possess the above-mentioned properties; it is now the most promising candidate for commercialization [173–176]. Electrolytes are the most important in batteries, as they carry ions between a pair of electrodes to complete a circuit of charges. The commercial electrolytes are the mixtures of carbonates and lithium salts; however, volatility and flammability are big problems to these kinds of liquid electrolytes [177–179]. To solve this, it is worth researching other materials to replace liquid electrolytes. The ILs have attracted lots of attention to be the suitable solution of these disadvantages due to their negligible flammability, broad electrochemical window, and vapor pressure, which are especially in demand of the electrolytes for the next-generation batteries [180–187]. Poly(ionic liquid)s (PILs), which are macromolecular analogs of ILs, have emerged as a new class of polymer electrolytes due to their combined properties emanating from the IL units and their intrinsic polymeric nature. For IL-based electrolytes, there is a balance between their ionic conductivities and mechanical properties depending on the composition ratio of the polymer phase and additive phase [188–193].

Hence, a membrane composed of a polymer matrix, lithium salt, and IL may have the potential to optimize the system of electrolyte. We demonstrated the preparation of new imidazolium-based ionogels through photopolymerization of 1-ethyl-3-vinylimidazolium bis(trifluoromethanesulfonylimide) with poly(ethylene glycol) diacrylate (PEGDA) ($M_n = 700$ g mol^{-1}), poly(ethylene glycol) methyl ether methacrylate (PEGME) ($M_n = 500$ g mol^{-1}), and lithium bis(trifluoromethanesulfonyl)imide (LiTFSI) in an IL to obtain mechanically robust membranes. And, their use as the electrolyte system for Li/LiFePO$_4$ cells was studied in particular. Additionally, the thermal stabilities and electrical properties, such as ionic conductivity, electrochemical stability as well as the charge/discharge test of the Li/LiFePO$_4$ cells based on these materials, will also be discussed in the main article.

Solar energy is one of the sources offering clean, sustainable, and stable energy. In recent years, Si-based solar cells have been widely used, but it requires a fairly complex process. Therefore, the various types of emerging solar cells and experimental syntheses have been proposed to fully present the essential properties, outstanding functionalities, and commercialized products of green energy materials. From various materials acquired for emerging-cell construction, the arrangement of PEDOT:PSS polymer and n-type crystalline Si was proposed as the preference in hybrid structures considering the superiorities in fabrication aspect, reliability, and sound efficiency. Moreover, many approaches to improve the performance of these hybrid solar cells were discovered as well [194,195].

The PEDOT:PSS material as the p-type layer in this type of cell arrangement surely to have an important impact in the junction section, however, a pristine PEDOT:PSS layer exhibits low electrical conductivity, with a value in the range of 0.2 to 10 S cm^{-1}, which particularly causes the poor device performance in photovoltaic applications. In addition, the deposition on hydrophobic substrates generally yields the nonuniform contact. Therefore, the low-cost, safety, and facile route to enhance the electrical also the adhesion capability of pristine PEDOT:PSS is still in progress. The chemical treatment is highly demanded which allows the low-cost and large-area production based on the solution-processing capability such as by using ethylene glycol (EG) and fluorosurfactant [196,197]. Up to now, there exist several well-established reports that cover the discussion of treated PEDOT:PSS on planar Si for cell design and proven to improve the photovoltaic performance [198,199]. However, the research for combining PEDOT:PSS thin film with nanowires (SiNWs) is still limited. SiNWs are known to have a light-trapping effect to enable the optimization of incoming light transmitted from the transparent PEDOT:PSS layer [200]. Therefore, PEDOT:PSS/ Sen has been considered an efficient way for the construction of high-performance hybrid solar cells. This study thoroughly explores the diversified physical, chemical, and material phenomena of fundamental properties and the unusual functionalities in PEDOT:PSS films and Si-nanowire-based solar cells. Furthermore, the relations between the theoretical and practical analyses are fully discussed. It is expected to be capable of providing very useful information about science fundamentals, integrated engineering, and real application.

The organization of this chapter is as follows. The beginning of the chapter includes the introduction, especially for the various developments in the theoretical and experimental researches. As for the chapter former, the synthesis, structure, and electrical properties of the as-prepared PEDOT:PSS layer is discussed in the following subchapter. The principles of "Baytron P" routes as the most fundamentally convenient PEDOT-based compound fabrication procedure are explained into consideration [201]. Moreover, related to structure and electrical properties, the PSS functions in commercial PEDOT:PSS complex clearly identified under a comprehensive framework [202,203]. The adjustment into higher conductivity values using diverse methods including physical treatment or chemical process is defined in the next discussion [204–206]. Furthermore, the experimental syntheses and measurements are covered in the third subchapter which is focused on how to enhance the pristine PEDOT:PSS conductivity in affordable routes followed by various analyses using four-point probe, Raman spectroscopy, XPS, and many more.

The addition of EG into precursors results in the optimum conductivity up to 10^3 order of magnitude for the reduction of sheet resistance. Concerning the application, the Si-nanowire-based hybrid cells are modified for the great enhancement of the photovoltaic performance under doping mechanisms, the conformational changes of PEDOT segment, the arrangement alteration of PEDOT:PSS, and the removal of excess insulating PSS chains from the PEDOT:PSS film achieving >14% can be realized.

REFERENCES

1. Wu X., Kang F., Duan W., Lia J., *Progress in Natural Science: Materials International*, 2019, 29, 247–255.
2. Kang J., Kim K. C., Jang S. S., *The Journal of Physical Chemistry C*, 2018, 122, 10675–10681.
3. Yang S.-J., Qin X.-Y., He R., Shen W., Li M., Zhao L.-B., *Physical Chemistry Chemical Physics*, 2017, 19, 12480–12489.
4. Urban A., Seo D.-H., Ceder G., *npj Computational Materials*, 2016, 2, 16002.
5. Meng Y. S., Dompablo M. E. A., *Energy and Environmental Science*, 2009, 2, 589–609.
6. Zhao Y., Li X., Yan B., Xiong D., Li D., Lawes S., Sun X., *Advanced Energy Materials*, 2016, 6(1–19), 1502175.

7. Deng D., *Energy Science and Engineering*, 2015, 3(5), 385–418.
8. Wu S., Xu R., Lu M., Ge R., Iocozzia J., Han C., Jiang B., Lin Z., *Advanced Energy Materials*, 2015, 5(21), 1–40.
9. Goriparti S., Miele E., Angelis F. D., Fabrizio E. D., Zaccaria R. P, Capiglia C., *Journal of Power Sources,* 2014, 257, 421–443
10. Chen J., *Materials,* 2013, 6, 156–183.
11. Etacheri V., Marom R., Elazari R., Salitra G., Aurbach D., *Energy and Environmental Science*, 2011, 4, 3243.
12. Zhao W., Yi J., He P., Zhou H., *Electrochemical Energy Reviews*, 2019, 2, 574–605.
13. Takami N., Sekino M., Ohsaki T., Kanda M., Yamamoto M., *Journal of Power Sources,* 2001, 97–98, 677–680.
14. Xu Z., Yang J., Li H., Nuli Y., Wang J., *Journal of Materials Chemistry A*, 2019, 7, 9432.
15. Ullah A., Majid A., Rani N., *Journal of Energy Chemistry*, 2018, 27, 219–237.
16. Haskins J. B., Lawson J. W., *The Journal of Chemical Physics,* 2016, 144, 184707.
17. Bhattacharya S., John S., *Scientific Reports,* 2019, 9, 12482.
18. Duan J., Xu H., Sha W. E. I., Zhao Y., Wang Y., Yang X., Tang Q., *Journal of Materials Chemistry A*, 2019, 7, 21036–21068.
19. Cheah S. K., et al., *Nano Letters*, 2009, 9, 3230.
20. Wang W., Tian M., Abdulagatov A., George S. M., Lee Y. C., Yang R., *Nano Letters*, 2012, 12, 655.
21. Ban C., Xie M., Sun X., Travis J. J., Wang G., Sun H., Dillon A. C., Lian J., George S. M., *Nanotechnology*, 2013, 24, 424002.
22. Kim S. W., Han T. H., Kim J., Gwon H., Moon H. S., Kang S. W., Kim S. O., Kang K., *ACS Nano*, 2009, 3, 1085.
23. Jung Y. S., Cavanagh A. S., Riley L. A., Kang S. H., Dillon A. C., Groner M. D., George S. M., Lee S. H., *Advanced Materials*, 2010, 22, 2172.
24. Lee M.-L., Su C.-Y., Lin Y.-H., Liao S.-C., Chen J.-M., Perng T.-P., Yeh J.-W., Shih H. C., *Journal of Power Sources*, 2013, 244, 410–416.
25. Wang H.-Y., Wang F.-M., *Journal of Power Sources*, 2013, 233, 1.
26. Goriparti S., Miele E., De Angelis F., Di Fabrizio E., Zaccaria R. P., Capiglia C., *Journal of Power Sources*, 2014, 257, 421–443.
27. Liu Y., Wang Y. M., Yakobson B. I., Wood B. C., *Physical Review Letters*, 2014, 113, 28304.
28. Raccichini R., Varzi A., Passerini S., Scrosati B., *Nature Materials*, 2015, 14, 271–279.
29. He Y., Yu X., Wang Y., Li H., Huang X., *Advanced Materials*, 2011, 23, 4938.
30. Xiao X., Lu P., Ahn D., *Advanced Materials*, 2011, 23, 3911.
31. Chen S., Chen P., Wang, Y., *Nanoscale*, 2011, 3(10), 4323.
32. Liu C., Li F., Ma L.-P., Cheng H.-M., *Advanced Materials*, 2010, 22, 28.
33. Yoon S.-H., Park C.-W., Yang H., Korai Y., Mochida I., Baker R. T. K., Rodriguez N. M., *Carbon*, 2004, 42, 21.
34. Yan X., Teng D., Jia X., Yu Y., Yang X., *Electrochimica Acta*, 2013, 108, 196.
35. Yue H., Li F., Yang Z., Tang J., Li X., He D., *Materials Letters*, 2014, 120, 39.
36. Kohandehghan A., Kalisvaart P., Cui K., Kupsta M., Memarzadeh E., Mitlin D., *Journal of Materials Chemistry A*, 2013, 1, 12850.
37. Lotfabad E. M., Kalisvaart P., Cui K., Kohandehghan A., Kupsta M., Olsen B., Mitlin D., *Physical Chemistry Chemical Physics*, 2013, 15 13646.
38. Lotfabad E. M., Kalisvaart P., Kohandehghan A., Cui K., Kupsta M., Farbod B., Mitlin D., *Journal of Materials Chemistry A*, 2014, 2, 2504.
39. Snyder M. Q., Trebukhova S. A., Ravdel B., Wheeler M. C., DiCarlo J., Tripp C. P., DeSisto W. J., *Journal of Power Sources*, 2007, 165, 379.
40. Liu J., Li X., Cai M., Li R., Sun X., *Electrochimica Acta*, 2013, 93, 195.
41. Ahn D., Xiao X., *Electrochemistry Communications*, 2011, 13, 796.

42. Riley L. A., Cavanagh A. S., George S. M., Jung Y. S., Yan Y., Lee S. H., Dillon A. C., *ChemPhysChem*, 2010, 11, 2124.
43. Wang D., Yang J., Liu J., Li X., Li R., Cai M., Sham T.-K., Sun X., *Journal of Materials Chemistry A*, 2014, 2, 2306.
44. Lee J.-H., Hon M.-H., Chung Y.-W., Leu I.-C., *Applied Physics A*, 2010, 102, 545.
45. Kang E., Jung Y. S., Cavanagh A. S., Kim G.-H., George S. M., Dillon A. C., Kim J. K., Lee J., *Advanced Functional Materials*, 2011, 21, 2430.
46. Lipson A. L., Puntambekar K., Comstock D. J., Meng X., Geier M. L., Elam J. W., Hersam M. C., *Chemistry of Materials*, 2014, 26, 935.
47. Meng X., Liu J., Li X., Banis M. N., Yang J., Li R., Sun X., *RSC Advances*, 2013, 3, 7285.
48. Meng X., Libera J. A., Fister T. T., Zhou H., Hedlund J. K., Fenter P., Elam J. W., *Chemistry of Materials*, 2014, 26, 1029.
49. Meng X., et al., *Advanced Functional Materials*, 2014, 24, 158.
50. Li S., Ping Liu P., Huang X. B., Tang Y. G., Wang, H.-Y., *Journal of Materials Chemistry A*, 2019, 7, 10988.
51. Dimesso L., et al., *Chemical Society Reviews*, 2012, 41, 5068.
52. Wang J., Sun X., *Energy and Environmental Science*, 2012, 5, 5163.
53. Liu D., Liu Y., Pan A., Nagle K. P., Seidler G. T., Jeong Y.-H., Cao G., *The Journal of Physical Chemistry C*, 2011, 115(11), 4959.
54. Karthik K., Phuruangrat A., Chowdhury Z. Z., Pradeeswari K., Kumar R. M., *Journals of Science*, 2019.
55. Donders M. E., Arnoldbik W. M., Knoops H. C. M., Kessels W. M. M., Notten P. H. L., *Journal of the Electrochemical Society*, 2013, 160, A3066.
56. Xie J., Zhao J., Liu Y., Wang H., Liu C., Wu T., Hsu P.-C., Lin D. C., Jin Y., Cui Y., *Nano Research*, 2017, 10(11), 3754.
57. Scott I D., Jung Y. S., Cavanagh A. S., Yan Y., Dillon A. C., George S. M., Lee S. H., *Nano Letters*, 2011, 11, 414.
58. Miikkulainen V., Ruud A., Østreng E., Nilsen O., Laitinen M., Sajavaara T., Fjellvåg H., *The Journal of Physical Chemistry C*, 2014, 118, 1258.
59. Seok Jung Y., Cavanagh A. S., Yan Y., George S. M., Manthiram A., *Journal of the Electrochemical Society*, 2011, 158, A1298.
60. Riley L. A., Van Atta S., Cavanagh A. S., Yan Y., George S. M., Liu P., Dillon A. C., Lee S.-H., *Journal of Power Sources*, 2011, 196, 3317.
61. Li X., Liu J., Banis M. N., Lushington A., Li R., Cai M., Sun X., *Energy and Environmental Science*, 2014, 7, 768.
62. Park J. S., Meng X., Elam J. W., Hao S., Wolverton C., Kim C., Cabana J., *Chemistry of Materials*, 2014, 26, 3128.
63. Xiao X., Ahn D., Liu Z., Kim J.-H., Lu P., *Electrochemistry Communications*, 2013, 32, 31.
64. Bettge M., Li Y., Sankaran B., Rago N. D., Spila T., Haasch R. T., Petrov I., Abraham D. P., *Journals of Power Sources*, 2013, 233, 346.
65. Zhang X., Belharouak I., Li L., Lei Y., Elam J. W., Nie A., Chen X., Yassar R. S., Axelbaum R. L., *Advanced Energy Materials*, 2013, 3, 1299.
66. Park J. S., Meng X., Elam J. W., Hao S., Wolverton C., Kim C., Cabana J., *Chemistry of Materials*, 2014, 26, 3128.
67. Gandrud K. B., Pettersen A., Nilsen O., Fjellvåg H., *Journal of Materials Chemistry A*, 2013, 1, 9054.
68. Liu J., Banis M. N., Sun Q., Lushington A., Li R., Sham T. K., Sun X., *Advanced Materials*, 2014, 26, 6472.
69. Zheng F., Kotobuki M., Songa S.-F., Lai M. O., Lu L., *Journal of Power Sources*, 2018, 389, 198.

70. Chen C., Xie S., Sperling E., Yang A., Henriksen G., Amine K., *Solid State Ionics*, 2004, 167, 263.
71. Aaltonen T., Alnes M., Nilsen O., Costelle L., Fjellvåg H., *Journal of Materials Chemistry*, 2010, 20, 2877.
72. Chen C., Amine K., *Solid State Ionics*, 2001, 144, 51.
73. Aatiq A., Ménétrier M., Croguennec L., Suard E., Delmas C., *Journal of Materials Chemistry*, 2002, 12, 2971.
74. Deng, Y. et al., *ACS Applied Materials and Interfaces*, 2017, 9, 7050.
75. Thangadurai V., Weppner W., *Ionics*, 2006, 12, 81.
76. Aaltonen T., Nilsen O., Magrasó A., Fjellvåg H., *Chemistry of Materials*, 2011, 23, 4669.
77. Comstock D. J., Elam J. W., *The Journal of Physical Chemistry C*, 2013, 117, 1677.
78. Hamalainen J., Holopainen J., Munnik F., Hatanpaa T., Heikkila M., Ritala M., Leskela M., *Journal of the Electrochemical Society*, 2012, 159, A259.
79. Hämäläinen J., Munnik F., Hatanpää T., Holopainen J., Ritala M., Leskelä M., *Journal of Vacuum Science and Technology A*, 2012, 30, 01A106.
80. Liu J., Banis M. N., Li X., Lushington A., Cai M., Li R., Sham T.-K., Sun X., *The Journal of Physical Chemistry C*, 2013, 117, 20260.
81. Østreng E., Sønsteby H. H., Sajavaara T., Nilsen O., Fjellvåg H., *Journal of Materials Chemistry C*, 2013, 1, 4283.
82. Østreng E., Vajeeston P., Nilsen O., Fjellvåg H., *RSC Advances*, 2012, 2, 6315.
83. Perng Y.-C., Cho J., Sun S. Y., Membreno D., Cirigliano N., Dunn B., Chang J. P., *Journal of Materials Chemistry A*, 2014, 2, 9566.
84. Kamaya N., Homma K., Yamakawa Y., Hirayama M., Kanno R., Yonemura M., Kamiyama T., Kato Y., Hama S., Kawamoto K., *Nature Materials*, 2011, 10, 682.
85. Boulineau S., Courty M., Tarascon J.-M., Viallet V., *Solid State Ionics*, 2012, 221, 1.
86. Zhao Y., Daemen L. L., *Journal of the American Chemical Society*, 2012, 134, 15042.
87. Tang T., Chen P., Luo W., Luo D., Wang Y., *Journal of Nuclear Materials*, 2012, 420, 31.
88. Ma S. G., Shen Y.-H., Kong X.-G., Gao T., Chen X.-J., Xiao C.-J., Lu T.-C., *Mater Design*, 2017, 118, 218.
89. Tada, K., Kitta, M., Ozaki, H., Tanaka, S. *Chemical Physics Letters*, 2019, 731, 136598.
90. Haregewoin, A. M., Wotango, A. S., Hwang, B., *Journal of Energy and Environmental Science*, 2016, 9(6), 1955.
91. Khan S. A., Shahid Ali S., Khalid Saeed S., Muhammad Usman M., Khan I., *Journal of Materials Chemistry A*, 2019, 7, 10159.
92. Lin S. Y., Tran N. T. T., Chang S. L., Chuu C. P., Lin M. F., *Structure- and Adatom-Enriched Essential Properties of Graphene Nanoribbons*, CRC Press, Boca Raton, FL, 2017.
93. Tran N. T. T., Lin S. Y., Lin C. Y., Lin M. F., *Geometric and Electronic Properties of Graphene-Related Systems: Chemical Bondings*, CRC Press, Boca Raton, FL, 2017.
94. Lin C. Y., Chen R. B., Ho Y. H., Lin M. F., *Electronic and Optical Properties of Graphite-Related Systems*, CRC Press, Boca Raton, FL, 2017.
95. Grotendorst J., Attig N., Blugel S., Marx D., *Multiscale Simulation Methods in Molecular Sciences*, NIC Series, Winter School Forschungszentrum Jülich March 2009, 42, pp. 2–6.
96. Nakayama T., Shima H., *Physical Review E*, 1998, 58(3), 3984.
97. Chen G. P., Voora V. K., Agee M. M., Balasubramani S. G., Furche F., *Annual Review of Physical Chemistry*, 2017, 68, 421.
98. Wu, B., Han, S., Shin, K. G., Lu, W., *Journal of Power Sources* 2018, 395, 128–136.
99. Sendek, A. D., Yang, Q., Cubuk, E. D., Duerloo, K. A. N., Cui, Y., Reed, E. J, *Energy and Environmental Science*, 2017, 10(1), 306–320.

100. Ahmad, Z., Xie, T., Maheshwari, C., Grossman, J. C., Viswanathan, V., *ACS Central Science*, 2018, 4(8), 996–1006.
101. Blomgren, G. E. *Journal of Electrochemical Society*, 2017, 164, A5019–A5025.
102. Zubi G., Dufo-López R., Carvalho M., Pasaoglu G., *Renewable and Sustainable Energy Reviews*, 2018, 89, 292–308.
103. Pistoia G. *Lithium-Ion Batteries: Advances and Applications*, 1st ed. Elsevier, Amsterdam, Netherlands, 2014.
104. Lin M.-C., Gong M., Lu B., Wu Y., Wang D.-Y., Guan M., Dai H., *Nature*, 2015, 520(7547), 324–328.
105. Das S. K., Mahapatra S., Lahan, H., *Journal of Materials Chemistry A*, 2017, 5(14), 6347–6367.
106. Ramaprabhu S., Kumar Saroja A. P. V., Samantaray, S. S., *Chemical Communications*, 2019, 55, 10416–10419.
107. Liang Y., Zhao C., Yuan H., Chen Y., Zhang W., Huang J., Zhang Q., *InfoMat*, 2019, 1(1), 6–32.
108. Pistoia, G. *Batteries for Portable Devices*, 1st ed. Elsevier Science, Amsterdam, Netherlands, 2005.
109. Gao Z., Lin Z., LaClair T. J., Liu C., Li J.-M., Birky A. K., Ward, J. *Energy*, 2017, 122, 588–600.
110. Gerssen-Gondelach S. J., Faaij A. P. C., *Journal of Power Sources*, 2012, 212, 111–129.
111. Liu Y., Gorgutsa S., Santato C., Skorobogatiy M., *Journal of the Electrochemical Society*, 2012, 159(4), A349–A356.
112. Liu Z., Deng H., Zhang S., Hu W., Gao F. *Physical Chemistry Chemical Physics*, 2018, 20, 22351–22358.
113. Morris R. S., Dixon B. G., Gennett T., Raffaelle R., Heben M. J., *Journal of Power Sources*, 2004, 138(1–2), 277–280.
114. Matsumura T., Kanno R., Inaba Y., Kawamoto Y., Takano, M. *Journal of the Electrochemical Society*, 2003, 149(12), A1509–A1513.
115. Qian J., Liu L., Yang J., Li S., Wang X., Zhuang H. L., Lu Y. *Nature Communications*, 2018, 9(1), 4918.
116. Chen H., Freeman C. L., Harding J. H., *Physical Review B*, 2011, 84(8), 085108.
117. Chen H., Dawson J. A., Harding J. H. *Journal of Materials Chemistry A*, 2014, 2(21), 7988.
118. Kuganathan N., Tsoukalas L. H., Chroneos, A. *Solid State Ionics*, 2019, 335, 61–66.
119. Lau J., DeBlock R. H., Butts D. M., Ashby D. S., Choi C. S., Dunn B. S. *Advanced Energy Materials*, 2018, 8, 1800933.
120. Sathiya M., Rousse G., Ramesha K., Laisa C. P., Vezin H., Sougrati M. T., Doublet M-L., Foix D., Gonbeau D., Walker W., Prakash A. S., Ben Hassine M., Dupont L., Tarascon J.-M. *Nature Materials*, 2013, 12(9), 827–835.
121. Zurek E., Jepsen O., Andersen O. K., *Physical Chemistry Chemical Physics*, 2005, 6, 1934.
122. Kashani H., Tian Y., Ito Y., Fujita J.-I., Oyama Y. *Nature Communications*, 2019, 10(1), 275.
123. Yao X., Huang B., Yin J., Peng G., Huang Z., Gao C., Liu D., Xu X., *Chinese Physics B*, 2016, 25(1), 018802.
124. Humana R. M., Ortiz M. G., Thomas J. E., Real S. G., Sedlarikova M., Vondrak J., Visintin A., *ECS Transactions*, 2014, 63(1), 91–97.
125. Zhang S.S., *Journal of Power Sources*, 2006, 162, 1379–139.
126. Yadroitsev I., Yadroitsava I., Le Roux S. G., *3D Printing and Additive Manufacturing*. doi: 10.1089/3dp.2018.0060.
127. Kuganathan N., Tsoukalas L. H., Chroneos A., *Solid State Ionics*, 2019, 335, 61–66.

128. Wei J., Diaconescu P. L., Diaconescu, P. L. *Macromolecules,* 2017, 505, 1847–1861.
129. Deng J., Lei Y., Wen S., Chen, Z., *International Journal of Mineral Processing,* 2015, 140, 43–49.
130. Zhang S. S., *Journal of Power Sources,* 2006, 162, 1379–1394.
131. Nair J. R., et al., *Electrochimica Acta,* 2016, 199, 172–179.
132. du Plessis, A., Yadroitsev, I., Yadroitsava, I., Le Roux S. G., *3D Printing and Additive Manufacturing.* doi: 10.1089/3dp.2018.0060.
133. GhaffarianHoseini A., Dahlan N. D., Berardi U., GhaffarianHoseini A., Makaremi N., GhaffarianHoseini M., *Renewable and Sustainable Energy Reviews,* 2013, 25, 1–17.
134. Greeley J., Markovic N. M., *Energy and Environmental Science,* 2012, 5, 9246–9256.
135. Zhang Q., Sun Y., Xu W., Zhu D. *Advanced Materials,* 2014, 26, 6829–6851.
136. Wang W. C., Yung Y. L., Lacis A. A., Mo T. A., Hansen, J. E. *Science,* 1976, 194, 685–690.
137. Orilall, M. C., Wiesner, U., *Chemical Society Reviews,* 2011, 40, 520–535.
138. Frey H. C., Rouphail N. M., Zhai H., Farias T. L., Gonçalves G. A. *Transportation Research Part D: Transport and Environment,* 2007, 12, 281–291.
139. Liu J., Barnett S. A. *Solid State Ionics,* 2003, 158, 11–16.
140. Prasad N. R., Ranade S. J., Hoang H. T., Phuc, N. H., December hydropower energy recovery (hyper) from water-flow systems in Vietnam. *In 2012 10th International Power and Energy Conference (IPEC),* Ho Chi Minh City, Vietnam, IEEE, 2012, pp. 92–97.
141. Morren J., De Haan S. W., Kling W. L., Ferreira, J. A., *IEEE Transactions on Power Systems,* 2006, 21, 433–434.
142. Yu Y., Gu L., Zhu C., Van Aken P. A., Maier J., *Journal of the American Chemical Society,* 2009, 131, 15984–15985.
143. Ambroz F., Macdonald T. J., Nann, T. *Advanced Energy Materials,* 2017, 7, 1602093.
144. Christiansen A. S., Johnsen R. E., Norby P., Frandsen C., Mørup S., Jensen S. H., Hansen K., Holtappels P., *Journal of The Electrochemical Society,* 2015, 162, A531–A537.
145. Zhang S. S. *Journal of Power Sources,* 2006, 162, 1379–1394.
146. Whittingham M. S. *Chemical Reviews,* 2004, 104, 4271–4302.
147. Stevens D. A., Dahn J. R. *Journal of the Electrochemical Society,* 2000, 147, 1271–1273.
148. Moura S. J., Argomedo F. B., Klein R., Mirtabatabaei A., Krstic, M., *IEEE Transactions on Control Systems Technology,* 2016, 25, 453–468.
149. Cheng Q., *Scientific Reports,* 2017, 7, 14782.
150. Lin C. Y., Wu J. Y., Chiu C. W., Lin M. F., *AB-Stacked Graphenes.* CRC Press, Boca Raton, FL, 2017. ISBN: 9780429277368.
151. Lin C. Y., Wu J. Y., Chiu C. W., Lin M. F. *ABC-Stacked Graphenes.* CRC Press, Boca Raton, FL, 2017. ISBN: 9780429277368.
152. Lin C. Y., Wu J. Y., Chiu C. W., Lin M. F. *AA-Stacked Graphenes.* CRC Press, Boca Raton, FL, 2017. ISBN: 9780429277368.
153. Li X., Zhi L. *Chemical Society Reviews,* 2018, 47, 3189–3216.
154. Xin H., Li W. *Applied Physics Reviews,* 2018, 5, 031105.
155. Randviir E. P., Brownson D. A. C., Banks, C. E., *Materials Today,* 2014, 17(9), 426–432.
156. Wu Z.-S., Ren W., Xu L., Li F., Cheng H.-M., *ACS Nano,* 2011, 5(7) 5463–5471.
157. Shu K., Wang C., Li S., Zhao C., Yang Y., Liu H., Wallace, G. *Journal of Materials Chemistry A,* 2015, 3(8), 4428–4434.
158. Han S., Wu D., Li S., Zhang F., Feng X., *Advanced Materials,* 2014 26(6), 849–864.
159. Dunn B., Kamath H., Tarascon J. M., *Science,* 2011, 334, 928–935.
160. Pan H., Hu Y.-S., Chen L., Dunn B., Kamath H., Tarascon J. M., Yang Z. G., *Energy and Environmental Science,* 2013, 6, 2338–2360.
161. Nishi Y., *Journal of Power Sources,* 2001, 100, 101–106.

162. Le P., Cointeaux L., Strobel P., Leprêtre J.-C., Judeinstein P., Alloin F., *The Journal of Physical Chemistry C*, 2012, 116, 7712–7718.
163. Le P., Alloin F., Strobel P., Leprêtre J.-C., Pérez del Valle C., Judeinstein P., *The Journal of Physical Chemistry B*, 2010, 114, 894–903.
164. Wongittharom N., Lee T.-C., Wang C.-H., Wang Y.-C., Chang J.-K., *Journal of Materials Chemistry A*, 2014, 2, 5655–5561.
165. Wang H., Yeh W., Wongittharom N., Wang C., Tseng J., Lee W., Chang S., Chang K., *Journal of Power Sources*, 2015, 274, 1016–1023.
166. Etacheri V., Marom R., Elazari R., Salitra G., Aurbach D., *Energy and Environmental Science*, 2011, 4, 3243–3262.
167. Yang Z., Zhang J., Kintner-Meyer M. C. W., Lu X., Choi D., Lemmon J. P., Liu J., *Chemical Reviews*, 2011, 111, 3577–3613.
168. Hong Y., Kim Y., Park Y., Choi A., Choi N.-S., Lee K. T., *Energy and Environmental Science*, 2013, 6, 2067–2081.
169. Palomares V., Serras P., Villaluenga I., Hueso K. B., Carretero-González J., Rojo T., *Energy and Environmental Science*, 2012, 5, 5884–5901.
170. Kim S. W., Seo D. H., Ma X., Ceder G., Kang K., *Advanced Energy Materials*, 2012, 2, 710–721.
171. Slater M. D., Kim D., Lee E., Johnson C. S., *Advanced Functional Materials*, 2013, 23, 947–958.
172. Palomares V., Casas-Cabanas M., Castillo-Martínez E., Han M. H., Rojo T., *Energy and Environmental Science*, 2013, 6, 2312–2337.
173. Armand M., Tarascon J. M., *Nature*, 2008, 451, 652–657.
174. Kang B., Ceder G., *Nature*, 2009, 458, 190–193.
175. Sun Y. K., Myung S. T., Park B. C., Prakash J., Belharouak I., Amine K., *Nature Materials*, 2009, 8, 320–324.
176. Scrosati B., Garche J., *Journal of Power Sources*, 2010, 195, 2419–2430.
177. Balakrishnan P. G., Ramesh R., Kumar T. P., *Journal of Power Sources*, 2006, 155, 401–414.
178. Goodenough J. B., Kim Y., *Chemistry of Materials*, 2010, 22, 587–603.
179. Lux S. F., Lucas I. T., Pollak E., Passerini S., Winter M., Kostecki R., *Electrochemistry Communications*, 2012, 14, 47–50.
180. Smiglak M., Reichert W. M., Holbrey J. D., Wilkes J. S., Sun L. Y., Thrasher J. S., Kirichenko K., Singh S., Katritzky A. R., Rogers R. D., *Chemical Communications*, 2006, 2554–2556. doi: 10.1039/b602086k,
181. Earle M. J., Esperanca J. M. S. S., Gilea M. A., Lopes J. N. C., Rebelo L. P. N., Magee J. W., Seddon K. R., Widegren J. A., *Nature*, 2006, 439, 831–834.
182. Nakagawa H., Fujino Y., Kozono S., Katayama Y., Nukuda T., Sakaebe H., Matsumoto H., Tatsumi K., *Journal of Power Sources*, 2007, 174, 1021–1026.
183. Jin Y. D., Fang S. H., Chai M., Yang L., Tachibana K., Hirano S., *Journal of Power Sources*, 2013, 226, 210–218.
184. Sun X. G., Liao C., Shao N., Bell J. R., Guo B. K., Luo H. M., Jiang D. E., Dai S., *Journal of Power Sources*, 2013, 237, 5–12.
185. Sakaebe H., Matsumoto H., Tatsumi K., *Electrochimica Acta*, 2007, 53, 1048–1054.
186. Fernicola A., Croce F., Scrosati B., Watanabe T., Ohno H., *Journal of Power Sources*, 2007, 174, 342–348.
187. Galinski M., Lewandowski A., Stepniak I., *Electrochim Acta*, 2006, 51, 5567–5580.
188. Yin K., Zhang Z. X., Li X. W., Yang L., Tachibana K., Hirano S. I., *Journal of Materials Chemistry A*, 2015, 3, 170–178.
189. Yin K., Zhang Z. X., Yang L., Hirano S. I., *Journal of Power Sources*, 2014, 258, 150–154.

190. Appetecchi G. B., Kim G. T., Montanina M., Carewska M., Marcilla R., Mecerreyes D., De Meatza I., *Journal of Power Sources*, 2010, 195, 3668–3675.

191. Li X. W., Zhang Z. X., Li S. J., Yang L., Hirano S., *Journal of Power Sources*, 2016, 307, 678–683.

192. Safa M., Chamaani A., Chawla N., El-Zahab B., *Electrochimica Acta*, 2016, 213, 587–593.

193. Zhang P. F., Li M. T., Yang B. L., Fang Y. X., Jiang X. G., Veith G. M., Sun X. G., Dai S., *Advanced Materials*, 2015, 27, 8088–8094.

194. Lu W., Wang C., Yue W., Chen L., *Nanoscale*, 2011, 3, 9, 3631–3634.

195. Chen C.-Y., Wei T.-C., Hsiao P.-H., Hung C.-H., *ACS Applied Energy Materials*, 2019, 2(7), 4873–4881.

196. Wichiansee W., Sirivat A., *Materials Science and Engineering C*, 2008, 29, 78–84.

197. Vosgueritchian M., Lipomi D. J., Bao Z., *Advanced Functional Materials*, 2012, 22, 2, 421–428.

198. Li Q., Yang J., Chen S., Zou J., Xie W., Zeng X., *Nanoscale Research Letters*, 2017, 12, 506.

199. Thomas J. P., Zhao L., McGillivray D., Leung K. T., *Journal of Materials Chemistry A*, 2014, 2(7), 2383.

200. Chen C.-Y., Wong C.-P., *Nanoscale*, 2015, 7(3), 1216–1223.

201. Groenendaal L., Jonas F., Freitag D., Pielartzik H., Reynolds J. R., *Advanced Materials*, 2000, 12(7), 481–494.

202. Ahonen H. J., Lukkari J., Kankare J., *Macromolecules*, 2000, 33(18), 6787–6793.

203. Ghosh S., Inganäs O., *Synthetic Metals*, 1999, 101(1), 413–416.

204. Kim Y., Ballantyne A. M., Nelson J., Bradley D. D. C., *Organic Electronics: Physics, Materials, Applications*, 2009, 10(1), 205–209.

205. Sun, K., Zhang, S., Li, P., Xia, Y., Zhang, X., Du, D., Ouyang, J., *Journal of Materials Science: Materials in Electronics*, 2015, 26(7), 4438–4462.

206. Zhang X., et al., *Solar Energy Materials and Solar Cells*, 2016, 144, 143–149.

2 Diverse Phenomena in Stage-*n* Graphite Alkali-Intercalation Compounds

Wei-bang Li and Ngoc Thanh Thuy Tran
National Cheng Kung University

Shih-Yang Lin
National Chung Cheng University

CONTENTS

2.1 INTRODUCTION

A bulk graphite,[1] which could be regarded as the layered graphene systems,[2] has attracted a lot of theoretical and experimental researches in the basic science,[3–5] engineering,[6] and application.[7] Each graphene layer is a pure carbon honeycomb lattice,[8] and the graphitic layers are attracted together through the weak, but significant van der Waals interactions.[9] There exist the very strong intralayer σ bondings of C-$(2s, 2p_x, 2p_y)$ orbitals and a significant intra- and interlayer C-$2p_z$ orbital hybridizations,[10] in which the former determine the most outstanding mechanical materials, and the latter dominate the low-lying energy bands and thus the essential physical properties. Very interesting, the 3D graphite systems might present the AA,[11] AB,[12] ABC,[13] and turbostratic stackings.[14] All of them belong to semimetals under the interlayer hopping integrals of C-$2p_z$ orbitals. In general, such condensed matter systems become the *n*- or *p*-type metals, depending on the kinds of intercalated atoms or molecules. For example, the intercalation of alkali atoms[15] and $FeCl_3$ into graphite, respectively, creates many free conduction electrons and valence holes, as observed in pure metal.

It is well known that graphite could serve as the best anode materials[16] in the commercial Li+-based batteries,[17] mainly owing to the lowest cost, the most stable structure for intercalation, and the outstanding ion transport under the charging and discharging processes. When the Li^+ ions are released from the cathode, they will transport through the electrolyte and then intercalate into the graphitic system. Most important, the flexible interlayer spacings between the graphene layers are capable of providing sufficient positions for the various intercalant concentrations.[18] Any intermediate states and meta-stable configurations, which are created during the ion/atom intercalation, could survive through the very strong σ bondings in graphitic sheets. As a result, graphite is rather suitable for studying the structural transformation in the chemical reactions. For example, the close relations between graphene and intercalant layers in lattice symmetries are expected to present the dramatic transformation before and after the intercalation/ de-intercalation processes.

Up to now, there are a lot of theoretical and experimental studies on the fundamental properties in graphite-related systems. The former covers the phenomenological models and number simulation methods. For example, the tight-binding model, the random-phase approximation, and the Kubo formula are, respectively, utilized to investigate their electronic properties and magnetic quantization behaviors,[19–22] Coulomb excitations[23–26] and impurity screenings,[27] and optical absorption spectra.[28] Furthermore, the first-principles calculations[29] are available in understanding the optimal stacking configurations; intercalant lattices, π, σ, and intercalation-induced energy bands; free conduction electrons/ valence holes; and density of states. However, certain critical physical quantities and pictures are absent in the previous studies, such as the atom-dominated band structures, the spatial charge densities between intercalant and graphene layers, the interlayer orbital hybridization of intercalant and carbon atom, the atom- and orbital-projected van Hove singularities, and the spin-dependent state degeneracy/ magnetic moment/density distribution.

The main focuses of this work are the features of geometric and electronic properties in stage-n graphite alkali-intercalation compounds using the first-principles method.[29] Very interesting, the dependences on the kinds of intercalants, their concentrations, and stacking configurations are thoroughly explored through the delicate calculations and analyses, especially for the lattice symmetries of intercalant layers, the dominances of valence and conduction states by carbon and/or alkali atoms, the blueshift of the Fermi level, the interlayer charge density variations after the alkali intercalations, and atom- and orbital-decomposed density of states. Whether the spin-induced Coulomb on-site interactions or the spin-orbital couplings play critical roles will be examined in detail. Most important, the orbital-hybridizations in chemical bonds and the atom-induced spin configurations are determined from the above-mentioned results. How to simulate VASP band structures by the tight-binding model is worthy of detailed discussions, since the interlayer hopping integrals are not very complicated under the concise primitive unit cells of layered graphite compounds. The predicted interlayer distance/intercalant lattice, band overlap, and unusual van Hove singularities could be detected using the high-resolution X-ray diffraction/low-energy electron diffraction,[30] angle-resolved photoemission spectroscopy (ARPES),[31] and scanning tunneling spectroscopy (STS),[32] respectively.

2.2 THEORETICAL CALCULATIONS

Both numerical simulations and phenomenological models are available in fully exploring the diversified physical, chemical, and material phenomena. For example, the former and the latter, respectively, cover the first-principles calculations[29]/molecular dynamics[33]/quantum Monte Carlo[34]and the generalized tight-binding model[5,35]/Kubo formul[36]/random-phase approximation.[37] Apparently, only the first method is able to investigate the lattice symmetries of the graphite-related systems with/without the chemical modifications. As for pure graphites and graphite alkali-intercalation compounds, the previous VASP results show that the stacking configurations and the intercalant lattice symmetries and concentrations play important roles in determining the semimetallic or metallic behaviors. However, the concise physical and chemical pictures, which are very useful in fully comprehending the fundamental properties, are required in the further studies, as done for 2D layered graphenes[38,39] and 1D graphene nanoribbons.[40–42]

It is well known that the Vienna ab initio simulation package is developed within the density functional theory.[43] The up-to-date VASP calculations can provide the sufficient information about the lattice symmetries, valence and conduction bands, charge density distributions/variations after chemical reactions, atom- and orbital-decomposed density of states,[44] being accompanied with the spin-dependent magnetic configurations. Very interesting, a lot of electron quasiparticles experience rather complicated interactions, when their wave packets propagate in any periodical material. Beyond the classical electrodynamics, the many-body electron-electron Coulomb interactions could be classified into the exchange and correlation energies according to the first order and the other orders, respectively. The Perdew–Burke–Ernzerhof formula, which depends on the local electron density, is utilized to deal with many-particle Coulomb effects. As for the frequent electron-crystal scatterings, they are characterized by the projector-augmented wave pseudopotentials. The electron Bloch wave functions are solved using the linear superposition of plane waves, with the maximum kinetic energy of $500\,eV$. The current study on stage-*n* graphite compounds shows that the first Brillouin zone is sampled by $9\times9\times9$ and $100\times100\times100$ k-point meshes within the Monkhorst–Pack scheme, respectively, for the optimal geometry and band structure. Moreover, the convergence condition of the ground-state energy is set to be $\sim10^{-5}\,eV$ between two consecutive evaluation steps, where the maximum HellmannFeynman force for each ion is below $0.01\,eV/\text{Å}$ during the atom relaxations.

The self-consistent calculations, which are closely related to the local carrier density,[45] account for the VASP simulations through the Kohn-Sham equations. According to the initial charge distribution, one evaluates the classical and many-particle exchange-correlation Coulomb interactions, and the electron-ion scattering potentials. After solving the second-order differential equations, the intermediate electron density and ground-state energy are obtained, and they are utilized to examine the convergence condition on the Feynman force. The significant physical quantities, the lattice symmetries/the various bond lengths in a unit cell, the atom-dominated valence and conduction bands, the spatial charge densities/their variations after the creation of chemical bonds, the atom- and orbital-projected van Hove singularities, the atom-induced spin distributions, the spin-split or degenerate states across the chemical potential, and the vanishing or nonvanishing magnetic moments, are available in making a final decision about the critical multi-/single-orbital hybridizations of chemical bonds and the atom-induced spin configurations in the stage-n graphite alkali-intercalation compounds. The above-mentioned theoretical framework, which is built from the first-principles results, presents the successful exploration on the diversified essential properties of 2D layered graphenes,[38,39] 1D graphene nanoribbons,[40–42] and 2D silicene-based systems.[18] Now, the developed viewpoints are suitable for the anode, cathode, and electrolyte materials of Li$^+$-based batteries.[46] For example, a detailed comparison between graphite and $Li_4Ti_5O_{12}$ is interesting and meaningful in terms of well-known anode compounds.

The direct linking between numerical simulations and phenomenological models is a very interesting topic. According to the up-to-date results of the low-energy band structures, the combination of the VASP calculations[18,46] and the tight-binding model[5,35] is rather successful in exploring layered graphene/graphitic systems. For example, both methods can account for the linear and gapless Dirac-cone band structure, two pairs of linear/parabolic valence and conduction bands, three vertical Dirac cones/two pairs of parabolic bands plus one separated Dirac cone/one pair of partially flat, sombrero-shape and linear energy dispersions/oscillatory, sombrero-shaped and parabolic bands, respectively, for the monolayer, bilayer AA-/AB-stacked, and trilayer AAA-/ABA-/ABC-/AAB-stacked graphene systems. Their results are consistent with each other, since the important interlayer atomic interactions are purely due to the carbon-$2p_z$ orbitals within the specific unit cells. That is to say, the single-orbital Hamiltonians, with the concise inter- and intralayer hopping integrals, could be expressed in the analytic forms. Furthermore, all the intrinsic interactions, being modified by the external fields, are further included in the generalized tight-binding model. This theoretical framework has been successful in creating and explaining the rich and unique magnetic quantization phenomena of few-layer group-IV systems, e.g., the unusual quantized Landau levels in graphene, silicone,[47] germanene,[48] and tinene.[49]

2.3 UNIQUE STACKING CONFIGURATIONS AND INTERCALANT DISTRIBUTIONS

The normal stacking configurations, which are formed by the layered graphene systems, cover AAA, ABA, ABC, and AAB. For example, according to the VASP calculations on the trilayer graphenes, the ground-state energy/the interlayer distance is lowest and highest/shortest and longest for the ABA and AAA stackings, respectively. The similar results are revealed in the AA-, AB-, and ABC-stacked graphites. Each honeycomb structure has two equivalent sublattices with a planar geometry, indicating the orthogonal relation between one $2p_z$ and three ($2s$, $2p_x$, $2p_y$) orbitals. The sp^2-bonding-based graphene systems maintain the flat sheets even under any interlayer couplings (chemical reactions). The significant van der Waals interactions, the interlayer single-orbital hybridizations of $2p_z$-$2p_z$, account for the stacking configurations and also noticed that the 2D silicene[50]/germanene[51]/tinene[52] systems possess the buckled structures, since the π and σ bondings are nonorthogonal to each other. Furthermore, the sp^3 bonding might play an important role in the inter-layer atomic interactions. Apparently, graphenes are quite different from silicenes/germanenes/tinenes in various essential properties.

The interlayer spacings of graphitic layers are able to provide a very suitable chemical environment for the alkali intercalations and de-intercalations, i.e., they are very useful during the charging and discharging processes. The chemical modifications hardly affect the planar honeycomb lattices and thus do not change the orthogonal feature of π and σ bondings. Very interesting, the optimal geometric properties are rather sensitive to the (x, y)-plane distribution and concentration of alkali atoms, e.g., the interlayer distances of neighboring planes and an enlarged primitive unit cell. According to the highest concentration, the graphite alkali-intercalation materials could be classified into two categories: lithium and non-lithium ones. There exists LiC_{6n} or AC_{8n} for the stage-n materials (Figure 2.1a-d), as examined from the experimental measurements and the theoretical predictions. The specific distance between carbon and intercalant layer is shortest/longest for the Li/Cs (Table 2.1), only directly reflecting the effective atomic radius. Although all the alkali atoms present the hollow-site optimal positions, their distribution symmetries are dominated by Li atoms or non-Li ones. The layered LiC_{6n} and AC_{8n} systems, respectively, possess the three- and four-time enhancement in the unit cell (Figure 2.2a and b); therefore, their reduced first Brillouin zones exhibit different high-symmetry points (Figure 2.3a). This is expected to have a strong effect on the initial π-electronic states of the Dirac-cone structure. More chemical bonds, which arise from the alkali-carbon atoms, are produced in graphite intercalation compounds, compared with pristine systems. Such bondings need to be taken into account for the diversified essential properties. In addition to the intercalant distribution, the stacking configuration of two neighboring graphitic sheets also affects the ground-state energy. The current study clearly shows that only the stage-1 LiC_6 presents the AA stacking, and the other compounds possess the AB stacking. That is to say, all the graphite alkali-intercalation materials exhibit the AB-stacked configurations except for the former system.

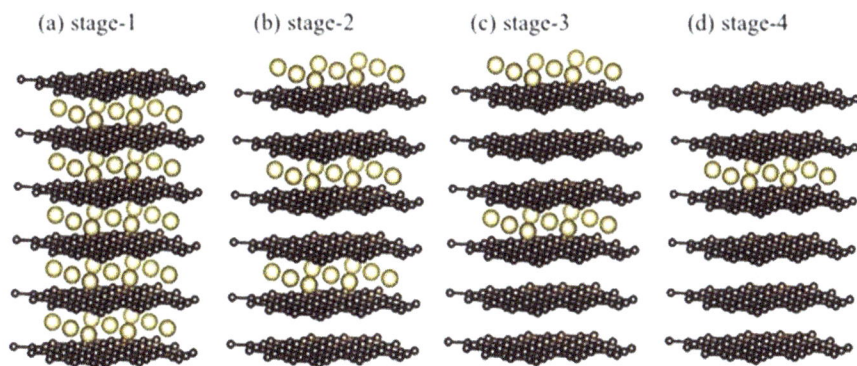

FIGURE 2.1 Stage-*n* graphite alkali-intercalation compounds: (a) $n=1$, (b) $n=2$, (c) $n=3$, and (d) $n=4$

TABLE 2.1

Stage-Dependent Ground State Energies, the Optimal Geometric Properties, the Blueshifts of the Fermi Levels, and the Redshifts of the Initial σ-Electronic States for the Stage-n Graphite Alkali-Intercalation Compounds

	Stage-1, alkali-doped graphite compounds				
	LiC6	NaC8	KC8	RbC8	CsC8
Distance between alkali metal and carbon layer (Å)	1.9075	2.3005	2.662	2.9127	3.016
Ground-state energy (eV)	−57.29	−74.96	−75.13	−75.04	−75.12
Blueshift of E_F in band structure (eV)	1.80	1.54	1.48	1.41	1.58
Blueshift of E_F in DOS	1.28−2.20	1.47−1.66	1.47−1.51	1.60−1.64	1.59−1.65
Redshifts of the initial σ-electronic (eV)	5	1.24	1.14	1.13	1.0

	Stage-2, alkali-doped graphite compounds				
	LiC12	NaC16	KC16	RbC16	CsC16
Distance between alkali metal and carbon layer (Å)	2.130/1.839	2.480/2.303	2.820/2.699	2.965/2.869	3.042/3.031
Ground-state energy (eV)	−112.52	−148.78	−148.93	−148.84	−148.94
Blueshift of E_F in band structure (eV)	1.37	1.23	1.17	1.12	1.10
Blueshift of E_F in DOS	1.34−1.38	1.19−1.20	1.15−1.16	1.08−1.09	1.06−1.06
Redshifts of the initial σ-electronic (eV)	1.22	0.99	0.93	0.86	0.82

	Stage-3, alkali-doped graphite compounds				
	LiC18	NaC24	KC24	RbC24	CsC24
Distance between alkali metal and carbon layer (Å)	1.932	2.439	2.680	3.028	3.029
Ground-state energy (eV)	−166.99	−222.64	−222.81	−222.69	−222.79
Blueshift of E_F in band structure (eV)	1.07	1.01	0.98	0.94	0.90
Blueshift of E_F in DOS	0.22−0.23	1.02−1.03	0.57−0.58	0.84−0.84	0.76−0.77
Redshifts of the initial σ-electronic (eV)	1.21	1.02	0.96	0.66	0.60

	Stage-4, alkali-doped graphite compounds				
	LiC24	NaC32	KC32	RbC32	CsC32
Distance between alkali metal and carbon layer (Å)	2.129/1.851	2.454/2.181	2.762/2.649	2.911/2.822	3.073/2.982
Ground-state energy (eV)	−223.15	−289.80	−296.55	−296.42	−296.48
Blueshift of E_F in band structure (eV)	0.89	0.83	0.80	0.77	0.78
Blueshift of E_F in DOS	0.31−0.33	0.63−0.64	0.63−0.63	0.51−0.52	0.41−0.41
Redshifts of the initial σ-electronic (eV)	1.23	0.90	0.88	0.70	0.58

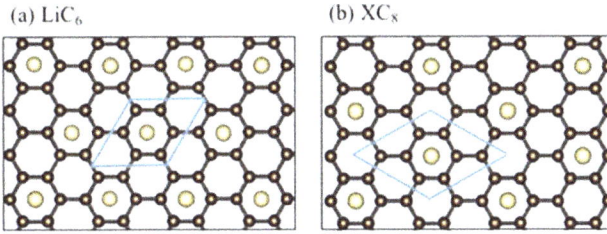

FIGURE 2.2   Planar structures of (a) LiC$_6$ and (b) XC$_8$ (X=Na, K, Rb, and Cs)

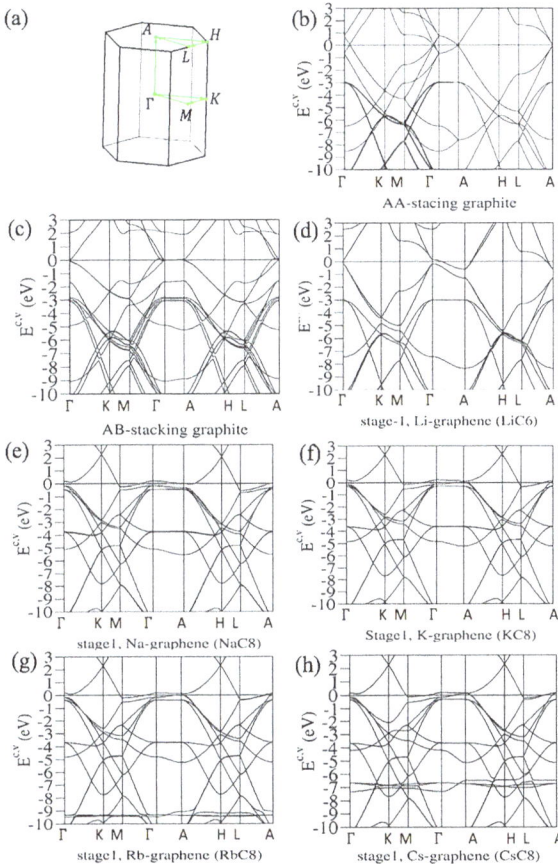

FIGURE 2.3   Rich band structures of the pristine graphite and stage-1 graphite alkali-intercalation compounds (a) within the corresponding first Brillouin zones: (b) AA and (c) AB stackings without intercalations, (d) LiC$_6$, (e) NaC$_8$, (f) KC$_8$, (g) RbC$_8$, and (h) CsC$_8$

The high-resolution experimental measurements of X-ray scatterings and low-energy electron diffractions[42] are able to examine the intercalant distribution, interlayer distance, and stacking configuration of graphite alkali-intercalation compounds; furthermore, those of scanning tunneling microscopy and tunneling electron microscopy are very suitable for the identifications of optimal geometric structures in few-layer graphene system.[49] The previous experiments have shown the AA/AB stacking of the stage-1 LiC_6/AC_8 material, while the predicted AB stacking in stage-2–4 compounds (LiC_{12}, LiC_{18}, and LiC_{24}) requires the further experimental verifications.

2.4 METALLIC AND SEMIMETALLIC BEHAVIORS

Without alkali atom intercalations, a pristine graphite exhibits the unique semimetallic behavior through the weak, but significant van der Waals interactions. It is well known that a monolayer graphitic sheet has the orthogonal π and σ bondings, in which the initial electronic states, respectively, come to exist at the Fermi level ($E_F=0$) and $E \sim -3.10$ eV. Obviously, the low-energy physical properties are mainly determined by the intralayer atomic interactions. Such hopping integrals, being accompanied with the uniform three nearest neighbors are responsible for a gapless and isotropic Dirac cone initiated from the K/K' valleys. The density of states is vanishing at E_F because of the specific Dirac points. A semiconducting zero-gap graphene is dramatically changed into a semimetallic system for any few-layer stackings, or infinite-layer graphites. For example, there exist the unusual overlaps of valence and conduction bands along the KH path (or the ΓA path under the reduced Brillouin zone in Figure 2.3a and b) in AA-, AB- and ABC-stacked graphites through the single-orbital interlayer atomic interactions. According to the VASP and tight-binding model calculations, the simple hexagonal/rhombohedral graphite (Figure 2.3a), with the highest/lowest stacking symmetry, presents the largest/lowest free carrier density of the conduction electrons and valence holes. In general, the π valence bands, which are purely related to the C-$2p_z$, show the initial states at the stable K valley, the saddle-point structure near the M point ($E^v \sim -3$ eV), and its termination in the Γ valley (the whole π-bandwidth more than 7 eV).

Very interesting, the band structures of lithium- and non-lithium graphite intercalation compounds sharply contrast with each other, such as those of stage-1 LiC_6 and XC_8 compounds shown in Figure 2.3d, e, and h, respectively. For the former, the Fermi level is transferred from the middle of the Dirac-cone structures into the conduction ones near the Γ valley after the intercalation of lithium atoms. That is to say, E_F presents a blueshift. Apparently, the free carriers are due to the outmost $2s$ orbitals of lithium atoms. The occupied and unoccupied states are highly asymmetric to each other about the Fermi level. The initial π/π^* valence/conduction states are mainly determined by the stacking configurations in the chemical intercalations (the distribution symmetry of Li-intercalants in Figure 2.2a), or the corresponding relation between the original and reduced first Brillouin zone. There exists an observable energy spacing of $E^c \sim 0.310.63$ eV along the Γ-A path in the modified valence and conduction Dirac-cone structures. The creation of discontinuous states in honeycomb lattices might arise from different ionization energies of Li-$2s$ and C-$2p_z$ orbitals (the distinct on-site Coulomb potential energies). The whole π valence-bandwidth of LiC_6, being created by the C-$2p_z$ orbitals, could be identified to about 7.32/7.40 eV from the electronic energy spectrum along the Γ-M-K-Γ/A-H-L-A (or Γ-K-M-Γ/A-L-H-A) paths. This result clearly illustrates the well-behaved π bondings in graphite alkali-intercalation compounds. Another σ bondings behave so, as indicated from the initial states at the Γ point of $E^v \sim -4.20$ eV. Their orthogonal relation remains the same after the chemical modification, being very useful in establishing the tight-binding model for graphite alkali-intercalation compounds.

The main features of band structure, as clearly indicated in Figure 2.3e–h, are dramatically changed under the other alkali-atom intercalations. Both stage-1 AC_8 and LiC_6 have totally different distribution configurations (Figure 2.2a and b), and so do the reduced first Brillouin zones (Figure 2.3a). For the former, the Dirac-cone structures of the π and π^* bands are initiated from the stable K/K' valleys, but not the Γ ones. The energy spacing of separated Dirac points is small or almost vanishing; furthermore, the Fermi level is situated above the conduction point about ~1.35–1.50 eV. In addition to E_F, whether the second conduction energy sub-band is partially occupied depends on the kinds of alkali atoms, such as the alkali-induced free electrons in the first and second sub-bands for $KC_8/RbC_8/CsC_8$ (Figure 2.3f–h). The low-energy bands belong to the single states, without the split double degeneracy (Figure 2.3d for LiC_6). The whole π-band energy spectra are identified from the K-Γ-K-M-Γ and H-A-H-L-A paths, leading to the width of ~7.31–7.41 eV. Very interesting, one pair of π-valence sub-bands come to exist near $E^v \sim -4.0$ eV, being accompanied with the initial pair of the σ valence bands. This further illustrates the zone-folding effects on band structure and the good separation of π and σ bondings.

Electronic energy spectra strongly depend on the n-stage of graphite intercalation compound, as clearly illustrated in Figure 2.4a–f. The stage-2 LiC_{12} and AC_{16} have different first Brillouin zones on the (k_x, k_y) plane, and their k_z-ranges are associated with the distances between two intercalant planes (Figure 2.1). Compared with stage-1 band structures (Figure 2.3d–h), the number of Dirac cones becomes double in the stage-2 cases, in which the further modifications cover the reduced blueshift of the Fermi level, the enhanced anisotropy, the induced energy spacings between valence and conduction Dirac points, and the diversified energy relations near the band-edge states. For example, four/two valence and conduction pairs come to exist from the Γ/K valley (or the A/H valley) for LiC_{12}/AC_{16} (Figure 2.4d–h). Furthermore, the whole π-bandwidths could be roughly estimated from the ΓKMΓ/KMKΓ path (or the ALHA/HLHA path). The low-energy essential properties are dominated by the π bondings of C-$2p_z$ orbitals. The above-mentioned obvious changes directly reflect the great enhancement of the interlayer atomic interactions due to the C-C bonds in two neighboring graphitic sheets/graphene-intercalant layers. Very interesting, all the stage-2 compounds exhibit a pair of σ bands at the Γ and A valleys, and the energy dispersions along Γ–A are negligible. These results indicate the mutual orthogonality of the planar σ and perpendicular π bondings. Such a phenomenon is expected to survive in any stage-*n* graphite alkali-intercalation compounds.

The blueshifts about the Fermi level and the free conduction electrons decline quickly in the increase in the *n* number. The alkali-concentration is greatly reduced, and do the charge transfer from their outmost *s*-orbitals to carbon $2p_z$-ones. Such results are clearly revealed in the stage-3 systems. Figure 2.5a–f shows the slight modifications of the Dirac-cone structures and the intersecting of the Fermi level with most of the conduction bands. LiC_{18} (Figure 2.5b) and AC_{24} (Figure 2.5c–f), respectively, possess six and three pairs of linear valence and conduction bands near the Γ–A and K–H valleys. Apparently, there exist the significant changes in the observable energy spacing between valence and conduction Dirac points, the anisotropic Fermi

(a)

(b)

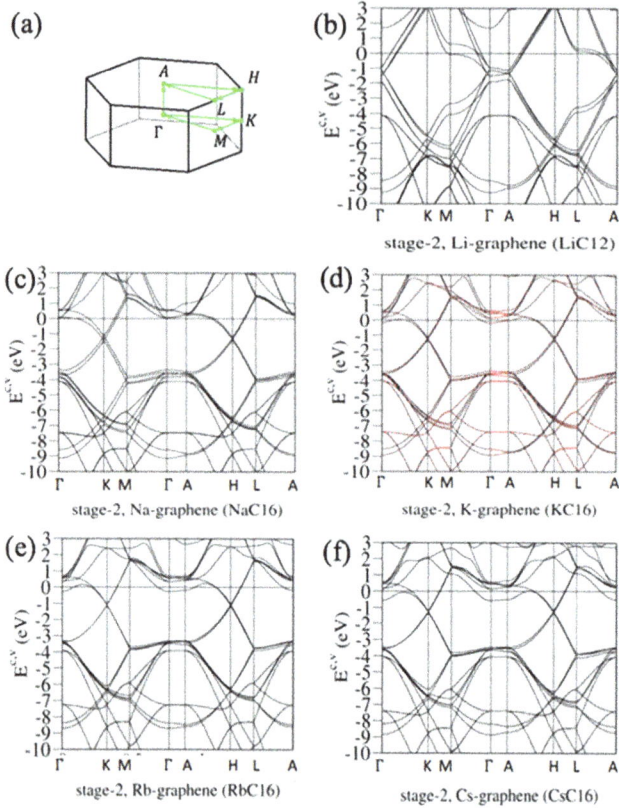

stage-2, Li-graphene (LiC12)

(c)

(d)

stage-2, Na-graphene (NaC16)

stage-2, K-graphene (KC16)

(e)

(f)

stage-2, Rb-graphene (RbC16)

stage-2, Cs-graphene (CsC16)

FIGURE 2.4 Unusual electronic energy spectra of the AB-stacked stage-2 graphite alkali-intercalation compounds: (a) the reduced first Brillouin zones of AB stacking, (b) LiC_{12}, (c) NaC_{16}, (d) KC_{16}, (e) RbC_{16}, and (f) CsC_{16}

velocities, and the distinct Fermi momenta slopes. The blueshifts of E_F are estimated to be 0.45, 0.58, 0.55, 0.53, and 0.5 eVs. Very interesting, the electronic energy spectra become more complicated under the stronger zone-folding effects. The main features of electronic properties are closely related to the weak, but important Li-C bonds, e.g., the minor contributions of alkali atoms on each of the electronic states. That is, the diversified n-type dopings are created by the critical Li-C and C-C bondings.

The conduction electron density, which is created by the alkali-atom doping, is worthy of a closer examination. It is mainly determined by the covered volume (area/length) in the wave-vector space between the lowest conduction state and the Fermi level for 3D graphite alkali-intercalation compounds[53] (2D alkali-adsorbed graphenes/1D graphene nanoribbons). In general, the VASP calculations are too heavy to obtain the delicate values because of the strong anisotropic energy dispersions (the nonlinear Dirac-cone structure at higher energies in Figures 2.3h and 2.4e). That a lot of 3D **k**-points are required for the accurate calculation of free electron density is almost impossible. The similar phenomenon is revealed in the alkali-adatom adsorptions on graphene. However, 1D graphene nanoribbons are very

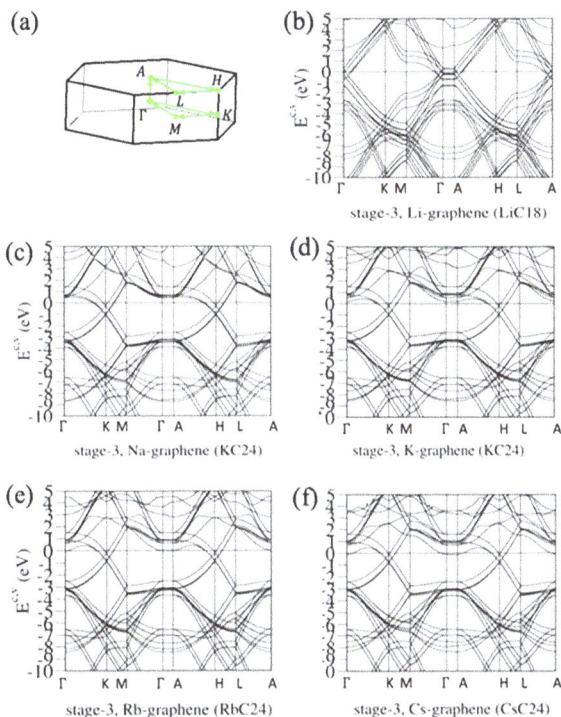

FIGURE 2.5 Rich band structures for the AB-stacked stage-3 graphite alkali-intercalation compounds: (a) the reduced first Brillouin zone, (b) LiC_{18}, (c) NaC_{24}, (d) KC_{24}, (e) RbC_{24}, and (f) CsC_{24}

suitable for the full exploration of charge transfer and conduction electron density, since the Fermi momenta are proportional to the latter. After examining the critical mechanisms, namely the zigzag/armchair edges, different widths, five kinds of alkali atoms, and their various distributions, the free electron density is almost identical to the adatom concentration.[54] It can be deduced that any one alkali adatom contributes the outmost half-occupied s-orbital as a conduction carrier, or it behaves the full charge transfer of $-e$ in the alkali-carbon bonds. Such conclusions might be suitable for 2D and 3D alkali-doped graphite compounds, being supported by the following result. According to the spatial charge distributions and their variations after the chemical modifications, the chemical bondings hardly depend on the distributions and concentrations of alkali adatoms.

Generally speaking, only the high-resolution ARPES is capable of detecting the wave-vector-dependent occupied energy spectra below the Fermi level. The up-to-date experimental measurements have successfully verified the diversified phenomena for the graphene-related sp^2-bonding systems, as clearly observed under the various dimensions, layer numbers, stacking configurations, substrates, and adatom/molecule chemisorptions. For example, there exist 1D parabolic dispersions with sufficiently wide energy spacings/band gaps in graphene nanoribbons, the linear Dirac cone for monolayer graphene, two parabolic valence bands in bilayer AB stacking, the linear and parabolic dispersions in twisted bilayer systems, one linear and two parabolic bands for tri-layer ABA stacking, the linear, partially flat and Sombrero-shaped dispersions in tri-layer ABC stacking, the substrate-created observable energy spacing between the separated valence and conduction bands in bilayer AB stacking/oscillatory bands in few-layer ABC stacking,[2] the semimetal-semiconductor/semimetal-metal transitions after the adatom/molecule adsorptions graphene surface, and the bilayer- and monolayer-like energy dispersions, respectively, at the K and H corner points (Figure 2.3a) for the Bernal graphite. The ARPES measurements are available in examining the main features of the occupied band structure in graphite alkali-intercalation compounds, namely the blueshift of the Fermi level, the initial K or Γ valley for the π and π^* bands, the separated or almost close Dirac-cone structures, their double or single degeneracy, the whole π band along the Γ-K-M-Γ A-L-H-A paths (or the K-Γ-K-M-Γ/H-A-H-L-A paths), the formation of σ sub-bands at $E \sim -4.0\,eV$ from the Γ valley, and the sensitive dependences on the alkali-atom distributions and concentrations. The experimental verifications could provide useful information about the zone-folding effects due to the alkali interaction, the single-orbital hybridizations in alkali-carbon bonds, and the almost full charge transfer (discussed in the previous paragraph).

The density of states at E (DOS(E)), the number of states within a very small energy range of dE, is available in thoroughly comprehending the valence and conduction energy spectra simultaneously. In general, its special structures are created by the band-edge states with the vanishing group velocities. The critical points in the energy-wave-vector space include the local extreme points (minima and maxima), the saddle points, and the partially flat energy dispersions. Apparently, the van Hove singularities are greatly diversified under different dimensions, e.g., the important differences in graphite- and graphene-related systems. Of course, the discrete energy levels in 0D quantum dots only present the delta-function-like peaks, as observed in the magnetically quantized Landau levels of the emergent layered materials (few-layer graphenes in Refs), and any-dimensional dispersionless energy bands. The 1D parabolic and linear bands, respectively, create the asymmetric divergent peaks in the square-root form and the plateau structures. As for the 2D condensed matter systems, the concave-/convex-form energy dispersions, the Dirac-cone structure (the linear valence and conduction bands without energy spacing), the saddle points, and the constant-energy loops, respectively, induce the broadening shoulders, V-shapes, the symmetric peaks related to the logarithmical divergence, and the square-root divergent peaks. Moreover, the higher dimension in 3D materials makes them only show the symmetric/asymmetric peaks and shoulder structures, mainly owing to the absence of quantum confinement. The latter come from the 3D parabolic energy dispersions.

The delicate calculations and analyses, which are conducted on the atom- and orbital-dependent density of states (Figure 2.6al), are capable of fully understanding the metallic behavior and the close relations among the different chemical bondings. Although band structures, with many band-edge states, become very complicated under the alkali-atom intercalations, the main features of van Hove singularities are sufficiently clear for the identifications of diversified phenomena through a suitable broadening factor (e.g., energy width of ~0.10 eV). This further illustrates a critical role of zone folding in the fundamental properties. Very interesting, the low-energy

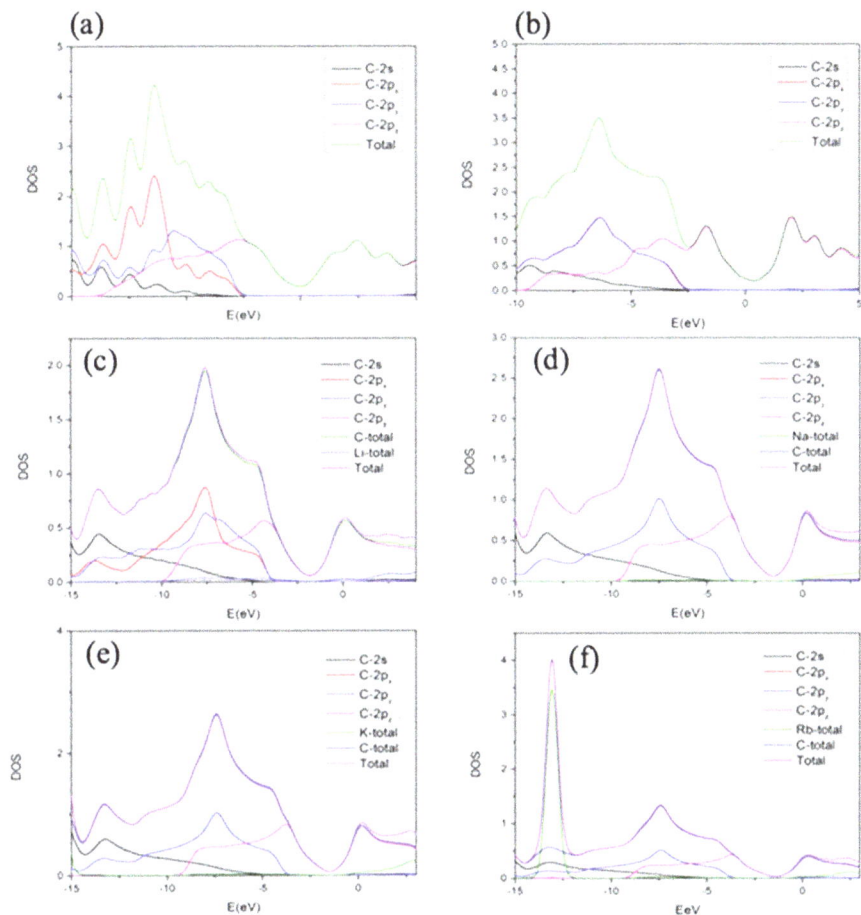

FIGURE 2.6 Atom- and orbital-decomposed density of states for the (a) AA- and (b) AB-stacked graphite systems, (c) LiC_6, (d) NaC_8, (e) KC_8, and (f) RbC_8. The black, red, lig and ht blue, pink, green, deep blue, and purple curves, respectively, C-2s, C-2p_x, C-2p_y, C-2p_z, alkali, carbon, and compound

DOSs in stage-1 and stage-2 graphite alkali-intercalation compounds present a prominent peak just at the Fermi level (E_F=0), regardless of the kind of alkali atoms. Furthermore, there exists a valley structure, with a minimum value, at its left-hand neighbor. When such characteristic is combined with the similar ones at E_F) in the pristine AA- and AB-stacked graphite systems (Figure 2.6a and b), the Fermi level is deduced to exhibit a blueshift. That is to say, E_F is situated at the conduction energy sub-bands (E_F roughly lies in the center of valence and conduction bands) roughly after (before) the alkali-atom intercalations. It should be noticed that the contributions due to the alkali atoms are weak, but rather important. The blueshifts of the stage-1 LiC_6, NaC_8, KC_8, RbC_8, and CsC_8 and stage-2 LiC_{12}, NaC_{16}, KC_{16}, RbC_{16} and CsC_{16} are, respectively, estimated to be 1.90, 1.80, 1.70, 1.60, 1.50, 1.30, 1.20, 1.10, 1.00, and 0.90 eVs.

The van Hove singularities in graphite intercalation compounds (Figure 2.6c–l), which survive in the specific energy ranges, are mainly determined by the carbon or alkali atoms and their orbitals. Most important, DOS in the critical energy range of -5.0 eV $< E < 3.0$ eV, being relatively easily examined from the experimental STS measurements, are dominated by the carbon-$2p_z$ orbitals (the pink curves). Furthermore, the minor contributions related to the outmost s-orbital of alkali atoms (the green curves), especially at the conduction energy spectrum, play a critical role in determining the blueshift of the Fermi level. As for the C-($2p_x$, $2p_y$) orbitals (the red and light blue curves), their contributions are initiated from $\sim E < -4.0$ eV, while they are absent in the opposite energy range. The redshift of the initial σ valence bands is about 1.0 eV, compared with those of the pristine simple hexagonal and Bernal graphites (Figure 2.6a and b). Only the LiC_6 and LiC_{12} cases (Figure 2.6c and h) exhibit the split contributions of $2p_x$ and $2p_y$ orbitals. This result directly reflects the anisotropic distribution configuration. Moreover, the C-$2s$ orbitals come to exist at the deeper energies of $\sim E < -5.0$ eV. Apparently, the above-mentioned features indicate the good separation of C-$2p_z$ and C-($2s$, $2p_x$, $2p_y$) orbital contributions and thus the normal perpendicular orbital hybridizations of π and σ chemical bonds. That such bonding behavior is strongly linked with the significant interlayer $2p_z$-orbitals due to the carbon-alkali bonds can account for the featured electronic properties, e.g., the main features of band structures and DOSs in Figures 2.3–2.7.

The clear identifications of stage-1 and stage-2 of graphite alkali-intercalation compounds could be achieved from the low-energy features of van Hove singularities, as indicated in Figure 2.6c–l, respectively. Compared with those of the former, the blueshifts of the Fermi levels are relatively small, in which they are, respectively, ~ 1.20, 1.15, 1.10, 1.06, and 1.03 eV for LiC_{12}, NaC_{16}, KC_{16}, RbC_{16}, and CsC_{16}. Most important, their densities of states at E_F do not belong to the local maxima. Such a result directly reflects whether the band-edge states of conduction bands are somewhat away from the Fermi level (Figure 2.4b–f). This significant difference between stage-2 and stage-1 systems further illustrates the intersecting of E_F and conduction bands, thus leading to the n-type doping cases after the alkali-atom intercalations. The experimental measurements could be utilized to verify the free conduction electrons due to the alkali-atom intercalations, e.g., the atom-dependent optical threshold absorption frequencies, and the doping-enhanced electrical conductivities.

The high-resolution STS measurements are the only method in examining the van Hove singularities due to the valence and conduction energy spectrum, especially for the semimetallic, metallic, or semiconducting behaviors near the Fermi level. Such examinations cover the form, energy, number, and intensity of special structures in the density of states. Apparently, they cannot identify the wave-vector-dependent energy dispersions. The up-to-date measured results have successfully verified the diverse electronic properties in graphene-related systems with the dominating sp^2 bondings, e.g., 1D graphene nanoribbons, carbon nanotubes, 2D few-layer graphenes, adatom-adsorbed graphenes, and 3D graphite. The main features of graphene nanoribbons, the width- and edge-dependent energy gaps and the square-root-divergent asymmetric peaks of 1D parabolic dispersions, are confirmed from the precisely defined boundary structures. The similar strong peaks are displayed in seamless carbon nanotubes,

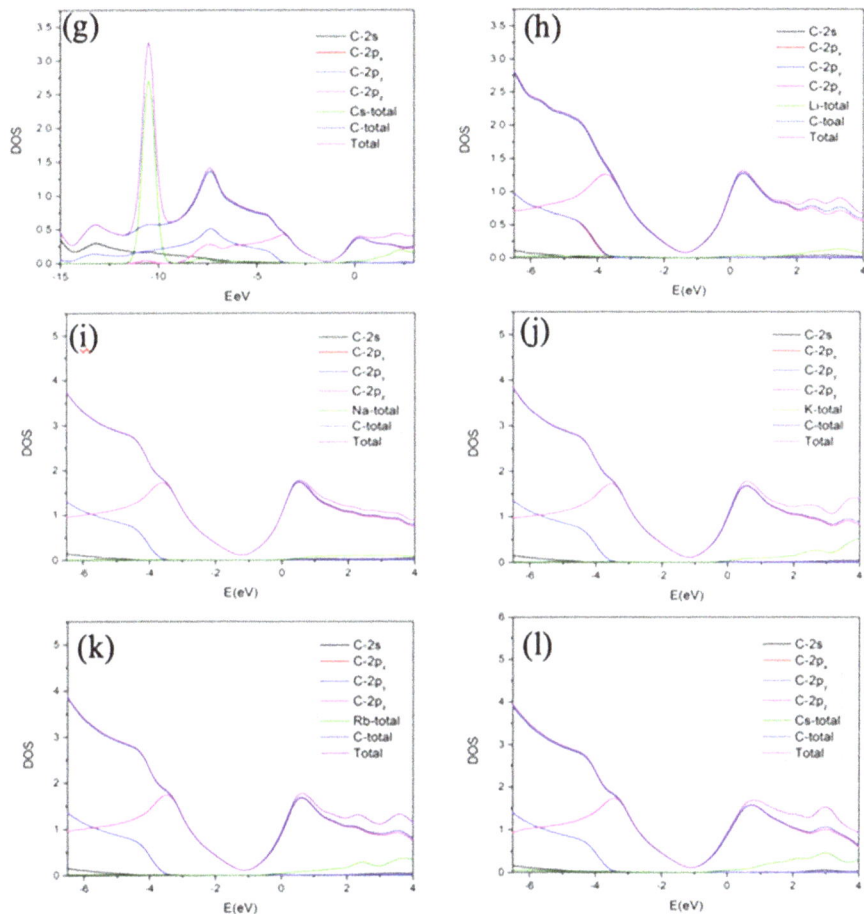

FIGURE 2.7 Atom- and orbital-decomposed density of states for the (g) CsC_8, (h) LiC_{12}, (i) NaC_{16}, (j) KC_{16}, (k) RbC_{16}, and (l) CsC_{16}. The black, red, light blue, pink, green, deep blue, and purple curves, respectively, C-2s, C-$2p_x$, C-$2p_y$, C-$2p_z$, alkali, carbon, and compound

in which these structures exhibit the chirality- and radius-dependent band gaps and energy spacings between two neighboring sub-bands. Even with the obvious curvature effects on a cylindrical surface (the misorientation of $2p_z$ orbitals and the significant hybridization of carbon four orbitals), armchair nanotubes belong to 1D metals with a finite density of states at the Fermi level (a sufficiently high free carrier density). A lot of STS measurements, which are made for few-layer and adatom-doped graphenes, clearly illustrate the low-lying characteristics of van Hove singularities: a V-shaped energy dependence vanishing at the Dirac point in monolayer system (a zero-gap semiconductor), the asymmetry-induced peak structures in the logarithmic form for twisted bilayer graphenes, a gate-voltage-created band gap in bilayer AB stacking and tri-layer ABC stacking, a delta-function-like peak centered about the Fermi level related to surface states of the partially flat bands in tri-layer and penta-layer ABC stackings, a sharp dip structure at E_F accompanied with a pair of asymmetric peaks in tri-layer AAB stacking (a narrow-gap semiconductor with the low-lying constant-energy loops), and a redshift of Dirac point arising from the n-type electron doping of Bi adatoms. The measured density of states in Bernal graphite is shown to be finite near E_F characteristic of the 3D semimetallic property and presents the splitting of π and π^* strong peaks at deeper/higher energies. It should be noticed that the well-defined π and σ bandwidths in graphene-related systems are worthy of the further STS examinations. In addition, there are no STS results for AA- and ABC-stacked graphites up to now, mainly owing to few content of natural graphite and the difficulties in sample growth. The experimental examinations on the current predictions, especially for the van Hove singularities near E_F and the π and σ bandwidths, are very useful in providing the detailed information about the n-type dopings and the orbital hybridizations of chemical bonds.

The close relations between the VASP simulations[30] and the tight-binding model/the effective-mass approximation[17,35] in graphite-/graphene-related systems are discussed in detail. The previous theoretical studies show that the electronic properties of pure carbon honeycomb lattices are well characterized by the phenomenological model. According to the calculated results of two different methods, the rich and unique band structures are created by the stacking configuration, layer number, and dimensionality, such as those due to monolayer graphene, bilayer AA/AB stackings,[55] trilayer AAA/ABA/ABC/AAB stackings,[56,57] 3D AA-/AB-/ABC-stacked graphites,[11–13] 1D achiral/chiral carbon nanotubes,[58,59] and graphene nanoribbons. Their outstanding consistence is deduced to be closely related to the good separations of π and σ chemical bondings and the interlayer single-$2p_z$-orbital hybridizations. Very interesting, the generalized tight-binding model,[5] which is combined with the static and dynamic Kubo formulas[60] and the modified random-phase approximation, is capable of fully exploring the diversified magnetic quantization. As for the emergent 2D group-IV and group-V materials, these theoretical frameworks are successful in studying the unusual magneto-electronic properties, the various magneto-optical selection rules, the unique quantum Hall conductivities, and complex magneto-plasmon modes and inter-Landau-level excitations.[61]

The low-energy physical properties of graphites and graphite alkali-intercalation compounds are dominated by the single-orbital interactions of C-$2p_z$ and A-s orbitals (the outmost one of alkali atom). The AA-, AB-, and ABC-stacked graphites[11–13] have two, four, and six carbon atoms in a unit cell, so their Hamiltonians are built from the corresponding tight-binding functions. After the diagonalization of Hermitian matrices, the π- and π^*-electronic energy spectra strongly depend on the stacking configurations, especially for the overlap of valence and conduction bands. The highest conduction electron/valence hole density is revealed in the first/third system. Apparently, the stacking-enriched interlayer hopping integrals of C-$2p_z$ orbitals are responsible for the diversified semimetallic properties. Most important, the above-mentioned phenomenological methods could cover the alkali intercalation effects, the enlarged unit cells, and the stacking configurations. For example, the modified Hamiltonians of graphite lithium-intercalation compounds include the ionization energy difference between C-$2p_z$ and Li-$2s$ orbitals, the intralayer nearest-neighboring interactions of $2p_z$-$2p_z$/$2s$-$2s$ orbital in C-C/Li-Li bonds, and the interlayer $2p_z$-$2s$/$2p_z$-$2p_z$ orbital hybridizations. Specifically, the third ones might be complicated and very sensitive to the change in the intercalant concentration and arrangement; therefore, they play a critical role in creating the diversified metallic behaviors. It should be noticed that the theoretical models, with the zone-folding effects, are absent up to now.

The X-ray diffraction[62] is available for thorough investigations of crystalline materials through the atomic lattices. The incident photons, which possess a wavelength comparable to/less than the lattice constants, experience the elastic scatterings with the periodic atom distribution and thus create the constructive and destructive patterns. In this process, the X-rays are scattered elastically by the carrier charge density and create the so-called Thomson scatterings. The valence electrons oscillate like a Hertz dipole at the frequency of incident photons and serve as a source of dipole radiation. The wavelength λ of X-rays is conserved during Thomson scatterings. The spectral intensity at distance R could be roughly expressed as $I(R) = I_0(r_e/R)(1 + \cos^2 2\theta/2)$ in the absence of detailed atomic nuclei, where $r_e = (\mu_0/4\pi)(e^2/m)$ is the classical radius of electron $(2.818 \times 10^{-15}\,\text{m})$,[63] and it also becomes an evidence why the scattering from atomic nuclei has not been considered in the detailed derivations. In fact, the intensity equation would hold even in the presence of specific scatterings from the atomic nuclei, since the nuclei components will only yield a $<10^{-6}$ contribution compared to that due to electrons. The significant difference simply lies in quasi-particle mass: a ratio, at least, of ~2,000, between an electron and a nucleus. The X-ray scatterings from nuclei could be neglected thoroughly. Figure 2.8 illustrates such an experiment where the crystal, e.g., a simple cubic system, is irradiated with monochromatic X-rays of wavelength λ. Under the case considered here, each atom is surrounded by six neighbor atoms at distance a and the angle related to two atomic bonds is always 90° or multiples of it. One is able to characterize the critical scattering vector $Q = K - K_0$, the wave vector difference between the incident and reflected photons, and then utilizes the geometry of scattering vector construction to derive $Q = 2k\sin\theta$. The magnitude of Q at $I(R) \to$ max is further obtained from three Laue conditions[64] by multiplying the inverse cell parameter $1/a$, adding the squares, and taking the square root. This generates the maximum-diffraction intensity corresponding to $|Q|/2\pi = \sqrt{h^2 + k^2 + l^2}/a$. The direct combination of two equations leads to the well-known Bragg diffraction condition.[65] The observable intensities are examined to only occur for certain orientations of momentum transfers. That is, the strongest diffraction appears when Q is equal to the reciprocal lattice vector. The high-resolution measurements of X-ray diffraction spectra are capable of providing two important points: the specific angle of θ and intensity of I. In general, there exist the single- and poly-crystal perfect structures, as identified in a lot of experimental observations. The former and the latter could be, respectively, verified from Laue/rotation-oscillation/Weissenberg/precession methods[66–69] and Debye-Scherrer/Guinier techniques.[58] Specifically, the up-to-date measurements of X-ray diffraction spectra on the geometric properties of stage-*n* alkali graphite intercalation compounds approximately agree with the theoretical predictions in Table 2.1, e.g., the measured and calculated interlayer distances in LiC_6, 3.706 and 3.815 Å[70].

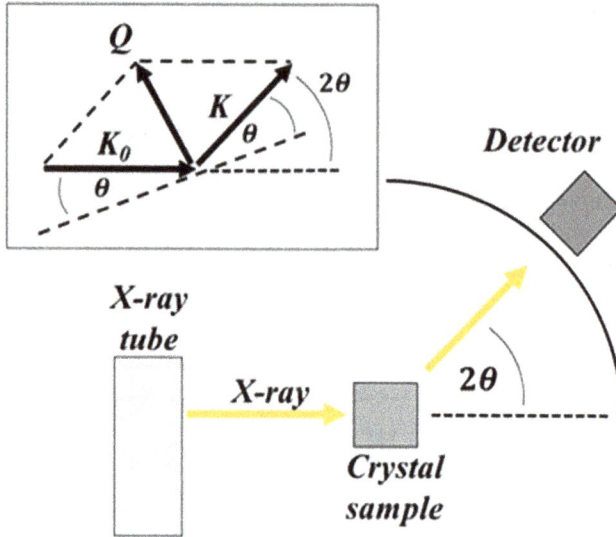

FIGURE 2.8 Scattering of X-ray by a crystallite of sample

The ARPES[71] is a direct experimental technique to observe the distribution of the occupied electronic states in the reciprocal space. This equipment, a refinement of ordinary photoemission spectroscopy, is available for studying photoemission of excited electrons from a sample usually achieved by illumination with soft X-rays. Figure 2.9 clearly displays the experimental process of ARPES measurements. Photoelectrons are stimulated by incident photons and escape from the material surface into the vacuum; furthermore, they are counted by an angle-resolved energy analyzer. A significant feature of the X-ray photoemission spectroscopy analysis is that this technique is one of the ultimate surface-sensitive methods for analyzing the elements that are present within 1–10 nm below the sample surface. The photoelectron

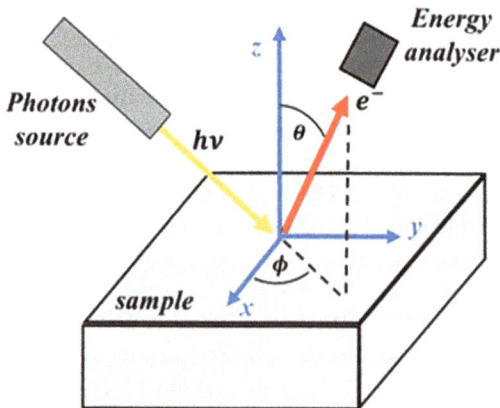

FIGURE 2.9 Angle-resolved photoemission spectroscopy for detecting the photoelectrons

momentum is characterized by $\sqrt{2mE_p}$, where m is the bare electron mass and E_p is its kinetic energy. The momentum components, which are parallel and perpendicular to the sample surface, are determined by the polar angle and azimuthal angle (θ and ϕ, respectively, in Figure 2.8). The total energy and the parallel-component momentum are conserved during the kinetic photoemission process, while the perpendicular-component momentum does not behave so. This is due to the breaking of translational symmetry along the normal direction. Generally speaking, the ARPES measurements are mainly focused on the occupied bands with energy dispersions along the high symmetry points in the (k_x, k_y)-space, e.g., the low-lying occupied states of graphene-related systems near the K valley. In addition, they provide energy widths of valence electrons and thus quasiparticle decay rates, such as the Coulomb and electron-phonon de-excitation scatterings.[72] Improvements in energy and angular resolutions are responsible for advancing this apparatus into a precision tool. Up to now, the highest resolutions for energy and angular distributions are, respectively, ~1 meV and 0.1° in the UV regime. ARPES is the most powerful technique in examining/identify/verifying the wave-vector-dependent electronic structures. The experimental measurements, which are conducted on graphene-related systems,[73–75] have successfully confirmed the feature-rich band structures under different dimensions, layer numbers, stacking configurations, and adatom/molecule chemical adsorptions and substitutions. The many earlier ARPES experiments delicately identify the important blueshifts of the Fermi level in alkali-atom graphite intercalation compounds, being consistent with the VASP calculations (Table 2.1). For example, the Fermi level above the Dirac point is ~1.48 and ~1.35 eV in KC_8[76,77] under the experimental and theoretical studies. The Fermi level is largely enhanced into the conduction Dirac cone, so there is a high 3D free conduction electron density. Such phenomena also indicate the almost full charge transfer from the outmost alkali *s*-orbital to the carbon $2p_z$ ones.[78,79]

In addition to geometric and electronic properties, free conduction electrons/valence holes in graphite intercalation compounds are able to create the rich and unique essential properties, such as optical absorption spectra,[28] single- and many-particle Coulomb excitations,[23–27] and electrical conductivities. Apparently, the Fermi level at the conduction/valence Dirac-cone is responsible for the optical threshold frequency of $2E_F$ under the vertical transitions of valence and conduction bands. Specifically, the many-body excitonic effects might come to exist under suitable conditions, as revealed in certain optical measurements. They would reduce the optical gap and greatly enhance the initial absorption structure. It is well known that free charge carriers have the outstanding Coulomb response under the dynamic and static perturbations of electron-electron interactions. As a result of the conduction and valence electrons (the π^* and π carriers), there exist the diverse (momentum, frequency) excitation phase diagrams. That is to say, certain intra- and interband electron-hole regions and different plasmon modes might appear simultaneously, strongly depending on the Fermi level/carrier density and the band structure. Moreover, the Fermi surface, which is formed by the various Fermi-momentum conduction/valence states, is predicted to generate the Friedel charge density oscillation at the long distance. Apparently, it has a significant effect on transport property, e.g., the residual resistivity related to the electron-impurity elastic scatterings.

The graphite-related materials quite differ from the 3D ternary lithium oxide systems in terms of the Li^+-based battery anode. There are certain important differences between the layered LiC_6/Li^+C_6 and $Li_4Ti_5O_{12}$ bulk compounds. The former are metals/semimetals, while the latter belongs to a sufficiently large direct-gap insulator with $E_g = 2.98$ eV at the Γ point. The interlayer spacings of graphitic sheets, which are determined by the van der Waals interactions of C-$2p_z$ orbitals, are available for the Li-atom/Li^+-ion intercalations and de-intercalations during the charging and discharging processes. Maybe, the heterojunction of graphite and electrolyte becomes an issue in greatly enhancing the performance of ion transport. This problem is under the current investigations. But for the latter, there are a lot of chemical bonds within a unit cell, in which the Li and TiO bond lengths exhibit the large modulation over 20%. It is deduced to be very easy to change the internal geometric structures (or create the intermediate configurations). The dramatic transformation between the quasistable and intermediate geometries is worthy of the further systematic studies. Of course, the boundary of two distinct lithium oxides needs to be examined thoroughly. Additionally, the semimetallic, metallic, and semiconducting properties, being associated with electrons and holes, might not be played important roles in ion-transport batteries.

In addition to the lithium-ion-based batteries, there exist the aluminum-chlorine-ion-induced ones.[80,81] The AB-stacked graphite in the second system could serve as the efficient cathode material under the anion transport. Very interesting, the coexistence of $AlCl_4^-$ and $Al_2Cl_7^-$ anions[82] will determine the optimal geometric symmetries between two neighboring graphitic layers and thus the low-lying energy valence and conduction bands. How to mix together for two kinds of large molecular ions could be thoroughly explored by the first-principles calculations, e.g., the diversified distribution configurations and concentrations associated with the specific relative ratios. As to the cathode and anode systems of graphite intercalation compounds, their fundamental properties are expected to be totally different from each other in terms of the geometric, electronic, magnetic, optical, and transport properties. The intercalations and de-intercalations, which are driven by the aluminum- and chloride-related anions, should be one of the non-negligible transport mechanisms in the cathode material of graphite. For example, such behaviors would be responsible for initiating the ionic currents and dominate the stationary ion flows.

2.5 CONCLUDING REMARKS

The fundamental properties of (Li, Na, K, Rb, and Cs)-intercalated graphite compounds under the distinct stage configurations have been investigated by means of the first-principles calculations. The weak, but significant van der Walls interactions, which arise from the interlayer $2p_z$-$2p_z$ and $2p_z$-s orbital hybridizations in C-C and C-A bonds, respectively, make the most important contributions to the low-lying π-electronic structure and thus dominate the essential physical properties. The dramatic changes cover the blueshift of the Fermi level/the redshift of the σ bands (the n-type doing behaviors), the greatly enhanced asymmetric electron and hole energy spectra, the obviously reduced conduction electron density for the dilute intercalant cases, the energy spacing between valence and conduction Dirac cones, the initial K/Γ valleys for the π-electronic state with the single/double degeneracy, the whole π

band along the K-Γ–K-M-Γ/Γ–K-M-Γ paths (or the H-A-H-L-A/A-L-H-A paths), and a pair of σ sub-bands due to carbon-($2p_x$, $2p_y$) orbitals at the Γ valley of $E \sim -4.0\,eV$. The stable layered structures and the metallic/semimetallic properties are responsible for the important differences with the Li^+-based battery anode of the insulating bulk $Li_4Ti_5O_{12}$. The alkali charge that transfers to the carbon sp^2 honeycomb lattices is deduced to be large, being comparable with that in 1D graphene nanoribbons.[65] That is to say, the free electron density is roughly identical to the alkali-atom concentration. Conduction electrons/valence holes have been predicted to strongly affect the other essential properties, e.g., the great enhancement of the optical threshold frequency, and the various (momentum, frequency) excitation phase diagrams of Coulomb excitations with the rich single- and many-particle modes.[35–39] Most important, the tight-binding model, which could be built from the low-lying VASP band structures under the simultaneous considerations of intralayer CC and AA bondings and interlayer C-C and A-C interactions, would be very useful in fully exploring the diversified phenomena, e.g., the rich magnetic quantization[31–34] in layered systems only available by this method.

This work is supported by the Hi-GEM Research Center and the Taiwan Ministry of Science and Technology under Grant No. MOST 108–2212-M-006–022-MY3.

REFERENCES

1. Lee, S.-M., Kang, D.-S & Roh, J.-S. Bulk graphite: Materials and manufacturing process. *Carbon Letters*, 2015, **16**, 135–146.
2. Tran, N. T. T., Lin, S. Y., Lin, C. Y. & Lin, M. F. *Geometric and Electronic Properties of Graphene-Related Systems: Chemical Bonding Schemes.* CRC Press, Boca Raton, FL, 2017. ISBN: 9781351368483.
3. Wang, Y., Panzik, J. E., Kiefer, B. & Lee, K. K. Crystal structure of graphite under room-temperature compression and decompression. *Scientific Reports*, 2012, **2**, 520.
4. Ho, J. H., Chang, C. P. & Lin, M. F. Electronic excitations of the multilayered graphite. *Physics Letters A*, 2006, **352**, 446–450.
5. Lin, C. Y., Chen, R. B., Ho, Y. H. & Lin, M. F. *Electronic and Optical Properties of Graphite-Related Systems.* CRC Press, Boca Raton, FL, 2017. ISBN: 9781138571068.
6. Zhang, Z., Huang, H., Yang, X. & Zang, L. Tailoring electronic properties of graphene by π-π stacking with aromatic molecules. *The Journal of Physical Chemistry Letters*, 2011, **2**, 2897–2905.
7. Cheng, Q. et al. Graphene-like-graphite as fast-chargeable and high-capacity anode materials for lithium ion batteries. *Scientific Reports*, 2017, **7**, 14782.
8. Mecklenburg, M. & Regan, B. Spin and the honeycomb lattice: Lessons from graphene. *Physical Review Letters*, 2011, **106**, 116803.
9. Gómez-Santos, G. Thermal van der Waals Interaction between graphene layers. *Physical Review B*, 2009, **80**, 245484.
10. Chung, D. D. L. Graphite review, *Journal of Materials Science*, 2002, **37**(8), 1475–1489.
11. Lin, C.-Y., Wu, J.-Y., Chiu, C.-W. & Lin, M.-F. *AA-Stacked Graphenes.* CRC Press, Boca Raton, FL, 2019, ISBN: 9780429277368.
12. Lin, C.-Y., Wu, J.-Y., Chiu, C.-W. & Lin, M.-F. *AB-Stacked Graphenes.* CRC Press, Boca Raton, FL, 2019, ISBN: 9780429277368.
13. Lin, C.-Y., Wu, J.-Y., Chiu, C.-W. & Lin, M.-F. *ABC-Stacked Graphenes.* CRC Press, Boca Raton, FL, 2019, ISBN: 9780429277368.

14. Negishi, R., Wei, C., Yao, Y., Ogawa, Y., Akabori, M., Kanai, Y., Matsumoto, K., Taniyasu, Y. & Kobayashi, Y. Turbostratic stacking effect in multilayer graphene on the electrical transport properties. *Physica Status Solidi*, 2019, **257**(2). doi: 10.1002/pssb.201900437.
15. Natori, A., Ohno, T., & Oshiyama, A. Work function of alkali-atom adsorbed graphite. *Journal of the Physical Society of Japan*, 1985, **54**, 3042–3050.
16. Mao, C. et al. Selecting the best graphite for long-life, high-energy Li-ion batteries. *Journal of the Electrochemical Society*, 2018, **165**, A1837–A1845.
17. Cheng, Q. Graphene-like-graphite as fast-chargeable and high-capacity anode materials for lithium ion batteries. *Scientific Reports*, 2017, **7**, 14782.
18. Lin, S.-Y., Tran, N. T. T., Chang, S.-L., Su, W.-P. & Lin, M. F. *Structure- and Adatom-Enriched Essential Properties of Graphene Nanoribbons*. CRC Press, Boca Raton, FL. ISBN: 9780367002299.
19. Ho, J. H., Lai, Y. H., Tsai, S. J., Hwang, J., Chang, C. & Lin, M. F. Magnetoelectronic properties of a single-layer graphite (condensed matter: Electronic structure and electrical, magnetic, and optical properties). *Journal of the Physical Society of Japan*, 2006, **75**, 21.
20. Huang, Y. K., Chen, S. C., Ho, Y. H., Lin, C. Y. & Lin, M. F. Feature-rich magnetic quantization in sliding bilayer graphenes. *Scientific Reports*, 2014, **4**, 7509.
21. Koshino, M. & McCann, E. Landau level spectra and the quantum Hall effect of multilayer graphene. *Physical Review B*, 2011, **83**, 165443.
22. Wang, Z. F., Liu, F. & Chou, M. Y. Fractal Landau-level spectra in twisted bilayer graphene. *Nano Letters*, 2012, **12**, 3833–3838.
23. Ho, J. H., Chang, C. P. & Lin, M. F. Electronic excitations of the multilayered graphite. *Physics Letters A*, 2006, **352**, 446–450.
24. Lin, M. F., Chuang, Y. C. & Wu, J. Y. Electrically tunable plasma excitations in AA-stacked multilayer graphene. *Physical Review B*, 2012, **86**, 125434.
25. Wu, J. Y., Chen, S. C., Roslyak, O., Gumbs, G. & Lin, M. F. Plasma excitations in graphene: Their spectral intensity and temperature dependence in magnetic field. *ACS Nano*, 2011, **5**, 1026–1032.
26. Lozovik, Y. E. & Sokolik, A. A. Influence of Landau level mixing on the properties of elementary excitations in graphene in strong magnetic field. *Nanoscale Research Letters*, 2012, **7**, 134.
27. Roth, L. M., Pratt, G. W., Jr. A many-body treatment of dielectric screening for impurity states and excitons in semiconductors. *Journal of Physics and Chemistry of Solids*, 1959, **8**, 47–49.
28. Ain, Q. T., Al-Modlej, A., Alshammari, A. & Anjum, M. N. Effect of solvents on optical band gap of silicon-doped graphene oxide. *Materials Research Express*, 2018, **5**, 035017.
29. Leenaerts, O., Partoens, B. & Peeters, F. M. Adsorption of H_2O, NH_3, CO, NO_2, and NO on graphene: A first-principles study. *Physical Review B*, 2008, **77**, 125416.
30. De Padova, P., Quaresima, C., Ottaviani, C., Sheverdyaeva, P. M., Moras, P., Carbone, C., Topwal, D., Olivieri, B., Kara, A. & Oughaddou, H. Evidence of graphene-like electronic signature in silicene nanoribbons. *Applied Physics Letters*, 2010, **96**, 261905.
31. Rufieux, P., Cai, J., Plumb, N. C., Patthey, L., Prezzi, D., Ferretti, A., Molinari, E., Feng, X., Mullen, K. & Pignedoli, C. A. Electronic structure of atomically precise graphene nanoribbons. *ACS Nano*, 2012, **6**, 6930–6935.
32. Kano, S., Tadaa, T. & Majima, Y. Nanoparticle characterization based on STM and STS. *Chemical Society Reviews* 2015, **44**, 970.

33. Berdiyorov, G., Neek-Amal, M., Peeters, F. & van Duin, A. C. Stabilized silicene within bilayer graphene: A proposal based on molecular dynamics and density-functional tight-binding calculations. *Physical Review B*, 2014, **89**, 024107.

34. Needs, R. The quantum Monte Carlo method: Electron correlation from random numbers. *Journal of Physics: Condensed Matter*, 2008, **20**, 6.

35. Liu, Z., Yu, G., Yao, H., Liu, L., Jiang, L. & Zheng, Y. A simple tight-binding model for typical graphyne structures. *New Journal of Physics*, 2012, **14**, 113007.

36. Wu, J. & Berciu, M. Kubo formula for open finite-size systems. *Europhysics Letters*, 2010, **92**, 3.

37. Paier, J., Ren, X., Rinke, P., Scuseria, G. E., Grüneis, A., Kresse, G. & Scheffler, M. Assessment of correlation energies based on the random-phase approximation. *New Journal of Physics*, 2012, **14**, 043002.

38. Novoselov, K. S. et al. Electric field effect in atomically thin carbon films. *Science*, 2004, **306**, 666–669.

39. Zhang, Y., Tan, Y. W., Stormer, H. L. & Kim, P. Experimental observation of the quantum Hall effect and Berry's phase in graphene. *Nature*, 2005, **438**, 201.

40. Jiao, L., Wang, X., Diankov, G., Wang, H. & Dai, H. Facile synthesis of high-quality graphene nanoribbons. *Nature Nanotechnology*, 2010, **5**, 321.

41. Kosynkin, D. V. et al. Longitudinal unzipping of carbon nanotubes to form graphene nanoribbons. *Nature*, 2009, **458**, 872.

42. Cai, J. et al. Atomically precise bottom-up fabrication of graphene nanoribbons. *Nature*, 2010, **466**, 470.

43. Medeiros, P. V., Mascarenhas, A. J., de Brito Mota, F. & de Castilho, C. M. C. A DFT study of halogen atoms adsorbed on graphene layers. *Nanotech*, 2010, **21**, 485701.

44. Kresse, G. & Furthmuller, J. Efficient iterative schemes for ab initio total-energy calculation using a plane-wave basis set. *Physical Review B*, 1996, **54**, 11169.

45. Kohn, W. & Sham, L. J. Self-consistent equations including exchange and correlation effects. *Physical Review*, 1965, **140**, A1133.

46. Lin, S.-Y., Li, W.-B., Tran, N. T. T., Hsu, W.-D., Liu, H.-Y. & Ming, M. F. Essential properties of Li/Li$^+$ graphite intercalation compounds. arXiv 2018:1810.11166.

47. Lin, S. Y., Chang, S. L., Tran, N. T. T., Yang, P. H. & Lin, M. F. H–Si bonding-induced unusual electronic properties of silicene: A method to identify hydrogen concentration. *Physical Chemistry Chemical Physics*, 2015, **17**, 26443–26450.

48. Li, S. S. et al. Tunable electronic and magnetic properties in germanene by alkali, alkaline-earth, group III and 3d transition metal atom adsorption. *Physical Chemistry Chemical Physics*, 2014, **16**, 15968–15978.

49. Chen, R. B., Chen, S. C., Chiu, C. W. & Lin, M. F. Optical properties of monolayer tinene in electric fields. *Scientific Reports*, 2017, **7**, 1849.

50. Sahin, H. & Peeters, F. M. Adsorption of alkali, alkaline-earth, and 3d transition metal atoms on silicene. *Physical Review B*, 2013, **87**, 085423.

51. Dávila, M. E., Xian, L., Cahangirov, S., Rubio, A. & Le Lay, G. Germanene: A novel two-dimensional germanium allotrope akin to graphene and silicene. *New Journal of Physics*, 2014, **16**, 095002.

52. Cai, B. et al. Tinene: A two-dimensional Dirac material with a 72 meV band gap. *Physical Chemistry Chemical Physics*, 2015, **17**, 12634–12638.

53. Lin, Y. T., Lin, S. Y., Chiu, Y. H. & Lin, M. F. Alkali-created rich properties in graphene nanoribbons: Chemical bondings. *Science Report*, 2017, **7**, 1722.

54. Tran, N. T. T., Lin, S. Y., Lin, C. Y. & Lin M. F. *Geometric and Electronic Properties of Graphene-Related Systems Chemical Bonding Schemes*, CRC Press, Baco Raton, FL, 2017. ISBN: 9781138556522.

55. Tran, N. T. T., Lin, S. Y., Glukhova, O. E. & Lin, M. F. Configuration-induced rich electronic properties of bilayer graphene. *Journal of Physical Chemistry C*, 2015, **119**, 10623–10630.

56. Redouani, I., Jellal, A., Bahaoui, A. & Bahlouli, H. Multibands tunneling in AAA-stacked trilayer graphene. *Superlattice and Microstructures*, 2018, **116**, 44–53.

57. Hattendorf, S., Georgi, A., Liebmann, M. & Morgenstern, M. Networks of ABA and ABC stacked graphene on mica observed by scanning tunneling microscopy. *Surface Science*, 2013, **610**, 53–58.

58. Ohta, T. et al. Interlayer interaction and electronic screening in multilayer graphene investigated with angle-resolved photoemission spectroscopy. *Physical Review Letters*, 2007, **98**, 206–802.

59. Shyu, F. L., Tsai, C. C., Lee, C. H. & Lin, M. F. Magnetoelectronic properties of chiral carbon nanotubes and tori. *Journal of Physics: Condensed Matter*, 2003, **18**, 8313–8324.

60. Do, T.-N., Chang, C.-P., Shih, P.-H., Wu, J.-Y. & Lin, M.-F. Stacking-enriched magneto-transport properties of few-layer graphenes. *Physical Chemistry Chemical Physics*, 2017, **19**, 29525–29533.

61. Wu, J.-Y., Chen, S.-C., Do, T.-N., Su, W.-P., Gumbs, G. & Lin, M.-F. The diverse magneto-optical selection rules in bilayer black phosphorus. *Scientific Reports*, 2018, **8**, 13303.

62. Bragg, W. L. The structure of some crystals as indicated by their diffraction of x-rays. *Proceedings of the Royal Society of London Series A-Containing Papers of a Mathematical and Physical Character*, 1913, **89**, 248–277.

63. Maslen, E. N., Fox, A. G. & O'Keefe, M. A. *International0 Tables for Crystallography*, vol. C. Kluwer Academic, Dordrecht, The Netherlands, 2004. ISBN: 978-1-4020-1900-5.

64. Kittel, C. *Introduction to Solid State Physics*. John Wiley & Sons, New York, 1976. ISBN: 0-471-49024-5.

65. Bragg, W. L. The structure of some crystals as indicated by their diffraction of X-rays. *Royal Society A*, 1913, **89**, 248–277.

66. Kittel, C. *Introduction to Solid State Physics*, John Wiley & Sons, Hoboken, NJ, 1976. ISBN: 0-471-49024-5.

67. Arndt, U. W., Champness, J. N., Phizackerley, R. P. & Wonacott, A. J. A single-crystal oscillation camera for large unit cells. *Journal of Applied Crystallography*, 1973, **6**, 457.

68. Santoro, A. & Zocchi, M. Multiple diffraction in the Weissenberg methods. *Acta Crystallographica*, 1966, **21**, 293.

69. Burger, M. J. *X-Ray Crystallography: An Introduction to the Investigation of Crystals by Their Diffraction of Monochromatic X-Radiation*. R. E. Krieger Pub. Co., 1980. ISBN: 978-0898741766.

70. Kganyago, K. R. & Ngoepe, P. E. Structural and electronic properties of lithium inter-calated graphite LiC_6. *Physical Review B*, 2003, **68**, 205111.

71. Damascelli, A. Probing the electronic structure of complex systems by ARPES. *Physica Scripta*, 2004, **109**, 61–74.

72. Cuk, T., Lu, D., Zhou, X., Shen, Z.-X., Devereaux, T. & Nagaosa, N. A. review of electron phonon coupling seen in the high-Tc superconductors by angle-resolved photo-emission studies (ARPES). *Physica Status Solidi (B)*, 2005, **242**, 11–29.

73. Coletti, C. et al. Revealing the electronic band structure of trilayer graphene on SiC: An angle-resolved photoemission study. *Physical Review B*, 2013, **88**, 155439.

74. Ohta, T., Bostwick, A., McChesney, J. L., Seyller, T., Horn, K. & Rotenberg, E. Interlayer interaction and electronic screening in multilayer graphene investigated with angle-resolved photoemission spectroscopy. *Physical Review Letter*, 2007, **98**, 206802.

75. Ohta, T., Bostwick, A., Seyller, T., Horn, K. & Rotenberg, E. Controlling the electronic structure of bilayer graphene. *Science*, 2006, **313**, 951954.

76. Eberhardt, W., McGovern, I. T., Plummer, E. W. & Fischer, J. E. Charge-transfer and non-rigid-band effects in the graphite compound LiC_6. *Physical Review Letter*, 1980, **44**, 200.

77. Chan, C. T., Kamitakahara, W. A., Ho, K. M. & Eklund, P. C. Charge-transfer effects in graphite intercalates: *Ab initio* calculations and neutron-diffraction experiment. *Physical Review Letter*, 1987, **58**, 1528.

78. Kamakura, N., Kubota, M., & Ono, K. Band dispersion and bonding character of potassium on graphite. *Surface Science*, 2008, **602**, 101.

79. Boesenberg, U., Sokaras, D., Nordlund, D., Weng, T.-C., Gorelov, E., Richardson, T. J., Kostecki, R. & Cabana, J. Electronic structure changes upon lithium intercalation into graphite – insights from ex situ and operando X-ray Raman spectroscopy. *Carbon*, 2019, **143**, 371–377.

80. Yang, K., Zhao, Z., Xin, X., Tian, Z., Peng, K. & Lai, Y. Graphitic carbon materials extracted from spent carbon cathode of aluminium reduction cell as anodes for lithium ion batteries: Converting the hazardous wastes into value-added materials. *Journal of Taiwan Institute of Chemical Engineers*, 2019, **104**, 201–209.

81. Narsimulu, D., Nagaraju, G., Sekhar, S. C., Ramulu, B. & Yu, J. S. Designed lamination of binder-free flexible iron oxide/carbon cloth as high capacity and stable anode material for lithium-ion batteries. *Applied Surface Science*, 2019, **497**, 143795.

82. Wang, Q., Zheng, D., He, L. & Ren, X. Cooperative effect in a graphite intercalation compound: Enhanced mobility of $AlCl_4$ in the graphite cathode of aluminum-ion batteries. *Physical Review Applied*, 2019, **12**, 044060.

3 Effect of Nitrogen Doping on the Li-Storage Capacity of Graphene Nanomaterials

A First-Principles Study

Chin-Lung Kuo and Yu-Jen Tsai
National Taiwan University

CONTENTS

3.1 INTRODUCTION

With the rise of global energy demand, searching an electrochemical storage device with high energy and power densities has become a critical issue nowadays. Among various energy storage devices, carbon-based rechargeable Li-ion batteries (LIBs) have gained lots of attention with the rapid development of portable electronics, hybrid electric vehicles, and the application to grid energy storage.[1] The commercial LIB system using bulk graphite as anode materials can only give rise to the theoretical specific capacity of 372 mA h g^{-1}, and there have been several efforts to increase the energy density and specific capacity by replacing the bulk graphite with other carbonaceous anode materials. Among them, graphene-based materials have been viewed as a prospective alternative of anode material mainly due to their high reversible capacity.[2-6] Furthermore, several researches have shown that graphene doped with nitrogen can give rise to higher reversible capacity compared to pristine graphene. Reddy et al.[7] have grown N-doped graphene layers by liquid precursor-based chemical vapor deposition (CVD) technique, and the measured reversible capacity is almost double compared

45

to pristine graphene. Wu et al.[8] carried out heat treatment of pristine graphene in a mixed gas of NH_3 and Ar to obtain N-doped graphene sheets with doping level of 3.06 at.%, which shows a high reversible capacity of 872 mA h g^{-1} after 30 cycles at a low current rate of 50 mA g^{-1} and retain a high capacity of 199 mA h g^{-1} at an extremely high rate of 25 A g^{-1}. Li et al.[9] annealed the synthesized graphene nanosheets in the presence of NH_3 gas and transferred into N-doped graphene nanosheets with a doping level of 2.8 at.%, which exhibit a superior specific capacity of 684 mA h g^{-1} in the 501st cycles. The N-doped graphene fabricated by Wang et al.[10] with a doping level of 3.9 at.% also exhibits a high capacity of 719 mA h g^{-1} after 20 cycles. These experimental reports suggest that N-doped graphene sheets can result in high reversible capacity mainly due to the fact that N-doping is effective in introducing defects into graphene. Previous density functional theory (DFT) calculations[11] verified that the substitutional N-dopants would increase the probability of point defect generation in graphene.

When a nitrogen atom is doped into graphene, three common bonding configurations within the carbon lattice are observed, including graphitic N (or quaternary N), pyridinic N, and pyrrolic N.[12] Generally, pyridinic N bonds with two C-atoms at the edges or defects of graphene and contributes one π electron to the π system. Pyrrolic N refers to N-atoms that contribute two π electrons to the π system, although unnecessarily bond into the five-membered ring. Graphitic refers to N-atoms that substitute for C-atoms in the hexagonal ring. These N-doping configurations have been suggested by X-ray photoelectron spectroscopy (XPS) and further directly observed by using scanning transmission electron microscopy (STEM).[13] Since it is hard to reveal which kinds of N-doped defects can effectively enhance the Li-storage capacity of N-doped graphene by experimental observations, several DFT studies have been performed to investigate the lithiation mechanism of graphene with different N-doped defects. Ma et al.[14] suggested that graphene with pyridinic N can result in high storage capacity, whereas graphene with graphitic N leads to the lowest storage capacity among three types of N. Besides, Yu[15] concluded that not all N-doped defects can improve the capacity of LIBs. His results indicated that Li-adsorption on graphitic N substituted from pristine graphene is energetically more unfavorable as compared to that on pristine graphene, while N-doped single and double vacancies can greatly improve the reversible capacity of the battery in comparison with the pristine graphene. Although the lithiation behavior for some of the N-doped defects has been examined, the reason why introducing some particular N-doped defects in graphene can lead to high Li-storage capacity remains unclear. Besides, whether increasing N-dopants in graphene must result in enhanced Li-storage capacity also remains unknown.

In this work, we have employed first-principles calculations based on DFT to investigate the adsorption of Li-atoms on various kinds of N-doped defect structures in graphene with different contents of graphitic and pyridinic N. Our calculated results first showed that introducing N-dopants in graphene can lower down the formation energy of defects and in turn increase the amount of N-doped defects in graphene, where the amount of N-doped MV can be more than N-doped DVs. Besides, the calculated reversible Li-storage capacity is not enhanced with increasing N-dopants near defect sites in graphene, which indicates that N-aggregation in defect sites may even lower down the Li-storage capacity compared to that of intrinsic defects. Although introducing too many N-dopants into graphene may lead to lower Li-storage capacity, an appropriate amount of N-dopants can effectively lift the migration energy barrier of vacancy defects in graphene, which can further suppress the loss of reversible capacity originated from vacancy aggregation.

3.2 COMPUTATIONAL DETAILS

All of our calculations in this study were performed within the DFT framework using generalized gradient approximation (GGA) with the parameterization of Perdew–Burke–Ernzerhof (PBE)[16,17] for exchange-correlation functional as implemented in the Vienna *ab initio* simulation package (VASP).[18–21] The projector augmented-wave (PAW) method was used to describe the core-electron interactions.[22,23] The cutoff energy of 550 eV was used for the expansion of plane-wave basis set. The Γ-centered k-point meshes with spacing <0.03 Å^{-1} were used for geometry optimization. For geometry optimization, all of the structures were fully relaxed until atomic forces were below 0.05 eV Å^{-1}.

For modeling of N-doped defective graphene, we employed 6 × 6 extension of the hexagonal unit cell of graphene to construct defect structures with either graphitic or pyridinic N atoms. For intrinsic defects, the most commonly observed defects in graphene are monovacancy (MV), divacancy (DV) with three variants, and Stone–Wales (SW) defects.[24] Herein we denoted MV as either MV or V1, and three DV variants with 5–8–5, 555–777, and 5555–6–7777 geometries are denoted as DV1 (V_2^1), DV2 (V_2^2), and DV3 (V_2^3), respectively. The relaxed structures of these intrinsic defects are presented in Figure 3.1. With the inclusion of N-dopants, the defect structures generated from x vacancies with y substituted N-atoms are denoted as V_xN_y. Moreover, the types ("p" for pyridinic; "g" for graphitic) as well as the number of N-atoms are also shown in the parentheses. For instance, DV1 with four substituted pyridinic N atoms is denoted as $V_2^1N_4$(4p). The relaxed N-doped defective graphene with corresponding denotations are presented in Figures 3.2 and 3.3. To avoid the spurious coupling effect between periodic graphene sheets, a vacuum separation of 18 Å along normal direction was set. Dipole corrections for the total energy of all the model systems were considered in our calculations.

The climbing image-nudged elastic band (CI-NEB) method[25,26] is used to search the saddle points and minimum energy path of vacancy migration in graphene. Nine images are employed between two endpoint structures. Each image is relaxed until the forces on the constituent atoms are <0.05 eV Å^{-1}.

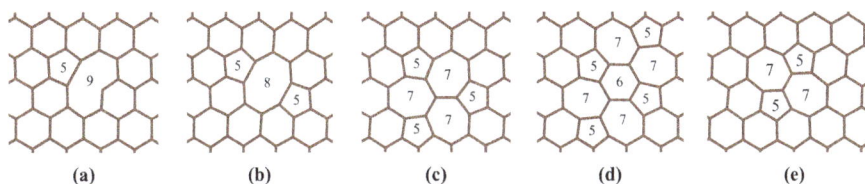

FIGURE 3.1 Graphene structures with intrinsic (a) MV, (b) DV1 (V_2^1), (c) DV2 (V_2^2), (d) DV3 (V_2^3), and (e) SW defects.

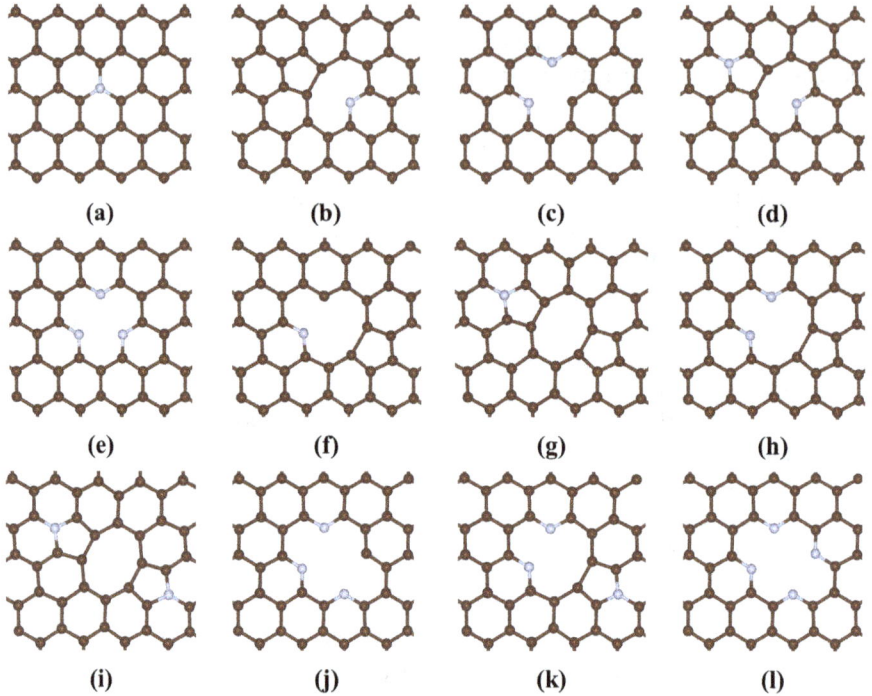

FIGURE 3.2 Structures of (a) graphitic N substituted from pristine graphene (N_1), and N-doped graphene with MV and DV1 structures considered in this work, including (b) V_1N_1(1p), (c) V_1N_2(2p), (d) V_1N_2(1p1g), (e) V_1N_3(3p), (f) $V_2^1N_1$(1p), (g) $V_2^1N_1$(1g), (h) $V_2^1N_2$(2p), (i) $V_2^1N_2$(2g), (j) $V_2^1N_3$(3p), (k) $V_2^1N_3$(2p1g), and (l) $V_2^1N_4$(4p).

3.3 RESULTS AND DISCUSSION

3.3.1 FORMATION ENERGY OF N-DOPED DEFECTS IN GRAPHENE

To assess the stability of N-doped defects in graphene, the formation energy was calculated by using the following definition:

$$E_f = E_{tot} - n_C\mu_C - n_N\mu_N$$

where E_{tot} is the total energy of N-doped graphene, n_C and n_N are the numbers of C- and N-atoms in the supercell, respectively, μ_C is the total energy of pristine graphene per atom, and μ_N is a half of the total energy of the isolated N_2 molecule. The calculated results are summarized in Table 3.1. In general, the formation energy of intrinsic defects is higher than that of N-doped defects, which indicates that the inclusion of N-dopants can reduce the formation energy of defects and thus increase the amount of N-doped defects in graphene. An exception occurs on $V_2^1 N_1$(1p) with the formation energy of 0.79 eV higher than that of intrinsic DV1; hence, this configuration was not considered for further calculations in this work. On the contrary, the calculated

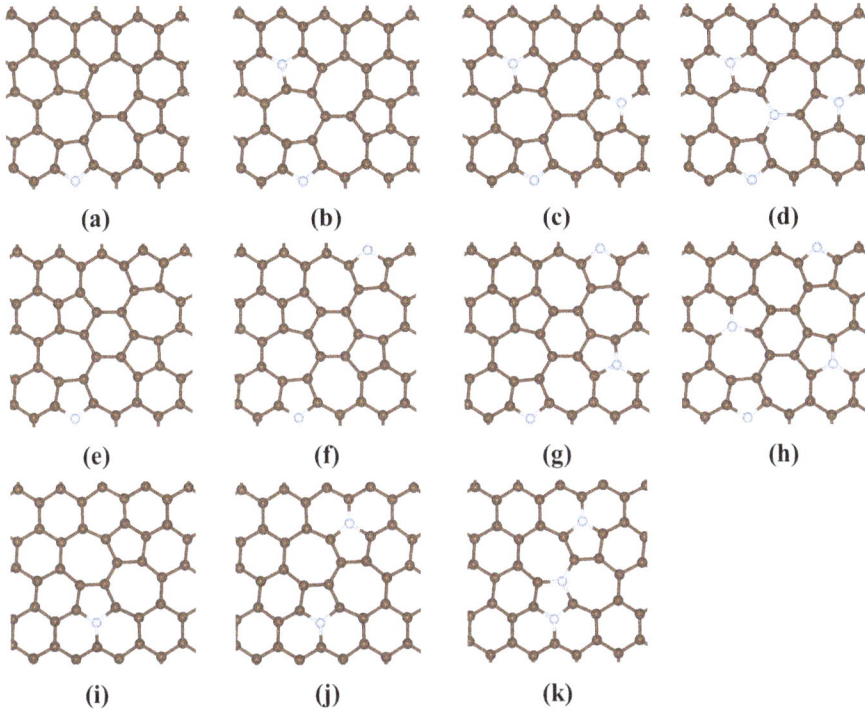

FIGURE 3.3 N-doped graphene with DV2 and DV3 structures considered in this work, including (a) $V_2^2N_1(1g)$, (b) $V_2^2N_2(2g)$, (c) $V_2^2N_3(3g)$, (d) $V_2^2N_4(4g)$, (e) $V_2^3N_1(1g)$, (f) $V_2^3N_2(2g)$, (g) $V_2^3N_3(3g)$, (h) $V_2^3N_4(4g)$, (i) SW + 1N, (j) SW + 2N, and (k) SW + 3N.

formation energy of intrinsic MV is higher than that of DVs, whereas the formation energy of N-doped MV at any N-content becomes lower than that of N-doped DVs. The results suggest that the amount of MV can be more than DVs in N-doped graphene, which is in contrast to the relative number of defects in graphene without N-dopants.

TABLE 3.1

Formation Energy of N-Doped Defective Graphene (Unit: eV)

Defect	no N	$n_N = 1$	$n_N = 2$	$n_N = 3$	$n_N = 4$
MV	7.45	$V_1N_1(1p)$: 5.19	$V_1N_2(2p)$: 4.82	$V_1N_3(3p)$: 3.49	
			$V_1N_2(1p1g)$: 4.41		
DV1	6.66	$V_2^1N_1(1p)$: 7.45	$V_2^1N_2(2p)$: 5.88	$V_2^1N_3(3p)$: 5.38	$V_2^1N_4(4p)$: 3.74
		$V_2^1N_1(1g)$: 6.08	$V_2^1N_2(2g)$: 5.29	$V_2^1N_3(2p1g)$: 5.69	
DV2	6.09	$V_2^2N_1(1g)$: 5.54	$V_2^2N_2(2g)$: 5.07	$V_2^2N_3(3g)$: 5.23	$V_2^2N_4(4g)$: 5.97
DV3	6.37	$V_2^3N_1(1g)$: 5.99	$V_2^3N_2(2g)$: 5.52	$V_2^3N_3(3g)$: 5.60	$V_2^3N_4(4g)$: 5.70
SW	4.31	4.24	3.86	4.47	

In the presence of MV, pyridinic N structures were found to be energetically favorable for most of the N-contents but for $n_N = 2$, where the configuration with one pyridinic N and one graphitic N is stable by 0.39 eV compared to that of MV with two pyridinic N. For DV1 with no more than two N-dopants, graphitic N structures were shown to be the most stable configurations, while pyridinic N structures turn to dominate as N-contents further increase. Both $V_1N_3(3p)$ and $V_2^1N_4(4p)$ exhibit fairly high structural stability compared to any other N-doped defects, implying that MV and DV1 can strongly induce aggregation of N-dopants. As for DV2, DV3, and SW defects, graphitic N structures appeared to be dominant at all of the N-contents and the formation energies are all shown to be the lowest at $n_N = 2$, indicating that all of these defects can aggregate two N-dopants at most. It is also worth noting that both intrinsic DV2 and N-doped DV2 with no more than three N-dopants are the most stable structures among three different DVs. However, N-doped DV1 becomes energetically more favorable than N-doped DV2 at $n_N = 4$, implying that N-dopants can help stabilize DV1 and suppress the transformation of DV1 into DV2.

3.3.2 SINGLE LI-ADSORPTION ON N-DOPED DEFECTS IN GRAPHENE

To investigate which kinds of N-doped defects in graphene can strongly bind Li, we first calculated the adsorption energy of single Li on various N-doped graphenes, which is defined as

$$E_{ads} = E\left(LiC_yN_z\right) - E\left(C_yN_z\right) - E\left(Li\right)$$

where $E(LiC_yN_z)$ and $E(C_yN_z)$ are the total energy of N-doped defective graphene with and without an adsorbed Li-ion, respectively, and μ_{Li} is the total energy of an isolated Li-atom. The calculated results summarized in Table 3.2 show that in general, the adsorption energy becomes stronger with an increasing amount of pyridinic N atoms in either MV or DVs but becomes weaker with an increasing amount of graphitic N atoms. Since nitrogen has bigger electronegativity than carbon, it could drag electrons

---

TABLE 3.2

Adsorption Energy of Single Li-Atom on N-Doped Defective Graphene (Unit: eV)

Defect	no N	$n_N = 1$	$n_N = 2$	$n_N = 3$	$n_N = 4$
MV	−2.70	V_1N_1(1p): −2.98	V_1N_2(2p): −3.62 V_1N_2(1p1g): −2.12	V_1N_3(3p): −4.66	
DV1	−2.28	$V_2^1N_1$(1g): −2.26	$V_2^1N_2$(2p): −3.61 $V_2^1N_2$(2g): −1.49	$V_2^1N_3$(3p): −4.40 $V_2^1N_3$(2p1g): −3.78	$V_2^1N_4$(4p): −5.23
DV2	−2.40	$V_2^2N_1$(1g): −2.26	$V_2^2N_2$(2g): −1.60	$V_2^2N_3$(3g): −1.52	$V_2^2N_4$(4g): −1.15
DV3	−2.10	$V_2^3N_1$(1g): −2.13	$V_2^3N_2$(2g): −1.88	$V_2^3N_3$(3g): −1.87	$V_2^3N_4$(4g): −1.38
SW	−1.62	−1.90	−1.31	−1.02	

from neighboring carbon atoms in N-doped graphene. Hence, it has been believed that the considerably high adsorption energy of Li on pyridinic N, such as V_1N_3 (3p) and V_2N_4(4p), can be attributed to the strong electrostatic attraction between Li-ion and N-dopants. However, the adsorption of Li on graphitic N substituted from pristine graphene (N_1) can be also relatively strong since there is also an electrostatic attraction between them, while the calculated adsorption energy is even weaker than that on pristine graphene. To explain this phenomenon, we have analyzed the electronic structure of these N-doped graphenes by plotting the total density of states (TDOS). Due to the fact that Li will transfer its 2s valence electron to the graphene sheet during Li-adsorption on graphene, the defective graphene structures with more number of states just above the Fermi level can result in less increase of energy for graphene. From the TDOS of N-doped graphene shown in Figure 3.4, we can see that there is still

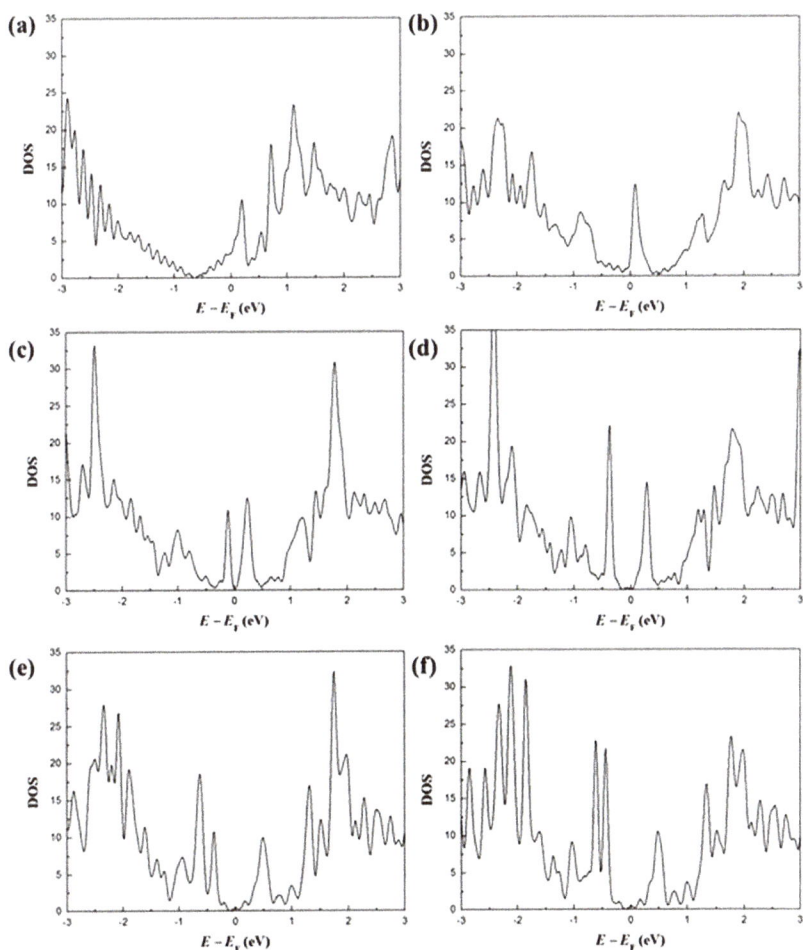

FIGURE 3.4 Total electronic density of states (TDOS) of (a) N_1, (b) V_2^1, (c) $V_2^1N_1$(1p), (d) V_2^1 N_2(2p), (e) $V_2^1N_3$(3p), and (f) $V_2^1N_4$(4p).

TABLE 3.3

Absolute Fermi Level Position (Fermi Level Related to Core Level) of N-Doped Defective Graphene

	N_1	V_2^1	$V_2^1 N_1(1p)$	$V_2^1 N_2(2p)$	$V_2^1 N_3(3p)$	$V_2^1 N_4(4p)$
E_{F_abs} (eV)	266.43	266.05	266.01	265.94	265.88	265.85

a considerable amount of states lying just above the Fermi level (n_{DOS}) for N_1 compared to DV1 with pyridinic N atoms. Besides, n_{DOS} of $V_2^1 N_4(4p)$ does not appear to be much more than that of $V_2^1 N_1(1p)$, implying that the adsorption energy of Li on N-doped defects is not mainly related to the amount of n_{DOS}.

The reason why Li-adsorption on pyridinic N is stronger compared to graphitic N can be explained by the position of the Fermi level for each N-doped graphene. Herein, we calculated the absolute Fermi level by first calculating the 1 s energy of C atom faraway from the defect site, denoted as E_{core}. Then, E_{core} is subtracted by the calculated Fermi level to obtain the position of absolute Fermi level, denoted as E_{F_abs}. The calculated E_{F_abs} of N-doped graphene is summarized in Table 3.3. It is found that N_1 has much higher E_{F_abs} than any other N-doped graphene, showing that it is more energetically unfavorable for graphitic N to accept electrons from Li as compared to pyridinic N. Besides, with the increasing amount of pyridinic N atoms, the value of E_{F_abs} becomes lower, indicating that increasing pyridinic N dopants near vacancy site can lower down the position of the Fermi level and induce a strong driving force to drag electron from Li.

3.3.3 Li-Storage Capacity of N-Doped Defective Graphene

The lithiation process and the Li-storage capacity for graphene with a variety of N-doped vacancy defects were systematically investigated. Here the achievable Li-capacity limit of N-doped defective graphene was determined by calculating its lithiation voltage with respect to Li/Li$^+$ at each Li-concentration and defined as follows:

$$V(x) = -\left[E\left(Li_x C_y N_z\right) - E\left(Li_{x-1} C_y N_z\right) - \mu_{Li} \right]$$

where $E(Li_x C_y N_z)$ is the total energy of N-doped defective graphene with x adsorbed Li-ions, and μ_{Li} is the total energy of single Li-atom in bulk BCC phase. As $V(x)$ becomes negative, it implies that xth Li-adsorption is thermodynamically unfavorable and starts to precipitate into Li-dendrites. Therefore, achievable Li-capacity limit, denoted as n_{Li}, was defined as the maximum amount of the Li-content that can reach before the lithiation voltage becomes negative. Since there are numerous configurations for Li-adsorption on N-doped defective graphene at given Li-content (x), we have extensively searched the most energetically favorable configuration for $Li_x C_y N_z$ compounds and then evaluated lithiation voltages at each lithiated composition of N-doped defective graphene.

Figures 3.5–3.7 present the calculated lithiation voltages of N-doped defective graphene, and the n_{Li} of N-doped defective graphenes with different defects is summarized in Table 3.4. For graphene with a substitutional N(N_1), the voltage of the first Li-adsorption

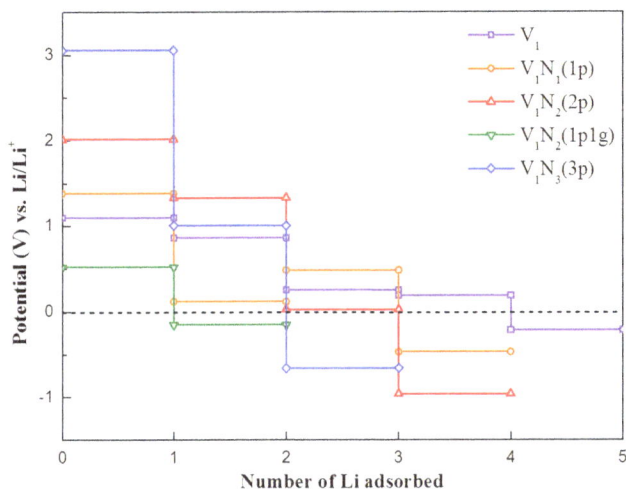

FIGURE 3.5   Calculated lithiation potential profiles of the graphene with intrinsic MV and N-doped MV at different N concentration.

is negative, indicating that a substitutional N cannot provide any Li-uptake. After introducing defect structures with N-dopants in graphene, our results show that most of these N-doped defects can strengthen the binding of Li and further enhance the Li-storage capacity. For N-doped MV defects, the maximum n_{Li} at each N-content (from zero to three N) is 4, 3, 3, and 2, respectively, indicating that n_{Li} gradually decreases with the increasing amount of N-dopants near vacancy site of MV. It should be noted that the voltage of the first Li-adsorption on MV becomes higher as more pyridinic N atoms aggregate to the vacancy site. However, considerably high lithiation voltage implies that the Li-atom could bind strongly on the defect sites, which would in turn lead to the irreversible Li-storage capacity and reduce the cycle performance of Li-ion battery. To focus on the effect of N-dopants on the reversible Li-storage capacity of graphene-based anode materials, herein we define that the adsorbed Li would contribute to irreversible Li-storage capacity if the voltage is higher than 2.0 V (a half of the voltage of the most commonly used cathode materials for LIBs [27]). Since the voltages of the first Li-adsorption for $V_1N_2(2p)$ and $V_1N_3(3p)$ are larger than 2.0 V, the maximum reversible n_{Li} of MV with N-content from zero to three N is 4, 3, 2, and 1, respectively, showing that more N-dopants in MV cannot enhance but even greatly lower down the reversible Li-capacity in comparison with intrinsic MV.

As for the Li-storage capacity of N-doped DVs, DV3 has the maximum n_{Li} of 4 among three intrinsic DV variants, which is in agreement with our previous work using a bigger supercell of graphene. At low N-contents ($n_N = 1$ and 2), DV3 with graphitic N atoms still contribute to the highest Li-uptake among three DV variants. It should be noted that although DV1 with two pyridinic N [$V_2^1 N_2(2p)$] has the same n_{Li} as DV3 with two graphitic N, the voltage of first Li-adsorption on $V_2^1 N_2(2p)$ is higher than 2.0 V, which would lead to the loss of reversible capacity. At high N-contents ($n_N = 3$ and 4), the n_{Li} of N-doped DV1 becomes higher than that of N-doped DV2 and DV3, and the n_{Li} of DV2 and DV3 with graphitic N greatly

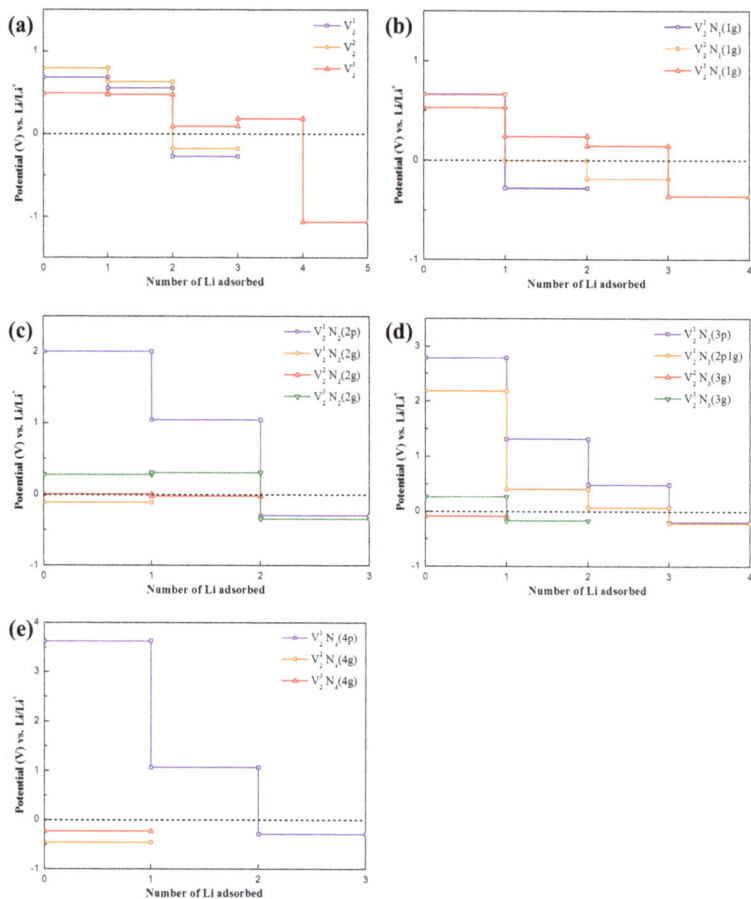

FIGURE 3.6 Calculated lithiation potential profiles of the graphene with (a) intrinsic DVs, and N-doped DVs with (b) one N-dopants, (c) two N-dopants, (d) three N-dopants, and (e) four N-dopants.

decreases and even becomes zero as the number of N-dopants reaches four. When not considering the irreversible capacity, the maximum reversible n_{Li} of N-doped DVs with N-content from zero to four is 4, 3, 2, 2, and 1, respectively. Accordingly, similar to the cases of N-doped MV, higher N-concentration does not enhance the reversible Li-storage capacity of N-doped DVs.

As for the Li-storage capacity of N-doped SW, the maximum n_{Li} decreases with the increasing N-dopants and becomes zero as the number of N-dopants reaches to two. It is worth noting that all of the N-dopants in N-doped DV2, DV3, and SW structures are all graphitic and induce n-type doping effects. When the effect of n-type doping is stronger, N-doped graphene with these defects becomes harder to accept electrons from Li-atoms, thereby resulting in the loss of Li-storage capacity. The results thus suggest that introducing many N-dopants near defect sites cannot enhance the reversible Li-storage capacity and may even lead to severe capacity loss.

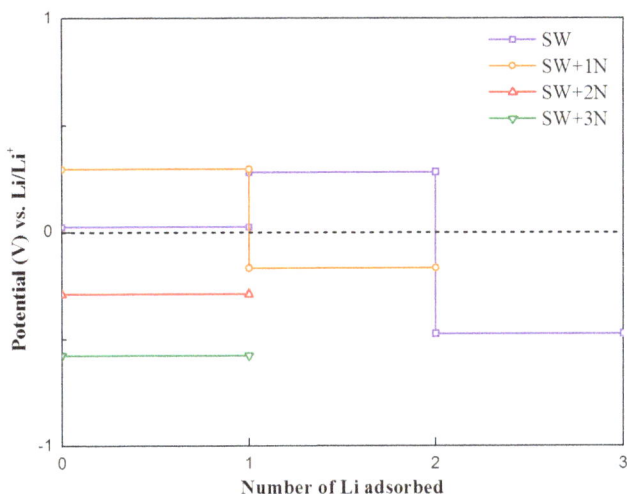

FIGURE 3.7 Calculated lithiation potential profiles of the graphene with intrinsic SW and N-doped SW at different N-concentration.

TABLE 3.4
Calculated n_{Li} of N-Doped Defective Graphene

Defect	no N	$n_N = 1$	$n_N = 2$	$n_N = 3$	$n_N = 4$
MV	4	$V_1N_1(1p)$: **3**	$V_1N_2(2p)$: **3** $V_1N_2(1p1g)$: 1	$V_1N_3(3p)$: **2**	
DV1	2	$V_2^1N_1(1g)$: 1	$V_2^1N_2(2p)$: 2 $V_2^1N_2(2g)$: 0	$V_2^1N_3(3p)$: **3** $V_2^1N_3(2p1g)$: **3**	$V_2^1N_4(4p)$: **2**
DV2	2	$V_2^2N_1(1g)$: 2	$V_2^2N_2(2g)$: 1	$V_2^2N_3(3g)$: 0	$V_2^2N_4(4g)$: 0
DV3	4	$V_2^3N_1(1g)$: 3	$V_2^3N_2(2g)$: 2	$V_2^3N_3(3g)$: 1	$V_2^3N_4(4g)$: 0
SW	2	1	0	0	

The values in boldface indicate that there is one Li-uptake that contributes to the irreversible capacity loss.

It is worth noting that these calculated reversible Li-storage capacity of graphene with N-doped defects can be used to compare with the experimental results. Wu et al.[8] synthesized N-doped graphene sheets with a doping level of 3.06 at.%, in which no graphitic N is formed at low doping temperature. The synthesized N-doped graphene electrodes exhibit a reversible capacity of 1,043 mA h g^{-1} in the first cycle at a low current rate of 50 mA g^{-1}, which is improved by 88 mA h g^{-1} as compared to the pristine graphene electrode. In this work, both V_1N_2 and V_2N_2 correspond to similar N-doping level (~2.8 at.%) as in Wu's work, and these two N-doped defect structures can contribute to an enhanced Li-storage capacity of ca. 64 mAh g^{-1}, which is in close agreement with the experimental value. In another experiment, Wang et al.[10] synthesized the N-doped graphene material with a doping level of ca. 3.9 at.%, at which the dominant binding types are pyridinic and pyrrolic N. The N-doped graphene electrodes exhibit ultrahigh discharge capacity of 1,284 mAh g^{-1} in the first cycle, which are higher than 1,080 mAh g^{-1} of the pristine graphene. In this work, we assume only pyridinic N defects exists in N-doped graphene, and thus, both V$_1$N$_3$ and V$_2$N$_3$ correspond to similar N-doping level (~4.2 at.%) as shown in Wang's work. From theoretical calculations, the reversible capacity will be improved by ca. 63 mAh g^{-1}, which is smaller than the reported value (204 mAh g^{-1}). The results indicate that reversible Li-storage capacity cannot be effectively enhanced only by N-aggregated defect structures. The additional Li-uptake can be attributed to the large number of topological defects on the graphene layer, which provide extra active sites for stabilizing the adsorption of Li-atoms.

3.3.4 MIGRATION ENERGY BARRIER OF N-DOPED DEFECTS IN GRAPHENE

Since vacancy defect graphene can migrate at elevated temperature as observed in experiments,[28] vacancy aggregation may occur during the synthesis procedure of N-doped graphene. To see whether the inclusion of N-dopants can influence the migration ability of vacancy defects, the migration energy barrier of vacancy defects with and without N-dopants in graphene has been evaluated by using CI-NEB method. In this work, we have considered the migration of V$_1$ and V$_1$N$_1$(1p) as well as the translational migration of V$_2^1$ and V$_2^1$N$_2$(2p), which is achieved by two times of bond rotations as proposed in previous work.[28] The migration pathways for these vacancy defects are displayed in Figure 3.8, and the calculated migration energy barrier for these migration pathways is summarized in Table 3.5. The results indicate that intrinsic MV has a relatively small migration energy barrier of 0.93 eV, which can be readily migrated on monolayer graphene. With the inclusion of single N-dopant near MV, the migration energy barrier greatly increases by 2.52 eV and the MV becomes more immobile. Similarly to the case of DV1, the translational migration energy barrier increases

(a)

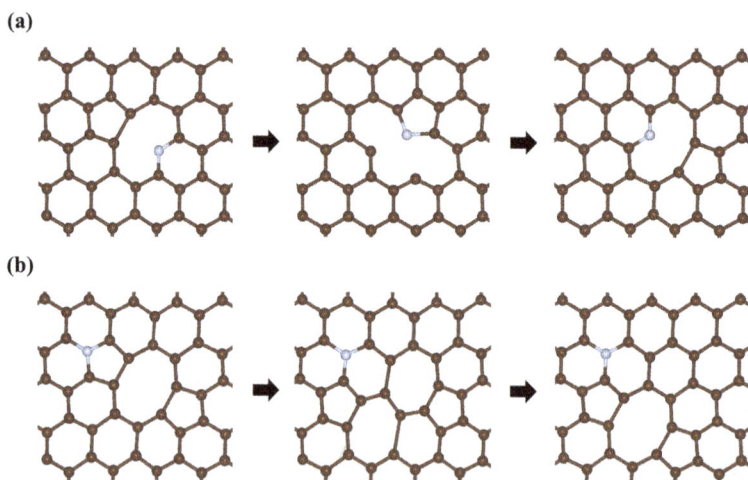

(b)

FIGURE 3.8 Migration pathways of (a) V$_1$N$_1$(1p) and (b) V$_2^1$N$_1$(1g) in graphene considered in this work.

TABLE 3.5

Migration Energy Barrier of MV and DV1 with and without Single N-Dopant

		DV migration	
$E_{barrier}$ (eV)	MV migration	(5–8-5) → (5–7-7-5)	(5–7-7-5) → (5–8-5)
no N	0.93	5.37 (1.37)	2.76 (6.77)
with N	3.45	6.39 (1.18)	2.22 (6.56)

The values in the parenthesis indicate the energy barrier of backward migration

by 1.02 eV after doping with N. The results thus imply that N-doping can suppress the migration ability of vacancy defects and further avoid the vacancy coalescence into a multivacancy. According to our previous work, vacancy aggregation (without N-dopants) in graphene will lead to the loss of Li-storage capacity. When N-dopants are involved in the graphene, considering that, for instance, two V_1N_1(1p) coalesce into either V_2N_2(2p) or V_2N_2(2g). The n_{Li} of the former is 6, whereas the latter decreases to 2 at most, showing that N-doped vacancy aggregation can also give rise to the loss of Li-storage capacity. Therefore, it concludes that with the inclusion of N-dopants, the vacancy defects can be stabilized and lower the probability of vacancy aggregation in graphene, which can in turn keep the high Li-storage capacity of N-doped graphene.

3.4 CONCLUSION

We have performed first-principles calculations based on the DFT to investigate the energetics of N-doped defects in graphene, Li-, and Na-storage capacity of N-doped defective graphene, and the migration energy barrier of N-doped vacancy defects in graphene. Our calculated results first show that introducing N-dopants can lower down the formation energy of defects in graphene and thus increase the amount of N-doped defects in graphene, which can provide more energetically favorable sites to bind Li- or Na-atoms. Besides, the formation energy of MV becomes lower than that of DVs when N-dopants are introduced in the graphene, suggesting that the amount of MV can be more than DVs in N-doped graphene. Moreover, MV with three pyridinic N and DV1 with four pyridinic N have the lowest formation energy among all N-doped defects considered, indicating that MV and DV1 can aggregate N-dopants to stabilize the defect structures.

The calculated reversible Li-/Na-storage capacity of N-doped defects was shown to decrease with the increasing N-dopants in graphene, implying that introducing many N-dopants near defect sites cannot enhance the reversible Li-/Na-storage capacity and may even lead to reversible capacity loss. Although the inclusion of N-dopants into the intrinsic defects could reduce the Li-/Na-uptake, the migration energy barrier of N-doped MV and N-doped DV1 were both found to become higher compared to that of intrinsic defects, which can effectively avoid the capacity loss due to the suppression of vacancy aggregation.

REFERENCES

1. J. B. Goodenough and K.-S. Park, *J. Am. Chem. Soc.*, 2013, **135**, 1167–1176.
2. E. Yoo, J. Kim, E. Hosono, H.-S. Zhou, T. Kudo and I. Honma, *Nano Lett.*, 2008, **8**, 2277–2282.
3. D. Pan, S. Wang, B. Zhao, M. Wu, H. Zhang, Y. Wang and Z. Jiao, *Chem. Mater.*, 2009, **21**, 3136–3142.
4. G. Wang, X. Shen, J. Yao and J. Park, *Carbon*, 2009, **47**, 2049–2053.
5. P. Lian, X. Zhu, S. Liang, Z. Li, W. Yang and H. Wang, *Electrochim. Acta*, 2010, **55**, 3909–3914.
6. S. Chen, P. Bao, L. Xiao and G. Wang, *Carbon*, 2013, **64**, 158–169.
7. A. L. M. Reddy, A. Srivastava, S. R. Gowda, H. Gullapalli, M. Dubey and P. M. Ajayan, *ACS Nano*, 2010, **4**, 6337–6342.
8. Z.-S. Wu, W. Ren, L. Xu, F. Li and H.-M. Cheng, *ACS Nano*, 2011, **5**, 5463–5471.
9. X. Li, D. Geng, Y. Zhang, X. Meng, R. Li and X. Sun, *Electrochem. Commun.*, 2011, **13**, 822–825.
10. X. Wang, Q. Weng, X. Liu, X. Wang, D.-M. Tang, W. Tian, C. Zhang, W. Yi, D. Liu, Y. Bando and D. Golberg, *Nano Lett.*, 2014, **14**, 1164–1171.
11. Z. Hou, X. Wang, T. Ikeda, K. Terakura, M. Oshima, M.-A. Kakimoto and S. Miyata, *Phys. Rev. B*, 2012, **85**, 165439.
12. H. Wang, T. Maiyalagan and X. Wang, *ACS Catal.*, 2012, **2**, 781–794.
13. Y.-C. Lin, P.-Y. Teng, C.-H. Yeh, M. Koshino, P.-W. Chiu and K. Suenaga, *Nano Lett.*, 2015, **15**, 7408–7413.
14. C. Ma, X. Shao and D. Cao, *J. Mater. Chem.*, 2012, **22**, 8911–8915.
15. Y.-X. Yu, *Phys. Chem. Chem. Phys.*, 2013, **15**, 16819–16827.
16. J. P. Perdew, K. Burke and M. Ernzerhof, *Phys. Rev. Lett.*, 1996, **77**, 3865–3868.
17. J. P. Perdew, K. Burke and M. Ernzerhof, *Phys. Rev. Lett.*, 1997, **78**, 1396–1396.
18. G. Kresse and J. Hafner, *Phys. Rev. B*, 1993, **47**, 558–561.
19. G. Kresse and J. Hafner, *Phys. Rev. B*, 1994, **49**, 14251–14269.
20. G. Kresse and J. Furthmüller, *Phys. Rev. B*, 1996, **54**, 11169–11186.
21. G. Kresse and J. Furthmüller, *Comp. Mater. Sci.*, 1996, **6**, 15–50.
22. P. E. Blöchl, *Phys. Rev. B*, 1994, **50**, 17953–17979.
23. G. Kresse and D. Joubert, *Phys. Rev. B*, 1999, **59**, 1758–1775.
24. F. Banhart, J. Kotakoski and A. V. Krasheninnikov, *ACS Nano*, 2011, **5**, 26–41.
25. G. Henkelman, B. P. Uberuaga and H. Jónsson, *J. Chem. Phys.*, 2000, **113**, 9901–9904.
26. G. Henkelman and H. Jónsson, *J. Chem. Phys.*, 2000, **113**, 9978–9985.
27. R. Koksbang, J. Barker, H. Shi and M. Y. Saïdi, *Solid State Ion.*, 1996, **84**, 1–21.
28. J. Kotakoski, A. V. Krasheninnikov, U. Kaiser and J. C. Meyer, *Phys. Rev. Lett.*, 2011, **106**, 105505.

4 Fundamental Properties of Li+-Based Battery Anode $Li_4Ti_5O_{12}$

*Thi Dieu Hien Nguyen, Hai Duong Pham,
Shih-Yang Lin, Ngoc Thanh Thuy Tran,
and Ming-Fa Lin*
National Cheng Kung University

CONTENTS

4.1 INTRODUCTION

The up-to-date lithium-ion batteries (LIBs; [1–4]) are frequently utilized in many electronic devices, such as laptops [2,3], cell phones [2,3], iPods [2], and so on, being mainly due to their high capacity [5,6], large output voltage [5,6], long-term stability [5,7], and friendly with the chemical environments [5,7]. The commercialized LIBs [6] principally consist of a cathode (positive electrode; [6]), an anode (negative electrode; [6]) and an electrolyte, in which the third component is closely related to the unusual transport of the positive lithium ions (Li+) between two electrodes. Very interesting, a separator membrane is designed to avoid the internal short circuit and only allow the lithium-ion to freely in/out of the cathode and anode. The crucial mechanisms of LIBs are characterized through the unique charging and discharging process that is based on the exchange of Li+-ions [8]. When the charging process comes into existence, the two electrodes are connected externally to an electrical supply. The electrons are released from the cathode material and move externally to the anode that creates the charge current [9]. Concurrently, the lithium ions move in the same direction internally from cathode to anode through the solid-/liquid-state electrolyte [7,9] to maintain the electric neutrality. By using this process, the external energy from the electrical supply is electrochemically stored in the battery, leading to the form of chemical energy. On the contrary, during the discharging process, electrons move in the opposite direction from anode to cathode through the external lead and thus can do the work; furthermore, lithium-ions transport back to the cathode via the specific electrolyte. The discharge-process energy is very useful for commercial purposes [10].

The experimental progress shows that two main types of negative electrodes in LIBs cover graphite and lithium titanium oxide. From the viewpoints of industry, the graphite-based anode material is very easy to be produced in LIBs, which leads to lowering the cost. However, this system has a serious disadvantage in the volume expansion during a lot of rapid charging and discharging processes [11,12]. While lithium titanium oxide serves as the negative electrode, it would be able to provide long life, rapid charging, high input/output power performance, excellent low-temperature operation, a wide effective state of charge range and overcome the drastic volume changes in the graphitic materials [12]. On the theoretical side, the graphite intercalation compounds of Li^+ ions/Li atoms are predicted to present the AA or AB stacking configurations [13], when the adion/adatom concentration is sufficiently high or low. Furthermore, the semimetallic/metallic behaviors, which are determined by the band overlap/the Fermi level at the conduction bands, come into existence in the Li^+/Li intercalation cases [13,14]. There are only a few first-principles studies [15] on LiTiO-related materials with unusual superconducting properties [15]. The thorough theoretical studies, which are conducted on their essential properties, are absent, even the pure numerical calculations [16], e.g., the absence of the significant orbital hybridizations in different chemical bonds. The critical physical/chemical/material pictures are one of the main focuses in this chapter.

The previous simulation methods [17], which are based on the first-principles calculations, can provide rich and unique phenomena, especially for the emergent 2D-layered materials. For example, the systematic studies have been done for the essential properties of few-layer graphene systems [18], 1D graphene nanoribbons [19], and silicene-related materials [20], with chemical absorptions [21,22] and substitutions [23]. Such investigations clearly illustrate that the quasiparticle charges/orbitals and spins dominate all the diversified phenomena. The delicate Vienna ab initio simulation package (VASP) results and thorough analyses are capable of proposing the significant mechanisms/pictures in fully understanding the geometric, electronic, and magnetic properties. The important multi-/single-orbital hybridizations in various chemical bonds are obtained from the optimal lattice symmetry, the atom-dominated band structures, the spatial charge densities and their variations after chemical modifications, and the atom- and orbital-decomposed density of states. Furthermore, the spin distribution configurations (non-, ferro-, and antiferromagnetic configurations), being associated with the host and/or guest atoms, are accurately identified from the spin-split/spin-degenerate energy bands, the spin density distributions, the net magnetic moments, and the spin-projected van Hove singularities. This developed framework, which is successfully conducted on the silicene- and graphene-related systems, could be generalized to other emergent materials, or it needs to be thoroughly tested in further investigations. It is thus expected to be very suitable for studying the rather complicated geometric and electronic properties of the mainstream Li^+-based batteries [24], mainly including the cathode [24], anode [24], and electrolyte materials [25].

In addition, the direct combinations of numerical simulations with phenomenological models would be very useful in understanding the unusual and diverse properties thoroughly, e.g., the linking of the VASP calculations [18,19,26] and the generalized tight-binding models [27] for the rich magnetic quantization [28].

This chapter is mainly focused on the geometric structure and electronic properties of the Li$_4$Ti$_5$O$_{12}$-related anode material in Li$^+$-based batteries [5,7]. The first principles calculate results covering the total ground-state energy, lattice symmetry, various Li-O and Ti-O, the atom-dominated band structure, the spatial charge density distribution, and the atom- and orbital-projected van Hove singularities. The spin-dependent behaviors, the spin-split electronic states, the net magnetic moment, and the spin density distributions, will be fully tested whether they could survive in the ternary transition-metal-atom compound. Such physical properties will play important roles to achieve the critical multi-orbital hybridizations in three kinds of chemical bonds. The analysis difficulties lie in the very complicated orbital-decomposed density of states, being supported by the electronic energy spectrum and carrier density. The theoretical predictions, which are conducted on the optimal geometry, the occupied valence state, and the energy gap and whole energy spectrum, could be verified from the high-resolution measurements of X-ray diffraction/ low-energy electron diffraction (LEED; [29]), angle-resolved photoemission spectroscopy (ARPES; [30]), and scanning tunneling microscopy (STM; [31]).

4.2 THEORETICAL SIMULATION METHODS

The rich and unique geometric structures and electronic properties of the 3D Li$_4$Ti$_5$O$_{12}$ compound are thoroughly investigated by the density functional theory (DFT; [32]) implemented by VASP [33]. The many-particle exchange and correlation energies, which mainly arise from the electron–electron Coulomb interactions, are calculated from the Perdew–Burke–Ernzerhof (PBE; [34]) functional under the generalized gradient approximation. Furthermore, the projector-augmented wave (PAW; [35,36]) pseudopotentials are able to characterize the electron–ion interactions. It is well known that these two kinds of intrinsic interactions have no exact formulas [37], i.e., it is very difficult to express the single- and many-particle Hamiltonian in an analytic form [37]. In general, plane waves, with the kinetic energy cutoff of 520 eV, are chosen as a complete set [37]; therefore, they are very reliable and suitable for evaluating Bloch wave functions and electronic energy spectra. The first Brillouin zone is sampled by $3 \times 3 \times 3$ and $12 \times 12 \times 12$ k-point meshes within the Monkhorst–Pack scheme for geometric optimizations and electronic structures, respectively. Such points are sufficient in obtaining the reliable orbital-projected density of states, spatial charge distributions, and spin density configurations. The convergence for the ground-state energy is 10^{-5} eV between two consecutive simulation steps, and the maximum Hellmann–Feynman force acting on each atom is <0.01 eV/A during the ionic relaxations.

The delicate VASP calculations and detailed analyses are conducted on certain physical quantities, such as [33] the atom-dominated band structures, the spatial charge densities, the atom- and orbital-decomposed van Hove singularities, the spin distribution configurations, the spin-split or spin-degenerate energy bands, and the net magnetic moments. As a result, the critical pictures, the multi- and/or single-orbital hybridizations in chemical bonds and the spin configurations due to different atoms, could be achieved under the concise scheme. Such viewpoints will be very useful in fully comprehending the diversified physical, chemical, and material phenomena. The theoretical framework has been successful in the systematic investigation of the geometric, electronic, and magnetic properties of few-layer 2D graphene systems [38], 1D graphene nanoribbons [39], and 2D silicene-related material. Very interestingly, the chemical modifications through the adatom chemisorptions and guest-atom substitutions can greatly diversify the various fundamental properties [26]. Apparently, the developed viewpoints are available in other condensed matter systems. For example, they should be suitable in thoroughly exploring the diverse phenomena of Li^+-based battery anode/cathode/electrolyte materials [13]. The numerical simulations might be available combining with the phenomenological models. For example, the first-principles calculations on band structures could be well fitted by the tight-binding model/the effective-mass approximation if the electronic energy spectra across the Fermi level are not so complicated [26]. This viewpoint has been very successful in fully exploring the diversified essential properties in few-layer graphene systems, e.g., the diverse magnetic quantization of AA-[40–42], AB-[42,43], ABC-[44], and AAB-stacked graphenes under the generalized tight-binding model with any external fields. Such a model is closely related to the parameterized Hamiltonian; that is, the various hopping integrals, due to different orbital hybridizations, need to be taken into account for its diagonalization. Apparently, they play critical roles in expressing suitable and reliable Hamiltonian. Specifically, the current material of $Li_4Ti_5O_{12}$ presents an extremely nonuniform chemical environment in a primitive unit cell (discussed later in Figure 4.1); therefore, it might be very difficult to achieve a reliable tight-binding model with various chemical bonds.

4.3   RICH GEOMETRIC SYMMETRIES OF 3D $LI_4TI_5O_{12}$ COMPOUND

The 3D $Li_4Ti_5O_{12}$ compound exhibits unusual geometric symmetries. One of the metastable configurations, which possesses the smallest unit cell, is chosen for a model study. This corresponds to a triclinic structure, as clearly illustrated in Figure 4.1. A primitive unit cell has 42 atoms (8-Li, 10-Ti, and 24-O atoms), the lattice constants of $a = 5.288$ Å, $b = 9.532$ Å and $c = 9.932$ Å, and the titled angles of $\alpha = 73.122$, $\beta = 78.042$, and $\gamma = 78.913$ about $\hat{x}$, $\hat{y}$ and $\hat{z}$, respectively. The space-group symmetries belong to Hermann Mauguin (P$\bar{1}$), Hall (−P1), and point group ($\bar{1}$). Apparently, there exists a highly anisotropic and extremely nonuniform chemical environment. The projections of this 3D material on different planes present diverse atom arrangements, such as those of (a) (100), (b) (010), (c) (001), (d) (110), (e) (011), (f) (101), and (g) (111) (Figure 4.2a–g), where (Li, Ti, O) atoms are, respectively, represented by blue, red, and green balls. The real-space lattice is available in getting the reciprocal lattice, where the band structure is done along the paths of high symmetry points: X-Γ-Y|L- Γ-Z|N- Γ-M|R- Γ (their definitions in Figure 4.3).

The unique chemical bonds, which survive in a primitive unit cell will determine the fundamental properties. They cover Li-O and Ti-O bonds, in which their total numbers are 48 and 52, respectively. According to the delicate first-principles calculations in Table 4.1, their bond lengths might lie in a wide range of 1.988–2.479 and 1.757–2.210

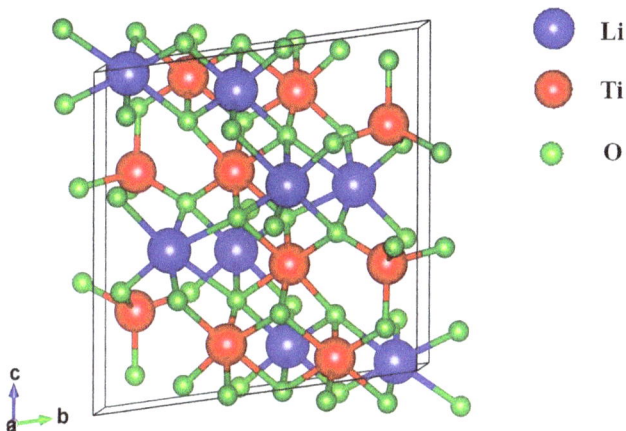

FIGURE 4.1   The optimal geometry of Li$_4$Ti$_5$O$_{12}$, with triclinic symmetry, where a primitive unit cell has 42 atoms, the lattice constants of $a = 5.288$ Å, $b = 9.532$ Å, and $c = 9.932$ Å and the titled angles of $\alpha = 73.122$, $\beta = 78.042$, and $\gamma = 78.91$

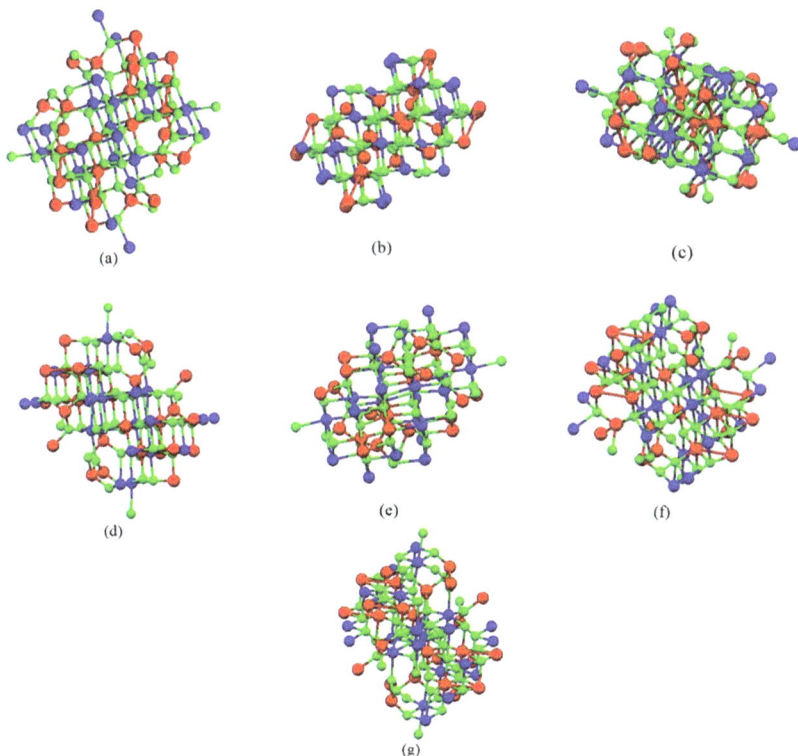

FIGURE 4.2   The geometric structure of Li$_4$Ti$_5$O$_{12}$ along different projections: (a) (100), (b) (010), (c) (001), (d) (110), (e) (011), (f) (101), and (g) (111), in which Li, Ti, and O atoms are, respectively, represented by the blue, red, and green balls

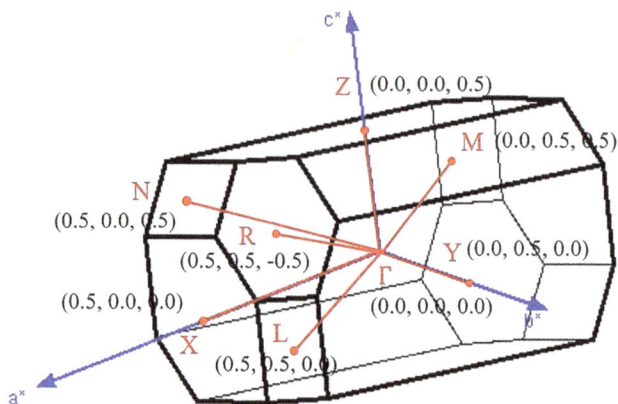

FIGURE 4.3   The first Brillouin zone with the paths of high symmetry points

Å. This diversified phenomenon will be directly reflected in the distribution width of the spatial charge density (Figure 4.5). Both Li-O and Ti-O bonds exhibit sufficiently large modulations, clearly indicating nonuniform bonding strengths. This might lead to the easy intercalation of the Li⁺ ions and thus the structural transformation to another metastable geometric configuration. Furthermore, all the chemical bonds are generated by the multi-orbital hybridizations (discussed later in orbital-dependent density of states by Figure 4.6). A lot of various orbital hybridizations come into existence in this anode material, and so do the hopping integrals in the tight-binding model [45].

The theoretical predictions on the optimal geometric structures could be verified from the experimental examinations. Both X-ray diffractions [46] and LEED [29,47] are suitable in measuring the 3D lattice symmetries, but not in STM [48] and tunneling electron microscopy (TEM; [31,48]) for the nanoscaled top- and side-view structures, respectively. For the 3D $Li_4Ti_5O_{12}$ material, they could be utilized to thoroughly explore the diversified lengths in LiO and TiO bonds. This will be very useful in indirectly identifying the complicated multi-orbital hybridizations in all chemical bonds.

4.4   RICH AND UNIQUE ELECTRONIC PROPERTIES

The 3D ternary compound of $Li_4Ti_5O_{12}$ exhibits unusual electronic states. The first-principles band structure, as indicated in Figure 4.4a, is very sensitive to the changes of wave vector. Consequently, it is highly anisotropic. Apparently, the occupied electron energy spectrum is asymmetric to the unoccupied hole one about the Fermi level ($E_F - 0$), mainly owing to different multi-orbital hybridizations in the nonuniform chemical bonds (Figure 4.1). The highest occupied valence-band state and the lowest unoccupied conduction-band one determine an energy gap of $E_g^d \sim 2.98$ eV (a red perpendicular arrow); furthermore, they come into existence at the same wave vector, leading to a direct-gap semiconductor. This band gap is just equal to the optical threshold absorption frequency; that is, the optical spectroscopies [49] are very suitable in examining the theoretical prediction of E_g^d. E_g is too wide to generate the spin-split electronic states across the Fermi level, where this phenomenon is supported by

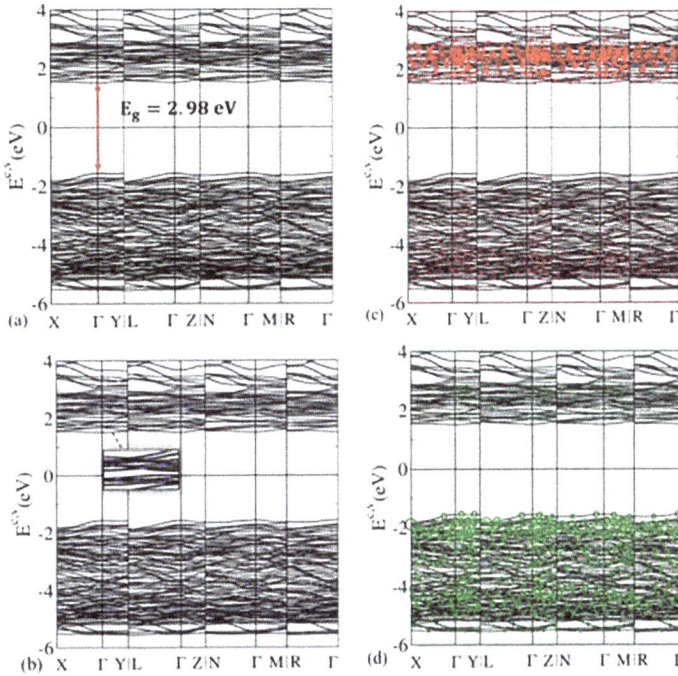

FIGURE 4.4 (a) Electronic energy spectrum for Li$_4$Ti$_5$O$_{12}$, along the high symmetry points in the first Brillouin zone within the range -6.0 eV $\leq E^{\nu} \leq 4.0$ eV for the specific, (b) lithium, (c) titanium, and (d) oxygen dominances (blue, red, and green balls, respectively).

the spin-degenerate valence and conduction bands, respectively, below and above E_F with a sufficiently great energy spacing. As a result, the net magnetic moment is zero and the spin-dependent interactions could be ignored in this ternary compound. There exist plenty of valence and conduction energy sub-bands because of many atoms and outer orbitals in a large unit cell. Generally speaking, the energy dispersions are weak, but significant. Their widths might lie in the range of ~0.08 to 0.31 eV. Different energy sub-bands would present the noncrossing, crossing, and anti-crossing behaviors. It should be noticed that the last phenomenon will appear when two neighboring sub-bands possess comparable independent components within the anti-crossing region [50], e.g., the almost competitive amplitude of the same mode in two anti-crossing Landau levels [50].

The atom dominances, which are represented by the ball radius, are very useful in understanding the critical roles of chemical bonding in the rich electronic states, such as the contributions to electronic states for the lithium, titanium, and oxygen atoms through the blue, red, and green balls, respectively. It is very difficult to observe the obvious contributions of the Li atoms, in which only small blue circles are revealed in the whole range of valence and conduction sub-bands (Figure 4.4b). This unique result clearly illustrates that the chemical bonding strengths (hopping integrals of neighboring atoms [27,51]) of the 48 Li-O bonds (Figures 4.2 and 4.3)

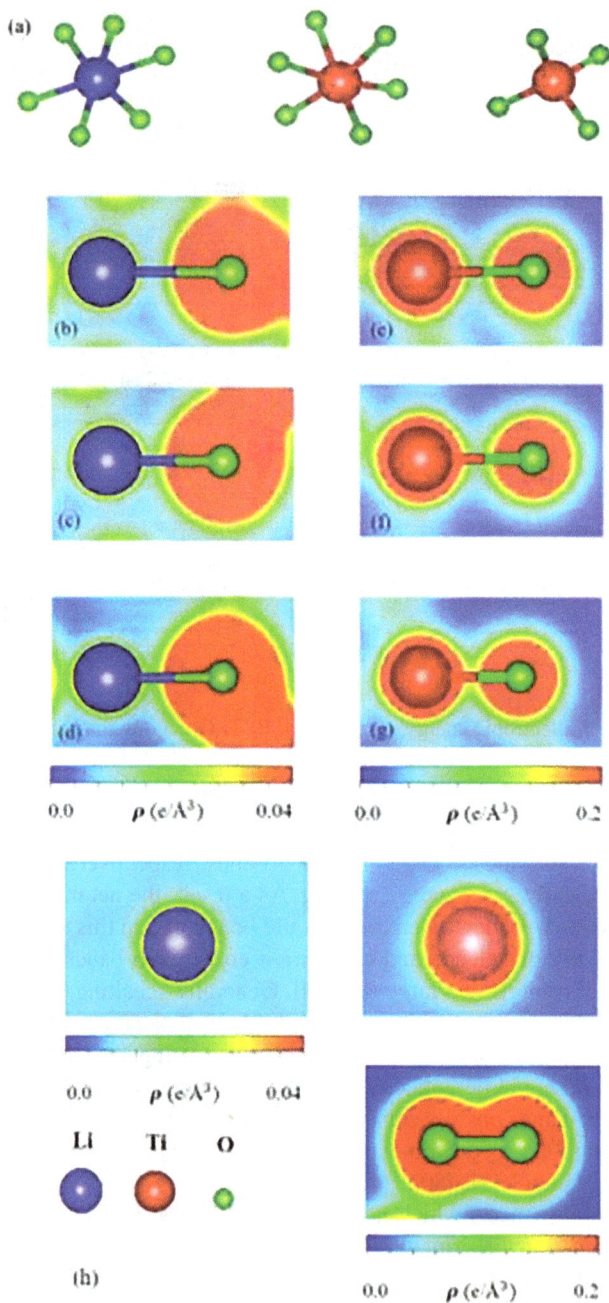

FIGURE 4.5 (a) Different LiO and TiO chemical bonds (LiO_6, TiO_4, TiO_6), in which the diversified spatial charge density distributions under the longest/middle/shortest for (b)/(c)/(d) the former and (e)/(f)/(g) the latter, respectively. Also shown in (h) are those of the isolated atoms

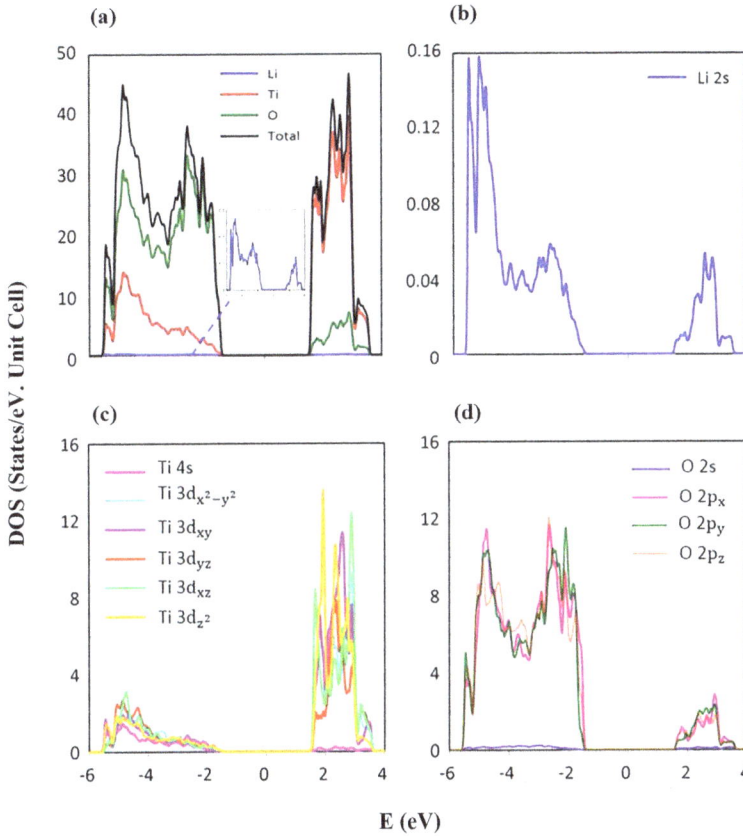

FIGURE 4.6 The atom- and orbital-decomposed density of states: those due to (a) Li, Ti, and O atoms (blue, red, and green curves), (b) Li-2s orbital (blue curve), (c) Ti-(4s, 3$d_{x^2-y^2}$, 3d_{xy}, 3d_{yz}, 3d_{zx}, 3d_{z^2}) orbitals (red, heavy red, purple, brown, light green, and light blue curves), and (d) O-(2s, 2p_x, 2p_y, 2p_z) orbitals (purple, pink, green, and orange curves).

should present a very wide range because of the large fluctuation of their lengths; furthermore, each Li-atom only makes the single-orbital contribution of 2s. On the contrary, the Ti-contributions are observable and very important within the valence- and conduction-band ranges of -5.50 eV $\le E^v \le 1.49$ eV and 1.49 eV $\le E^c \le 4.0$ eV, respectively, especially for the latter. Apparently, the unoccupied electronic states are dominated by the titanium atoms, but not the oxygen ones. The opposite is true for oxygen contributions. Such a phenomenon might be associated with the simultaneous existence of many O-related chemical bonds (100), with deeper (2s, 2p_x, 2p_y, 2p_z) orbital energies. In short, the atom-dominated electronic structure could provide partial information about the critical multi-orbital hybridizations.

The theoretical predictions, which are conducted on the wave-vector-dependent band structures below E_F, could be verified from the high-resolution ARPES [30]. The many previous measurements have clearly identified the low-energy valence bands of layered graphene systems, being only initiated from the K/K' valleys [52], such as the linear Dirac-cone structure in twisted bilayer graphenes (Moire superlattices with very large unit cells [53]) and monolayer graphene [54], parabolic/parabolic and linear dispersions in bilayer/trilayer AB stackings [54,55], linear, partially at and Sombrero-shaped bands in trilayer ABC stacking [55,56], and semimetallic property in bulk Bernal graphite [55]. These diversified electronic energy spectra are identified to only arise from the pure and unique interlayer $2p_z$-$2p_z$ orbital hybridizations in the normal/enlarged honeycomb lattices, according to a good consistence between the first-principles method [51] and the tight-binding model [27,57]. The similar experimental measurements are available in examining the main features of the 3D ternary $Li_4Ti_5O_{12}$ compound, such as the sensitive dependence on wave vector, the large band gap, the highly asymmetric electron and hole energy spectra, the weak, but significant energy dispersions, and the frequent noncrossing/crossing/anticrossing behaviors. Their verifications are very helpful in solving the critical orbital hybridizations of chemical bonds.

The spatial charge distributions on different chemical bonds, as clearly indicated in Figure 4.5a–h, are capable of providing the first-step orbital hybridizations, being further examined by the delicate atom- and orbital-decomposed van Hove singularities (discussed later in Figure 4.6). The optimal geometry, with a triclinic unit cell (Figure 4.2), shows the LiO_6, TiO_4, and TiO_6 chemical bonds, illustrating the highly nonuniform chemical environment. Their distinct bond lengths (Table 4.1) only direct the diversified chemical bonding strengths (diverse charge density distributions). That is to say, the spatial carrier densities, being closely related to electron orbitals of each atom, are very sensitive to the changes of the nearest-neighbor distances. As for the shortest Li-O bonds (1.988 Å in Figure 4.5d), only the $2s$ orbital only provides a dilute charge density around the lithium atom (Figure 4.5d), in which the effective distribution range is about 0.52 Å as measured from the deep blue region of the Li^+-ion core to the light one of the extended $2s$ orbital. Apparently, the two $1s$ orbitals are independent of the critical orbital hybridizations in the Li-O bonds. There exists an obvious overlap of distinct orbital charges in the range of 1.20–1.60 Å in the distance with the O-core. Furthermore, the significant O orbitals, being indicated by the light green and yellow colors, are deduced to arise from the $(2p_x, 2p_y, 2p_z)$ ones. However, the O-dependent $2s$ orbitals are far away from that of the lithium atom; therefore, their contributions to the chemical bonds are almost negligible. With the increasing of Li-O length (1.987 and 2.479 Å in Figure 4.6c and d, respectively), the charge overlap behavior between Li and O atoms is obviously reduced. Very interestingly, the multi-orbital hybridizations of $2s$-$(2p_x, 2p_y, 2p_z)$ in Li-O bonds present the diverse hopping integrals [57].

There are more Ti-O chemical bonds and quite strong bonding strengths (Figure 4.5d–f), compared with those of Li-O bonds. The effective distribution range of Ti atom is somewhat higher than that of O one because of the large atomic number. Most importantly, it could be classified into heavy red, light red, and yellow-green regions, which, respectively, corresponds to $(3s, 3p_x, 3p_y, 3p_z)$, $(4s)$, and $(3d_{x^2-y^2}, 3d_{xy}, 3d_{yz}, 3d_{zx}, 3d_{z^2})$. Apparently, the first ones do not take part in the important multi-orbital hybridizations in the Ti-O bonds. The deformed carrier distributions between Ti and O atoms, as shown in the distinct bond-length cases, clearly illustrate the diversified charge transfers from the former to the latter. The extremely nonuniform chemical bonds even for similar bonds will induce the high barriers in generating

TABLE 4.1
Various Chemical Bond Lengths in Li$_4$Ti$_5$O$_{12}$, in Which the Total Number of Li-O and Ti-O Bonds Is 100

No.	Atom	Atom	Bond Length	No.	Atom	Atom	Bond Length
1	Li0001	O0001	2.0474	25	Li0005	O0007	2.4002
2	Li0001	O0013	2.1712	26	Li0005	O0013	2.3493
3	Li0001	O0015	2.249	27	Li0005	O0016	1.9995
4	Li0001	O0022	2.2568	28	Li0005	O0019	2.1456
5	Li0001	O0004	2.4791	29	Li0005	O0021	2.4662
6	Li0001	O0014	2.0096	30	Li0005	O0002	2.0744
7	Li0002	O0002	2.0474	31	Li0006	O0008	2.4002
8	Li0002	O0014	2.1712	32	Li0006	O0014	2.3493
9	Li0002	O0016	2.249	33	Li0006	O0015	1.9995
10	Li0002	O0021	2.2568	34	Li0006	O0020	2.1456
11	Li0002	O0003	2.4791	35	Li0006	O0022	2.4662
12	Li0002	O0013	2.0096	36	Li0006	O0001	2.0744
13	Li0003	O0023	2.4256	37	Li0007	O0011	2.1386
14	Li0003	O0005	2.3937	38	Li0007	O0022	2.2864
15	Li0003	O0006	1.9903	39	Li0007	O0023	2.1019
16	Li0003	O0009	2.2102	40	Li0007	O0004	2.0344
17	Li0003	O0017	2.257	41	Li0007	O0009	2.2124
18	Li0003	O0018	1.9877	42	Li0007	O0012	2.2576
19	Li0004	O0024	2.4256	43	Li0008	O0012	2.1386
20	Li0004	O0005	1.9903	44	Li0008	O0021	2.2864
21	Li0004	O0006	2.3937	45	Li0008	O0024	2.1019
22	Li0004	O0010	2.2102	46	Li0008	O0003	2.0344
23	Li0004	O0017	1.9877	47	Li0008	O0010	2.2124
24	Li0004	O0018	2.257	48	Li0008	O0011	2.2576

No.	Atom	Atom	Bond Length	No.	Atom	Atom	Bond Length
1	Ti0002	O0001	2.1529	27	Ti0006	O0009	1.8607
2	Ti0002	O0011	2.09	28	Ti0006	O0012	1.9487
3	Ti0002	O0012	2.1281	29	Ti0006	O0018	2.1414
4	Ti0003	O0005	1.7735	30	Ti0007	O0001	1.7859
5	Ti0003	O0015	1.8475	31	Ti0007	O0017	1.7749
6	Ti0003	O0008	1.8822	32	Ti0007	O0020	1.9086
7	Ti0003	O0019	1.8413	33	Ti0007	O0007	1.8958
8	Ti0004	O0006	1.7735	34	Ti0008	O0002	1.7859
9	Ti0004	O0016	1.8475	35	Ti0008	O0018	1.7749
10	Ti0004	O0007	1.8822	36	Ti0008	O0019	1.9086
11	Ti0004	O0020	1.8413	37	Ti0008	O0008	1.8958
12	Ti0005	O0003	2.1221	38	Ti0009	O0003	2.0313
13	Ti0005	O0005	1.9673	39	Ti0009	O0013	2.2098
14	Ti0005	O0024	1.8371	40	Ti0009	O0014	1.7614

(Continued)

No.	Atom	Atom	Bond Length	No.	Atom	Atom	Bond Length
15	Ti0005	O0010	1.8607	41	Ti0009	O0015	2.0196
16	Ti0005	O0011	1.9487	42	Ti0009	O0021	1.948
17	Ti0005	O0017	2.1414	43	Ti0009	O0024	2.0203
18	Ti0006	O0004	2.1221	44	Ti0010	O0004	2.0313
19	Ti0006	O0006	1.9673	45	Ti0010	O0013	1.7614
20	Ti0006	O0023	1.8371	46	Ti0010	O0014	2.2098
21	Ti0001	O0002	2.1529	47	Ti0010	O0016	2.0196
22	Ti0001	O0011	2.1281	48	Ti0010	O0022	1.948
23	Ti0001	O0012	2.09	49	Ti0010	O0023	2.0203
24	Ti0002	O0004	2.0336	50	Ti0001	O0009	1.8139
25	Ti0002	O0010	1.8139	51	Ti0001	O0003	2.0336
26	Ti0002	O0022	1.7571	52	Ti0001	O0021	1.7571

suitable hopping integrals for the phenomenological models, e.g., the reliable parameters in the tight-binding model [51,57].

The atom- and orbital-projected density states, as clearly indicated in Figure 4.6a–d, are able to provide the full information on the significant multi-orbital hybridizations on the Li-O and Ti-O chemical bonds. Apparently, the density of states per unit cell vanishes with a large band gap of $E_g^d \sim 2.98$ eV (Figure 4.4a–d), and its valence and conduction spectra are highly asymmetric to each other. Whether a very wide gap will induce the difficulties of experimental observations could be tested in further STS measurements (discusses later; [58,59]). There exist a lot of shoulders and asymmetric/symmetric peaks. Such van Hove singularities principally originate from the band-edge states of energy sub-bands, with the local minimum, maximum, saddle, and almost dispersionless points in the energy wave-vector spaces. They might appear in between the high symmetry points. It is well known that the 3D parabolic energy dispersions can create the square-root dependences [60]. Generally speaking, the titanium and oxygen atoms, respectively, dominate the density of states in the energy range of -5.50 eV $\leq E^v \leq 1.49$ eV and 1.49 eV $\leq E^c \leq 4.0$ eV (red and green curves). Furthermore, the lithium atoms only make minor contributions in the whole valence and conduction energy spectrum (blue curve in Figure 4.6b) through only the $2s$ orbital, but not two core-level 1s orbitals.

The important contributions of various orbitals, which mainly come from (I) Li-$2s$ orbital (blue curve in Figure 4.6b), (II) Ti-($4s$, $3d_{x^2-y^2}$, $3d_{xy}$, $3d_{yz}$, $3d_{zx}$, $3d_{z^2}$) orbitals (red, heavy red, purple, brown, light green, and light blue curves in Figure 4.6c), and (III) O-($2s$, $2p_x$, $2p_y$, $2p_z$) orbitals (purple, pink, green, and orange curves in Figure 4.6d) are worthy of a closer examination. The enlarged spectral scale, which is done for lithium atoms, is able to provide clear van Hove singularities, i.e., their number, energies, intensities, and forms are clearly illustrated in Figure 4.6b. These main features are similar to those of ($2p_x$, $2p_y$, $2p_z$)-decomposed density of states for oxygen atoms (Figure 4.6d). The similarities, which cover the whole energy spectrum (e.g., -6.0 eV $\leq E \leq 4.0$ eV), only reflect the very important 48 Li-O chemical bonds with the multi-orbital hybridizations (Figures 4.1 and 4.2). It should be noticed that the O-$2s$ orbitals make almost zero contributions in the specific energy range. The rincipal reason might be that they belong to the fully occupied states under the spin-up and spin-down configurations; therefore, any dangling bonds/chemical bonds are forbidden. The effective width of energy spectrum, being closely relate to the Li-O Chemical bonds, is more than 10.0 eV. This unusual characteristic is attributed to the complicated $2s$-$(2p_x, 2p_y, 2p_z)$ chemical bonds (multi-orbital hopping integrals and on-site Coulomb potentials; [57]) and the large bond-length modulations (1.988–2.479 Å; the easily modulated hopping integrals).

In addition to the Li-O bonds, the oxygen atoms also make significant contributions to the Ti-O bonds, especially for the valence-state energy spectrum. The obvious evidences are revealed in the similar van Hove singularities in terms of the orbital-decomposed special forms, energies, and numbers. Concerning the transition-metal titanium atoms there exist unique six orbitals ($4s, 3d_{x^2-y^2}, 3d_{xy}, 3d_{yz}$, $3d_{zx}, 3d_{z^2}$) that take part in the chemical bonding of oxide compounds, as clearly illustrated in Figure 4.6c. The orbital-dependent contributions are almost comparable in the whole energy spectrum except for the small $4s$ orbital density of state with the conduction-band range (pink curve). Very interestingly, the initial conduction-/valence-state energy spectrum is mainly determined by the titanium and oxygen atoms. In short, the Li-O and Ti-O chemical bonds are deduced to have multi-orbital hybridizations of $2s$-$(2p_x, 2p_y, 2p_z)$, $(4s, 3d_{x^2-y^2}, 3d_{xy}, 3d_{yz}$ and $3d_{zx}, 3d_{z^2})$-$(2p_x, 2p_y, 2p_z)$, respectively.

The high-resolution STS measurement is the most efficient technique in examining the various van Hove singularities due to the band-edge states. Up to date, a rather weak tunneling quantum current could be measured in a very accurate way; furthermore, its differential conductance of dI/dV is deduced to be approximately proportional to the density of states. For example, such experiments have been successful in identifying the significant coupling effects in few-layer graphene systems, e.g., an almost symmetric V-shape structure vanishing at the Fermi level for monolayer (a zero-gap semiconductor; [61]), the logarithmically symmetric peaks in twisted bilayer graphenes [62], a gate-voltage-induced energy gap for bilayer AB and trilayer ABC stackings [55,62], a delta-function-like peak centered about E_F in trilayer and penta-layer ABC stackings [63,64], a sharp dip structure near E_F combined with a pair of square-root peaks under trilayer AAB stacking (a narrow-gap semiconductor with constant-energy loops; [65]). Further experimental examinations, which are very suitable for the 3D ternary Li₄Ti₅O₁₂ compound, are required to detect a large band gap of $E_g^d \sim 2.98$ eV, a lot of asymmetric/symmetric peaks and broadening shoulders, the high asymmetry of electron and hole energy spectrum, and their different widths. The measured results, being combined with the theoretical predictions on van Hove singularities, might be very useful in comprehending the complicated multi-orbital hybridizations of Li-O and Ti-O.

The first-principles results could be utilized to establish the phenomenological model, the tight-binding model, as discussed in Section 4.3 But for the 3D Li₄Ti₅O₁₂ compound [57], there exist a lot of different chemical bonds in a primitive unit cell (100 Li-O and Ti-O bonds Figures 4.1 and 4.2), very complicated valence and conduction sub-bands with a large band gap of $E_g \sim 2.98$ eV (Figure 4.4a–d), nonhomogeneous spatial charge densities in diverse chemical bonds (Figure 4.5a–g), and three kinds of multi-orbital hybridizations in van Hove singularities (Figure 4.6a–d). Such critical factors are almost impossible to be included in the parameterized tight-binding model.

That is to say, the extremely nonuniform hopping integrals, which are associated with various orbital hybridizations, are rather difficult to be achieved in order to the main features of the first-principles band structure. Apparently, how to obtain a reliable tight-binding model in Li^+-based anode [66], cathode [66], and electrolyte materials [66] will become an open issue, since it is relatively easy in understanding the essential physical, material, and chemical properties [67] from the concise pictures.

By using VASP based on the first-principles calculations, the optimal geometry structures and band structures have been achieved, being consistent with the high-resolution experimental measurements [68–72]. The delicate X-ray diffraction analysis is conducted in the range of 2θ (θ is known as Bragg angle, causing by the incident beam hits the parallel planes at a certain angle), and its pattern is utilized to compare with the standard diffraction peaks of spinel compounds. Based on recent researches, $Li_4Ti_5O_{12}$ under the $Fd\overline{3}m$ space group is well fixed between experimental and theoretical data through the Raman spectra [69,72]. For example, equilibrium lattice parameter of $Li_4Ti_5O_{12}$ is 8.42 Å, which is close to the experimental value of 8.38 Å [68]. However, $Li_4Ti_5O_{12}$ of the $\overline{P}$ group still lacks experimental examinations. Besides the X-ray diffraction, the LEED [70] is available in investigating the 3D lattice symmetries. Using this technique, a collimated beam of low-energy electrons (in the range of 20–200 eV) will incident on a single-crystalline material. As a result, the diffracted electrons as spots on a fluorescent screen could be observed to determine the surface morphology. Scanning electron microscopy (SEM; Refs) is able to provide the information about surface properties, as revealed from defects [69, 72]. These drawbacks in 3D compounds might play an important role in controlling and manipulating the fundamental material properties. As to band gap, the predicted value of 2.98 eV is almost identical to the experimental result of 2.95 eV by using spectroscope ellipsometry (SE) and UV–vis spectrophotometer data, in which the absorption edge is observed at ~400–420 nm [71]. Moreover, the band energy dispersion along the synthetic dimension needs to be identified from the ARPES [30,73] which is a powerful experimental technique measuring directly the single-particle spectral function $A(k, \omega)$. In APRES experiments, electrons are probed inside solids in the energy and momentum space and are able to provide some physical information, such as Fermi surface, energy gap, and many-body interactions (e.g., electron–phonon, electron–electron interactions). As an advanced technique, high-resolution ARPES microscopy using a well-focused light source has recently attracted many interests based on its potential to achieve local electronic information at micro- or nanoscale [73].

4.5 CONCLUDING REMARKS

The theoretical framework, which is based on the first-principles calculations [26], is developed for the essential properties of the 3D ternary $Li_4Ti_5O_{12}$ compound. The critical multi-orbital hybridizations in Li-O and Ti-O chemical bonds are delicately identified from the atom-dominated valence and conduction bands, the spatial charge density, the atom- and orbital-decomposed density of states. Their highly anisotropic and nonuniform characteristics in a large unit cell clearly indicate that the reliable tight-binding model, with various hopping integrals and on-site Coulomb potentials, is almost impossible to achieve for the simulation of the VASP band structure.

Similar developments could be generalized to the Li$^+$-based battery cathode [74,75], anode [74,75], and electrolyte materials [76]. Especially, the rich oxide compounds are worthy of systematic investigations for the diversified physical, chemical, and material science phenomena.

The current anode material, with the smallest unit cell of 42 atoms, is a triclinic structure in the titled three axes. There exist 100 chemical bonds, in which the very strong covalent bonds make this system presents a large direct gap of $E_g^d \sim 2.98$ eV. This gap is equal to the optical threshold absorption frequency. The high-resolution measurements of optical spectroscopies [77] are expected to be very efficient in examining the semiconducting behavior. A lot of valence and conduction bands are highly asymmetric to each other about the Fermi level, in which their energy dispersions are weak along any directions. Generally speaking, the whole band structure, being related to the critical chemical bonds, lies in the range of -6.0 eV $\leq E^v \leq 4.0$ eV. It also reveals the frequent sub-band noncrossings, crossings, and anti-crossings [50]. Moreover, the band-edge states, the critical points in the energy wave-vector spaces, appear as the van Hove singularities in the density of states. The atom- and orbital-dependent special structures, asymmetric/symmetric peaks and broadening shoulders, are successful in identifying the important multi-orbital hybridizations, such as the $2s$-$(2p_x, 2p_y, 2p_z)$, and $(4s, 3d_{x^2-y^2}, 3d_{xy}, 3d_{yz}, 3d_{zx}, 3d_{z^2})$-$(2p_x, 2p_y, 2p_z)$ in Li-O and Ti-O bonds, respectively. The critical mechanisms are also supported by the spatial charge density distribution. The diverse covalent bonds are partially supported by the atom-created energy bands and charge density distributions. Very interestingly, in addition to Li$_4$Ti$_5$O$_{12}$ compound, there exist other 3D LiTiO-related ternary materials, e.g., LiTi$_2$O$_4$ [78], Li$_2$Ti$_2$O$_4$ [78,79], and Li$_7$Ti$_5$O$_{12}$ [78,80]. Such unusual condensed matter systems are examined to have diverse geometric symmetries [78,80] and thus exhibit quite different electronic [78], optical [78], and superconducting properties. This current study clearly shows the very large variations of chemical bond lengths. The relatively easy modulations on chemical bonds might be helpful in solving the open issue: the structural transformation between two metastable configurations during the charging or discharging processes of Li$^+$-based batteries.

REFERENCES

1. Li, M.; Lu, J.; Chen, Z.; Amine, K. 30 years of lithium-ion batteries. *Advanced Materials* **2018**, 30, 1800561.
2. Pistoia, G. *Lithium-Ion Batteries: Advances and Applications*, 1st edition, Elsevier, Amaterdam, Netherlan, **2014**.
3. Deng, D. Li-ion batteries: Basics, progress, and challenges. *Energy Science and Engineering* **2015**, 3(5), 385–418.
4. Nitta, N.; Wu, F.; Lee, J. T.; Yushin, G. Li-ion battery materials present and future. *Materials Today* **2015**, 18, 1369–7021.
5. Gwon, H.; Hong, J.; Kim, H.; Seo, D.-H.; Jeon, S.; Kang, K. Recent progress on flexible lithium rechargeable batteries. *Energy and Environmental Science* **2014**, 7(2), 538–551.
6. Choi, S.; Wang, G. Advanced lithium-ion batteries for practical applications: Technology, development, and future perspectives. *Advanced Material Technology* **2018**, 1700376.
7. Zhou, G.; Li, F.; Cheng, H.-M. Progress in flexible lithium batteries and future prospects. *Energy Environmental Science* **2014**, 7(4), 1307–1338.

8. Yu, M.; Patrick, H.; Annette, V. J.; Alexandre, Y. Current Li-ion battery technologies in electric vehicles and opportunities for advancements. *Energies* **2019**, 12, 1074.
9. Armand, M.; Tarascon, J. M. Building better batteries. *Nature* **2008**, 451(7179), 652–657.
10. Ouyang, D.; Chen, M.; Liu, J.; Wei, R.; Weng, J.; Wang, J. Investigation of a commercial lithium-ion battery under overcharge/over-discharge failure conditions. *RSC Advances* **2018**, 8(58), 33414–33424.
11. Wang, X.; Sone, Y.; Segami, G.; Naito, H.; Yamada, C.; Kibe, K. Understanding volume change in lithium-ion cells during charging and discharging using in situ measurements. *Journal of the Electrochemical Society* **2007**, 154(1), A14.
12. Schweidler, S.; Biasi, L. D.; Schiele, A.; Hartmann, P.; Brezesinski, T.; Janek, J. Volume changes of graphite anodes revisited: A combined operando X-ray diffraction and in situ pressure analysis study. *The Journal of Physical Chemistry C* **2018**, 122, 8829–8835.
13. Lin, S. Y.; Li, W. B.; Tran, N. T. T.; Schiele, A.; Hartmann, P.; Brezesinski, T.; Janek, J. Essential properties of Li/Li+ graphite intercalation compounds. arXiv: *Materials Science* **2018**, 37–63.
14. Ji, K.; Han, J.; Hirata, A.; Fujita, T.; Shen, Y.; Ning, S.; Liu, P.; Kashani H.; Tian, Y.; Ito, Y.; Fujita, J.; Oyama, Y. Lithium intercalation into bilayer graphene. *Nature Communications* **2019**, 10(1), 1–10.
15. Tanaka, Y.; Ikeda, M.; Sumita, M.; Ohno, T.; Takada, K. First-principles analysis on role of spinel (111) phase boundaries in $Li_4 + 3_x Ti_5 O_{12}$ Li-ion battery anodes. *Physical Chemistry Chemical Physics* **2016**, 18(33), 23383–23388.
16. Zahn, S.; Janek, J.; Mollenhauer, D. A Simple ansatz to predict the structure of $Li_4 Ti_5 O_{12}$. *Journal of the Electrochemical Society* **2016**, 164(2), A221–A225.
17. Ohta, T.; Bostwick, A.; Seyller, T.; Horn, K.; Rotenberg, E. Controlling the electronic structure of bilayer graphene. *Science* **2006**, 313, 951–954.
18. Park, H. J.; Meyer, J.; Roth, S.; Skakalova, V. Growth and properties of few-layer graphene prepared by chemical vapor deposition. *Carbon* **2010**, 48(4), 1088–1094.
19. Celis, A.; Nair, M. N.; Taleb-Ibrahimi, A.; Conrad, E. H.; Berger, C.; de Heer, W.; Tejeda, A. Graphene nanoribbons: Fabrication, properties and devices. *Journal of Physics D: Applied Physics* **2016**, 49(14), 143001.
20. Jose, D.; Datta, A. Structures and electronic properties of silicene clusters: A promising material for FET and hydrogen storage. *Physical Chemistry Chemical Physics* **2011**, 13(16), 7304.
21. Nguyen, D. K.; Tran, N. T. T.; Chiu, Y.-H.; Lin, M. F. Concentration-diversified magnetic and electronic properties of halogen-adsorbed silicene. *Scientific Reports* **2019**, 9(1), 13746.
22. Sivek, J.; Sahin, H.; Partoens, B.; Peeters, F. M. Adsorption and absorption of boron, nitrogen, aluminum, and phosphorus on silicene: Stability and electronic and phonon properties. *Physical Review B* **2013**, 87, 085444.
23. Teshome Ayan, T.; Orcid, D. Effect of doping in controlling the structure, reactivity, and electronic properties of pristine and Ca (II)-intercalated layered silicene. *Journals of Physical Chemistry C* **2017**, 121(28), 15169–15180.
24. Bhatt, M. D.; O'Dwyer, C. Recent progress in theoretical and computational investigations of Li-ion battery materials and electrolytes. *Physical Chemistry Chemical Physics* **2015**, 17(7), 4799–4844.
25. Wu, X.; Pan, K.; Jia, M.; Ren, Y.; He, H.; Zhang, L.; Zhang, S. Electrolyte for lithium protection: From liquid to solid. *Green Energy and Environment* **2015**, 3(5), 385–418.
26. Lin, S. Y.; Chang, S. L.; Chen, H. H.; Su, S. H.; Huang, J.-C.; Lin, M.-F. Substrate-induced structures of bismuth adsorption on graphene: A first principles study. *Physical Chemistry Chemical Physics* **2016**, 18(28), 18978–18984.

27. Nielsen, E.; Rahman, R.; Muller, R. P. A many-electron tight binding method for the analysis of quantum dot systems. *Journal of Applied Physics* **2012**, 112(11), 114304.
28. Huang, Y. K.; Chen, S. C.; Ho, Y. H.; Lin, C. Y.; Lin, M. F. Feature-rich magnetic quantization in sliding bilayer graphenes. *Scientific Reports* **2014**, 4(1), 7509.
29. Jona, F.; Strozier, J. A.; Yang, W. S. Low-energy electron diffraction for surface structure analysis. *Reports on Progress in Physics* **1982**, 45(5), 527–585.
30. Lv, B.; Qian, T.; Ding, H. Angle-resolved photoemission spectroscopy and its application to topological materials. *Nature Reviews Physics* **2019**, 1, 609–626.
31. Binnig, G.; Rohrer, H. Scanning tunneling microscopy. *Surface Science* **1985**, 152/153, 17–26.
32. Argaman, N.; Makov, G. Density functional theory: An introduction. *American Journal of Physics* **2000**, 68(1), 69–79.
33. Kresse, G.; Furthmuller, J. Efficiency of ab-initio total energy calculations for metals and semiconductors using a plane-wave basis set. *Computational Materials Science* **1996**, 6(1), 15–50.
34. Perdew, J. P.; Burke, K.; Ernzerhof, M. Generalized gradient approximation made simple. *Physical Review Letters* **1996**, 77(18), 3865.
35. Blochl, P.E. Projector augmented-wave method. *Physical Review B* **1994**, 50, 17953–17979.
36. Rangel, T.; Caliste, D.; Genovese, L.; Torrent, M. A wavelet-based Projector Augmented-Wave (PAW) method: Reaching frozen-core all-electron precision with a systematic, adaptive and localized wavelet basis set. *Computer Physics Communications* **2016**, 208, 1–8.
37. Booth, G. H.; Tsatsoulis, T.; Chan, G. K.-L.; Gruneis, A. From plane waves to local Gaussians for the simulation of correlated periodic systems. *The Journal of Chemical Physics* **2016**, 145(8), 084111.
38. Warner, J. H.; Rummeli, M. H.; Gemming, T.; Buchner, B.; Briggs, G. A. D. Direct imaging of rotational stacking faults in few layer graphene. *Nano Letters* **2008**, 9, 102–106.
39. Lin, S. Y.; Tran, N. T. T.; Chang, S. L.; Su, W. P.; Lin, M.-F. *Structure- and Adatom-Enriched Essential Properties of Graphene Nanoribbons*, 1st edition. CRC Press, Boca Raton, FL, **2018**.
40. Ho, Y.-H.; Wu, J.-Y.; Chen, R.-B.; Chiu, Y.-H.; Lin, M.-F. Optical transitions between Landau levels: AA-stacked bilayer graphene. *Applied Physics Letters* 2010, 97, 101905.
41. Lee, J. K.; Lee, S. C.; Ahn, J. P.; Kim, S. C.; Wilson, J. I.; John, P. The growth of AA graphite on (111) diamond. *The Journal of Chemical Physics* **2008**, 129, 234709.
42. Rozhkova, A. V.; Sboychakova, A. O.; Rakhmanova, A. L.; Noria, F. Electronic properties of graphene-based bilayer systems. *Physical Chemistry Chemical Physics* **2015**, 17(7), 4799–4844.
43. Lu, C. L.; Chang, C. P.; Huang, Y. C.; Chen, R. B.; Lin, M. L. Influence of an electric field on the optical properties of few-layer graphene with AB stacking. *Physical Review B* **2006**, 73, 144427.
44. Yankowitz, M.; Joel, I.; Wang, J.; Birdwell, A. G.; Chen, Y.-A.; Watanabe, K.; Taniguchi, T.; Jacquod, P.; San-Jose, P.; Jarillo-Herrero, P. Electric control of soliton motion and stacking in trilayer graphene. *Nature Materials* **2014**, 13, 786.
45. Wang, C.-Z.; Lu, W.-C.; Yao, Y.-X; Li, J.; Yip, S.; Ho, K.-M. Tight-binding Hamiltonian from first-principles calculations. *Science Model Simulation* **2008**, 15, 81–95.
46. Sharma, R.; Bisen, D. P.; Shukla, U.; Sharma, B. G. X-ray diffraction: A powerful method of characterizing nanomaterials. *Recent Research in Science and Technology* **2012**, 4(8), 77–79.
47. Clarke, L. J. *Surface Crystallography: An Introduction to Low Energy Electron Diffraction*. John Wiley & Sons Ltd, Hoboken, NJ, **1985**.

48. Lauer, P.; Emtsev, K. V.; Graupner, R.; Seyller, T.; Ley, L.; Reshanov, S. A.; Weber, H. B. Atomic and electronic structure of few-layer graphene on SiC (0001) studied with scanning tunneling microscopy and spectroscopy. *Physical Review Letters B* **2008**, 77, 155426.

49. Jeong, T. Y.; Kim, H.; Choi, S.-J.; Watanabe, K.; Taniguchi, T.; Yee, K. J.; Kim, Y. S.; Jung, S. Spectroscopic studies of atomic defects and bandgap renormalization in semi-conducting monolayer transition metal dichalcogenides. *Nature Communications* **2019**, 10(1), 3825.

50. Zhang, X. C.; Martin, I.; Jiang, H. W. Landau level anti-crossing manifestations in the phase-diagram topology of a two-subband system. *Physical Review B* **2006**, 74(7), 073301.

51. Koskinen, P.; Makinen, V. Density-functional tight-binding for beginners. *Computational Materials Science* **2009**, 47, 237–253.

52. Koshino, M.; McCann, E. Parity and valley degeneracy in multilayer graphene. *Physical Review B* **2010**, 81(11), 115315.

53. Ryu Cho, Y. K.; Frisenda, R.; Castellanos-Gomez, A. Superlattices based on van der Waals 2D materials. *Chemical Communications* **2019**, 55, 11498.

54. Castro Neto, A. H.; Guinea, F.; Peres, N. M. R.; Novoselov, K. S.; Geim, A. K. The electronic properties of graphene, *Reviews of Modern Physics* **2009**, 81, 109.

55. Coletti, C.; Forti, S.; Principi, A.; Emtsev, K. V.; Zakharov, A. A.; Daniels, K. M.; Daas, B. K.; Chandrashekhar, M.; Ouisse, T.; Chaussende, D. Revealing the electronic band structure of trilayer graphene on SiC: An angle-resolved photoemission study. *Physical Review B* **2013**, 88, 155439.

56. Hattendorf, S.; Georgi, A.; Liebmann, M., Morgenstern, M. Networks of ABA and ABC stacked graphene on mica observed by scanning tunneling microscopy. *Surface Science* **2013**, 610, 53–58.

57. Zolyomi, V.; Wallbank, J. R.; Fal'ko, V. I. Silicane and germanane: Tight-binding and first-principles studies. *2D Materials* **2014**, 1(1), 011005.

58. Kano, S.; Tadaa, T.; Majima, Y. Nanoparticle characterization based on STM and STS. *Chemical Society Reviews* **2015**, 44, 970.

59. Li, G.; Luican, A.; Andrei, E. Y. Scanning tunneling spectroscopy of graphene on graphite. *Physical Review Letters* **2009**, 102, 176804.

60. Tang, S.; Dresselhaus, M. S. Anisotropic transport for parabolic, non-parabolic, and linear bands of different dimensions. *Applied Physics Letters* **2014**, 105(3), 033907.

61. Chung, H. C.; Chang, C. P.; Lin, C. Y.; Lin, M. F. Electronic and optical properties of graphene nanoribbons in external fields. *Physical Chemistry Chemical Physics* **2016**, 18(11), 7573–7616.

62. Wang, Z. F.; Liu, F.; Chou, M. Y. Fractal Landau-level spectra in twisted bilayer graphene. *Nano Letters* **2012**, 12(7), 3833–3838.

63. Ho, C. H.; Tsai, S. J.; Chen, R. B.; Chiu, Y. H.; Lin, M. F. Low-energy Landau level spectrum in ABC-stacked trilayer graphene. *Journal of Nanoscience and Nanotechnology* **2011**, 11(6), 4938–4947.

64. Que, Y.; Xiao, W.; Chen, H.; Wang, D.; Du, S.; Gao, H.-J. Stacking-dependent electronic property of trilayer graphene epitaxially grown on Ru (0001). *Applied Physics Letters* **2015**, 107(26), 263101.

65. Lin, C.-Y.; Huang, B.-L.; Ho, C.-H.; Gumbs, G.; Lin, M.-F. Geometry-diversified Coulomb excitations in trilayer AAB stacking graphene. *Physical Review B* **2018**, 98(19), 195442.

66. Chou, C.-P.; Sakti, A. W.; Nishimura, Y.; Nakai, H. Development of divide-and-conquer density-functional tight-binding method for theoretical research on Li-ion battery. *The Chemical Record* **2018**, 18, 1–13.

67. Lee, M. H.; Chung, H. C.; Lu, J. M.; Chang, C. P.; Lin, M.-F. Electronic and optical properties in graphene. *Philosophical Magazine* **2015**, 95(24), 2717–2730.

68. Wang, Y.; Zhang, Y.; Yang, W.-J.; Jiang, S.; Hou, X.; Guo, R., Liu, W.; Ping Huang, P.; Lu, J. -C.; Gu, H.-T., Xie, J. Enhanced rate performance of $Li_4Ti_5O_{12}$ anode for advanced lithium batteries. *Journal of the Electrochemical Society* **2018**, 166(3), A5014–A5018.

69. Vikram Babu, B.; Vijaya Babu, K.; Tewodros Aregai, G.; Seeta Devi, L.; Madhavi Latha, B.; Sushma Reddi, M.; Samatha, K.; Veeraiah, V. Structural and electrical properties of $Li_4Ti_5O_{12}$ anode material for lithium-ion batteries. *Results in Physics* **2018**, 9, 284–289.

70. *The Surface Science Society of Japan, Compendium of Surface and Interface Analysis*. Springer, Singapore, **2018**, 349–354.

71. Özen, S.; Şenay, V.; Pat, S.; Korkmaz, Ş. Optical, morphological properties and surface energy of the transparent $Li_4Ti_5O_{12}$(LTO) thin film as anode material for secondary type batteries. *Journal of Physics D: Applied Physics* 2016, 49(10), 105303.

72. Chen, W.; Jiang, H., Hu, Y.; Dai, Y.; Li, C. Mesoporous single crystals $Li_4Ti_5O_{12}$ grown on rGO as high-rate anode materials for lithium-ion batteries. *Chemical Communications* **2014**, 50(64), 8856–8859.

73. Iwasawa, H.; Takita, H.; Goto, K.; Mansuer, W.; Miyashita, T.; Schwier, E. F.; Ino, A.; Shimada, K.; Aiura, Y. Accurate and efficient data acquisition methods for high-resolution angle-resolved photoemission microscopy. *Scientific Reports* **2018**, 8(1), 17431.

74. Blomgren, G. E. The development and future of lithium ion batteries. *Journal of the Electrochemical Society* **2016**, 164(1), A5019–A5025.

75. Lu, J.; Chen, Z.; Pan, F.; Cui, Y.; Amine, K. High-performance anode materials for rechargeable lithium-ion batteries. *Electrochemical Energy Reviews* **2018**, 1(1), 35–53.

76. Yamada, Y.; Wang, J.; Ko, S.; Watanabe, E.; Yamada, A. Advances and issues in developing salt-concentrated battery electrolytes. *Nature Energy* **2019**, 4(4), 269–280.

77. Wan, N. H.; Meng, F.; Schroder, T.; Shiue, R.-J.; Chen, E. H.; Englund, D. High-resolution optical spectroscopy using multimode interference in a compact tapered fibre. *Nature Communications* **2015**, 6(1), 7762.

78. Liu, Y.; Lian, J.; Sun, Z.; Zhao, M.; Shi, Y.; Song, H. The first principles study for the novel optical properties of $LiTi_2O_4$, $Li_4Ti_5O_{12}$, $Li_2Ti_2O_4$, $Li_7Ti_5O_{12}$, *Chemical Physics Letters* **2017**, 677, 114–119.

79. Yoshimatsu, K.; Niwa, M.; Mashiko, H.; Oshima, T.; Ohtomo, A. Reversible superconductor insulator transition in $LiTi_2O_4$ induced by Li-ion electrochemical reaction. *Scientific Reports* 2015, 6, 16325.

80. Tanaka, S.; Kitta, M; Tamura, T.; Maeda, Y.; Akita, T., Kohyama, M. Atomic and electronic structures of $Li_4Ti_5O_{12}$/$Li_7Ti_5O_{12}$ (001) interfaces by first-principles calculations. *Journal Materials Science* **2014**, 49(11), 4032–4037.

5 Diversified Properties in 3D Ternary Oxide Compound Li_2SiO_3

Nguyen Thi Han
National Cheng Kung University
Thai Nguyen University of Education

Ngoc Thanh Thuy Tran and Vo Khuong Dien
National Cheng Kung University

Duy Khanh Nguyen
Ton Duc Thang University

Ming-Fa Lin
National Cheng Kung University

CONTENTS

5.1 INTRODUCTION

Up to date, the Li^+-based batteries (LIBs) have become one of the mainstream systems in the material basic science, engineering, and applications, mainly owing to the diversified geometric, electronic, and transport properties [1–5]. They are principally composed of the electrolyte [6–8], cathode [9], and anode materials [10]. Generally speaking, each component possesses the unusual geometric symmetry

79

with a very large primitive unit cell, directly reflecting the complicated chemical bondings due to a lot of chemical bonds. The greatly modulated bond lengths could be regarded as the most important common characteristics, therefore. They should be the critical condition in searching for the optimal match of three kinds of core components. Apparently, how to achieve the best LIBs, with the highest performance, belongs to a unified engineering issue [11]. It is well known that LIBs are widely utilized in a lot of electronic devices, e.g., cell phones [12], laptops [13], iPods [14], cars/buses [15], and radios [16]; their main features cover the high capacity [17], large output voltage [18], long-term stability [19], and friendly chemical environment [20]. This work only investigates the essential properties of LiSiO-related electrolyte materials through the first-principles simulations [21].

Very interestingly, the Li^+-ion transports occur at any time during the charging/discharging processes by the path of cathode $\rightarrow$ electrolyte $\rightarrow$ anode/anode $\rightarrow$ electrolyte $\rightarrow$ cathode [1–3]. Specifically, a separator membrane is inserted to avoid the internal short circuit and only accept the smallest Li^+ to freely pass the positive and negative electrodes [22]. The two electrodes are linked externally to an electric supply after the initial charging process, in which electron carriers rapidly escape from the cathode and transport by the external lead to the anode, leading to the creation of charge current [23]. In order to keep the electric neutrality, Li^+-ions rapidly transport along the parallel direction internally from cathode to anode by the solid-/liquid-/gluon-state electrolytes [1–5]. With this efficient process, the external energy from the electrical supply is stored in the battery using the chemical energy. The opposite chemical process, in which electrons move from anode to cathode through the external lead and Li^+-ions transport back to the cathode via the specific electrolyte, can provide the electric power and thus do works on electronic devices [1–5].

In general, the cathode and anode systems of the LIBs belong to the solid-state materials, such as the 3D ternary LiFe/Co/NiO [24] and LiTiO/graphite compounds [25], respectively. Very interestingly, the various electrolytes could be classified into the solid and liquid states [6–8]; furthermore, the layers are frequently utilized in the commercialized products. For example, the representative systems, respectively, cover LiSiO/LiGeO/LiSnO [26]. A lot of theoretical and experimental studies have been conducted on the optimal electrolytes [27], in which the critical conditions are closely related to the rather high Li^+-ion conductivity, the almost vanishing electron transport through them, the large electrochemical potential window, and the very stable electrode-electrolyte boundary. Compared with the liquid ones [28], all the solid-state secondary LIBs might present the following merits: the highly intrinsic, large energy density safety, comparable power density, long cycle lifetime. However, certain drawbacks need to be overcome in the near-future studies, such as the too high cost [29] and the optimal match with cathode and anode materials [9,10]. Among the whole solid-state electrolytes, the Li_2SiO_3 is chosen for fully understanding it.

Generally speaking, the theoretical investigations could be classified into two different categories, the numerical simulations and the phenomenological models. For example, the former and the latter, respectively, correspond to the first-principles calculations/Monte Carlo methods/molecular dynamics simulations [30] and the generalized tight-binding models/effective-mass approximations/Kubo formulas/ random-phase approximations [31]. Obviously, the many previous predictions show that the first-principles method is very efficient and reliable in determined the rich essential properties [32], the geometric [33], electronic [34], magnetic [35], and optical properties [36]. However, it cannot deal with the unique magnetic quantization because of a very large Moire superlattice due to the vector-potential-dependent Peierls phase (details) in [37]. Very interestingly, the VASP-calculated results (discussed later in Section 5.2) are sufficient in developing the theoretical framework for the thorough understandings of the diversified physical, chemical, and material phenomena [38]. The systematic studies have been successfully conducted on the layered 2D graphenes [39], 1D graphene nanoribbons [40], and 2D silicenes with/without significant chemical modifications (absorptions and substitutions). Through the delicate and thorough analyses can achieve the critical mechanisms and the concise pictures in fully comprehending the geometric, electronic, and magnetic properties. The significant multi-/single-orbital hybridizations in various chemical bonds are gotten from the optimal lattice symmetry, the atom-created valence and conduction bands, the spatial charge densities and their changes after chemical modifications, and the atom- and orbital-decomposed density of states (DOS). Furthermore, the spin distribution configurations, non-, ferro-, and antiferro-magnetic ones, which are generated by the host and/or guest atoms, are accurately examined through the spin-split/spin-degenerate energy bands near the Fermi level, the spin density distributions, the net magnetic moments, and the spin-projected DOS. Now, this theoretical framework could be further developed for the emergent materials (would be thoroughly tested in further investigations). It might be very suitable for exploring the diverse fundamental properties in a lot of complicated oxide compounds, e.g., the electrolyte, cathode, and anode materials, LiSiO [41], LiFe/Co/NiO [24], and $Li_4Ti_5O_{12}$, in the mainstream LIBs [1–5]. In addition, whether the direct combinations of numerical simulations with phenomenological models are reliable needs to be clarified under the independent cases, since this linking is very important in the full exploration of various properties, such as the useful combination of the VASP calculations [42] and the generalized tight-binding models [43] in the rich magnetic quantization [44]. The main focuses of this chapter are the geometric symmetries and electronic properties of the three-dimensional ternary compound of Li_2SiO_3 (electrolyte material of LIBs) as shown in Figure 5.1. The first-principles method is available in delicately calculating the total ground-state energy, lattice symmetry, distinct LiO and SiO bond lengths, the atom-dominated valence and conduction bands, the spatial charge density, and the atom and orbital-projected DOS. The spin-created phenomena, the spin-split band structure across the Fermi level, the net magnetic moment, and the spin density distributions, will be thoroughly examined whether they could survive in this emergent material. These physical quantities are very important in achieving the critical multi-orbital hybridizations of two kinds of chemical bonds and the spin-dependent magnetic configuration. Most of the analysis difficulties arise from the very complicated orbital-projected van Hove singularities, being partially supported by the electronic structure and charge density distribution. The theoretical

All-solid-state battery

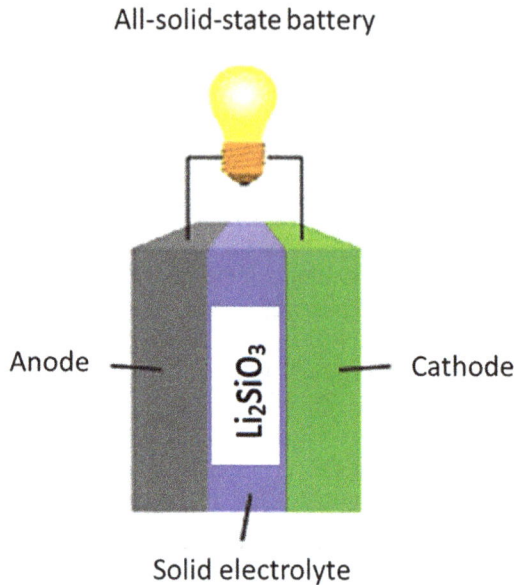

Anode — Li$_2$SiO$_3$ — Cathode

Solid electrolyte

FIGURE 5.1 All-solid-state Li$^+$-based battery with the electrolyte of 3D ternary Li$_2$SiO$_3$ compound.

predictions on the optimal geometry, the occupied electronic states and the band gap and whole energy spectrum, could be tested from high-resolution measurements of X-ray diffraction/low-energy electron diffraction (LEED) [45], angle-resolved photoemission spectroscopy (ARPES) [46], and scanning tunneling microscopy (STM) [47], respectively. In addition, the close relation between the numerical VASP calculations and the tight-binding model is discussed in detail.

5.2 NUMERICAL SIMULATIONS

Generally speaking, the theoretical studies could be classified into the numerical simulation methods and the phenomenological models. The former and the latter, respectively, cover the first-principles calculations/quantum Monte Carlo/molecular dynamics simulations [48] and the tight-binding model/static and dynamic Kubo formulas/random-phase approximation/self-energy method [49]. Only the first one is very suitable for the current investigations on the rich and unique geometric and electronic properties of the 3D ternary Li$_2$SiO$_3$ material. All the delicate evaluations, which are based on the density functional theory (DFT) [50], are done through the Vienna ab initio simulation package (VASP) [21].

The many-body exchange and correlation energies, being due to the electron–electron Coulomb interactions, are investigated from the Perdew–Burke–Ernzerhof (PBE) functional [51] within the generalized gradient approximation. Moreover, the projector-augmented wave (PAW) [52] pseudopotentials are available in characterizing the significant electron-ion scatterings. Of course, such two critical interactions have no exact solutions in the analytic forms. Therefore, it is difficult to finish the accurate diagonalization of the many-particle Hamiltonian. That the plane waves, with the kinetic energy cutoff of 500 eV, are chosen as a complete set would be more convenient and reliable to solve Bloch wave functions and band structures. The first Brillouin zone is sampled by $18 \times 10 \times 8$ and $24 \times 16 \times 14$ k-point meshes within the Monkhorst–Pack scheme [53] for geometric optimizations and electronic energy spectrum, respectively. These wave-vector points should be sufficient in finishing the suitable orbital-projected van Hove singularities, spatial charge distributions, and spin density configurations. Moreover, the convergence condition of the ground-state energy is set to be $\approx 10^{-5}$ eV between two consecutive simulation steps; furthermore, the maximum Hellmann–Feynman force acting on each atom is smaller than 0.01 eV under the ionic relaxations. The delicate VASP calculations are very useful in thoroughly exploring certain physical quantities, e.g., the atom-induced valence and conduction bands, the spatial charge densities after the chemical bondings, the atom- and orbital-projected DOS, the atom-dependent spin configurations, the spin-split or degenerate states across the Fermi level, and the finite or vanishing magnetic moments. By the detailed analyses, the critical physical and chemical pictures, the multi- and/or single-orbital hybridizations of chemical bonds and the spin configurations related to distinct atomic configurations, could be obtained under the concise form. According to such reliable viewpoints, the theoretical frameworks are responsible for the rich and unique phenomena. That is, the up-to-date theories are successful in systematic researches on the geometric structures, electronic properties, and magnetic configurations of 2D layered graphenes [39], 1D graphene nanoribbons, and 2D silicene-based systems [54]. Apparently, they have shown that the chemical modifications, being achieved by the adatom adsorptions guest-atom substitutions can greatly diversify the various essential properties.

Also, the developed viewpoints could be generalized to other condensed matter systems, such as the diverse phenomena in electrolyte/cathode/anode materials of LIBs [3–5]. The direct combinations of numerical simulations with phenomenological models would be very powerful in fully comprehending a lot of unusual properties. It is well known that the VASP energy bands could be obtained through the good fitting of the parameterized tight-binding model/the effective-mass approximation [55] when the electron and hole energy spectra do not present the complex dispersion relations (e.g., the nonmonotonous or oscillatory ones). Up to now, such a viewpoint is very successful in fully exploring the diversified phenomena of few-layer graphene systems. Since the weak, but significant interlayer van der Waals interactions only arise from the single-orbital hybridizations of $2p_z$-$2p_z$. For example, the diversified magnetic quantization phenomena come to exist in AA-, AB-, ABC-, and AAB-stacked [56,57], twisted, sliding, and guest-atom-substituted graphene systems through the generalized tight-binding model with any external field strengths. This model lies in the parameterized Hamiltonian, in which the various magnetic hopping integrals due to the different interlayer orbital hybridizations should be considered during its diagonalization. That is, they play important roles in expressing a lot of the magnetic Hamiltonian matrix elements [58].

5.3   RESULTS AND DISCUSSION

5.3.1   GEOMETRIC STRUCTURES

In this work, the 3D ternary Li_2SiO_3 compound is predicted to display an extremely nonuniform chemical environment in a primitive unit cell (discussed later) as shown in Figure 5.1, so a suitable tight-binding model, with the various chemical bondings, might be very difficult to obtain in fitting the first-principles band structure, unique lattice symmetries of 3D Li_2SiO_3 compound. The ternary 3D Li_2SiO_3 material possesses the unique lattice symmetries, according to the delicate first-principles calculations on the optimal geometric structures. We have chosen one of the metastable systems for studying the rich and unusual phenomena. Such a material, as clearly illustrated in Figure 5.1, directly corresponds to an orthorhombic structure, with a primitive unit cell of 24 atoms (8-Li, 4-Si, and 12-O atoms). The total ground-state energy is -10.0 eV per unit cell, in which the spin-dependent interactions do not make any contributions. Very interestingly, the lattice constants are $a = 9.467$ Å, $b = 5.440$ Å, and $c = 4.719$ Å about the a, b, and c axes, respectively, and very close to the previous experimental and other work results (Table 5.1). The space group symmetry arises from Cmc21 [59]. Obviously, the physical, chemical, and material environments are highly anisotropic and extremely nonuniform, and the other essential properties are expected to behave similarly. The diversified atomic arrangements are easily observed under the distinct plane projection, e.g., the geometric structures for (a) (100), (b) (010), (c) (001), (d) (110), (e) (011), (f) (101), and (g) (111) as illustrated in Figure 5.2a–g. Li-, Si-, and O-atoms are, respectively, denoted as the green, blue, and red balls (here and therein). The above-mentioned real-space lattice creates the orthorhombic first Brillouin zone shown in Figure 5.3, in which the complex high symmetry points are very useful in characterizing the electronic energy spectra and states. They cover Γ(0.00, 0.00, 0.00), X(0.33, 0.33, 0.00), S(0.00, 0.50, 0.00), Y(-0.50, 0.50, 0.00), T(0.50, 0.50, 0.00), Z(0.00, 0.00, 0.50), A(0.33, 0.33, 0.5), and R(0.00, 0.50, 0.00). Various chemical bonds, which come to exist in an orthogonal primitive unit cell (Figure 5.1), dominate all the fundamental properties. There exist only 32 Li-O and 16 Si-O bonds, and the other kinds of chemical bondings are thoroughly absent in the 3D ternary Li_2SiO_3 material. Most importantly, the optimal

TABLE 5.1

Lattice Constant of 3D Li_2SiO_3 Compound

Lattice constant (Å)	Exp[a]	Other work (DFT)[b]	Other work (MD)[c]	This work (DFT)
A	9.396	9.487	9.396	9.467
B	5.396	5.450	5.390	5.44
C	4.661	4.713	4.661	4.719

[a] Reference: [109].
[b] Reference: [110].
[c] Reference: [111].

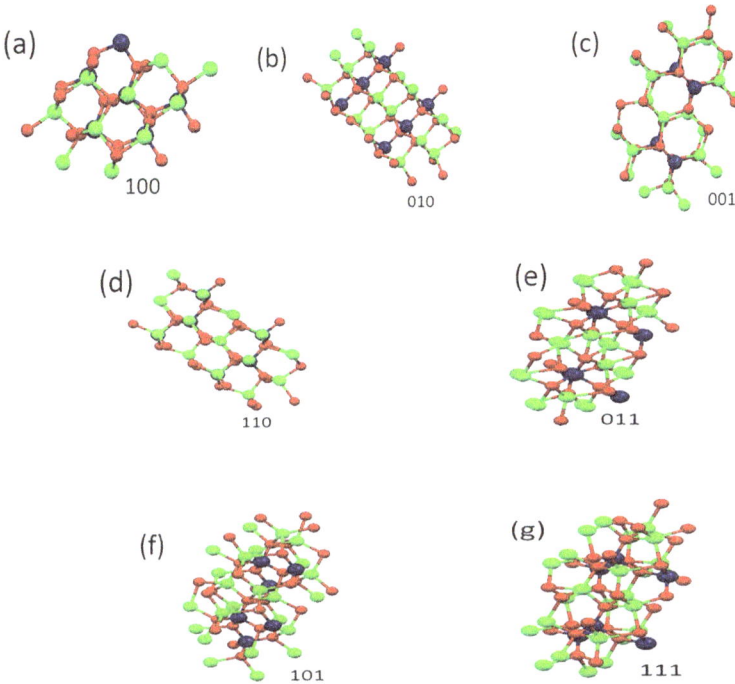

FIGURE 5.2 Geometric projections of Li$_2$SiO$_3$ on the distinct planes: (a) (100), (b) (010), (c) (001), (d) (110), (e) (011), (f) (101), and (g) (111), where Li, Si, and O atoms, respectively, correspond the green, blue, and red balls.

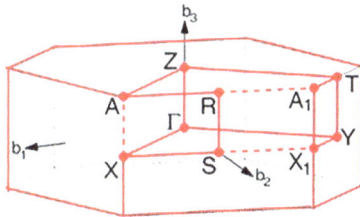

FIGURE 5.3 First Brillouin zone with the high symmetry points within three orthogonal axes.

geometric parameters in Table 5.2 clearly illustrate that their bond lengths possess an observable range of 1.943–2.291 and 1.612–1.702 Å. The fluctuation percentages are over 13% and 5.6%, indicating the possible structural transformation between two metastable systems with distinct geometric symmetries. This might be responsible for the outstanding electrolyte role of Li$_2$SiO$_3$- in Li$^+$-ion batteries. It is well known that the bond's lengths directly determine the spatial charge density distributions/the strengths of chemical bondings (Figure 5.6), so their large changes indicate the

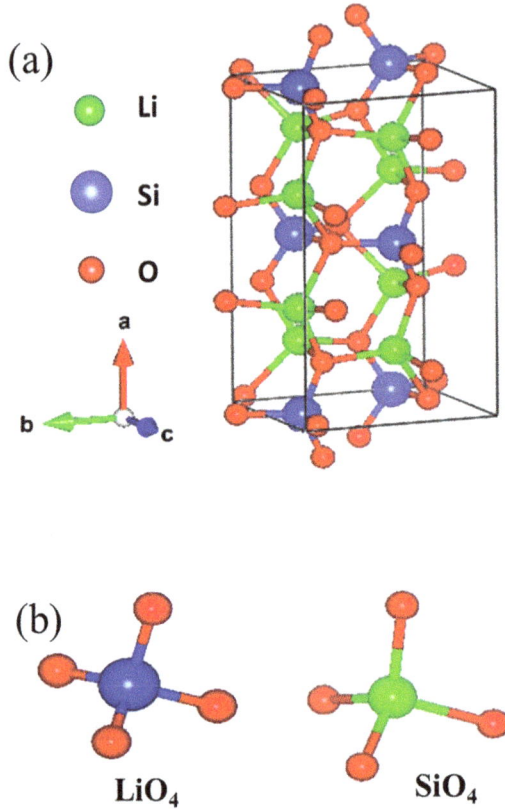

FIGURE 5.4 (a) Optimal lattice structure for Li_2SiO_3 under the orthorhombic symmetry with 24 atoms in a unit cell of the lattice constants $a = 9.467$ Å, $b = 5.440$ Å, and $c = 4.719$ Å; (b) also shows two kinds of chemical bondings: Li-O and Si-O.

highly nonuniform multi-orbital hybridizations (the various hopping integrals in the phenomenological models). Such complicated behaviors might induce extra theoretical barriers in exploring the other fundamental properties, e.g., the rich magnetic quantization phenomena in LiSiO-related compounds [60].

The high-resolution X-ray elastic scatterings (HR-XES) and low-energy electron diffractions (LEED) in [45] are very suitable for verifying the 3D lattice symmetries of LiSiO-based compounds, while the opposite is true for the scanning tunneling microscopy (STM) (the nanoscale top views in [61]) and tunneling electron microscopy (TEM) (the side-view structures [62]). Whether the wide modulations of the Li-O and Si-O bond lengths can be delicately identified from the measured data is worthy of systematic investigations. This is closely related to the very complex multi-orbital hybridizations in all the chemical bonds. The similar examinations could be generalized to the other metastable or intermediate configurations, being very useful in understanding the transformation paths between them.

TABLE 5.2

Different Chemical Bond Lengths in Li$_2$SiO$_3$ with the Total Number of 32 Li–O and 16 Si–O Bonds

	Bond lengths Li–O (Å)		Bond lengths Li–O (Å)		Bond lengths Si–O (Å)
1	Li$_1$–O$_4$ = 1.9433	17	Li$_8$–O$_8$ = 1.9710	1	Si$_1$–O$_1$ = 1.6115
2	Li$_4$–O$_7$ = 1.9485	18	Li$_8$–O$_3$ = 1.9485	2	Si$_1$–O$_3$ = 1.6115
3	Li$_4$–O$_{10}$ = 2.2012	19	Li$_8$–O$_{12}$ = 2.2012	3	Si$_1$–O$_{10}$ = 1.7016
4	Li$_5$–O$_5$ = 1.9710	20	Li$_1$–O$_9$ = 2.2012	4	Si$_1$–O$_9$ = 1.6978
5	Li$_5$–O$_{11}$ = 2.2012	21	Li$_3$–O$_8$ = 1.9485	5	Si$_2$–O$_2$ = 1.6115
6	Li$_5$–O$_2$ = 1.9485	22	Li$_3$–O$_2$ = 1.9433	6	Si$_2$–O$_4$ = 1.6115
7	Li$_5$–O$_8$ = 1.9433	23	Li$_2$–O$_3$ = 1.9433	7	Si$_2$–O$_9$ = 1.7016
8	Li$_6$–O$_6$ = 1.9710	24	Li$_2$–O$_5$ = 1.9485	8	Si$_2$–O$_{10}$ = 1.6978
9	Li$_6$–O$_1$ = 1.9485	25	Li$_3$–O$_3$ = 1.9710	9	Si$_3$–O$_5$ = 1.6115
10	Li$_6$–O$_7$ = 1.9433	26	Li$_2$–O$_{10}$ = 2.2012	10	Si$_3$–O$_7$ = 1.6115
11	Li$_1$–O$_6$ = 1.9485	27	Li$_3$–O$_9$ = 2.2012	11	Si$_3$–O$_{12}$ = 1.7016
12	Li$_6$–O$_{12}$ = 2.2012	28	Li$_4$–O$_1$ = 1.9433	12	Si$_3$–O$_{11}$ = 1.6978
13	Li$_7$–O$_7$ = 1.9710	29	Li$_3$–O$_9$ = 2.2012	13	Si$_4$–O$_6$ = 1.6115
14	Li$_7$–O$_{11}$ = 2.2012	30	Li$_4$–O$_4$ = 1.9710	14	Si$_4$–O$_8$ = 1.6115
15	Li$_7$–O$_6$ = 1.9433	31	O$_2$–Li$_2$ = 1.9710	15	Si$_4$–O$_{11}$ = 1.7016
16	Li$_1$–O$_1$ = 1.9710	32	O$_5$–Li$_8$ = 1.9433	16	Si$_4$–O$_{12}$ = 1.6978

5.3.2 RICH ELECTRONIC PROPERTIES

Very interestingly, the solid-state electrolyte in LIBs, the 3D ternary Li$_2$SiO$_3$ compound (Figure 5.1), presents the rich and unique band structure. The electronic energy spectrum, as clearly illustrated in Figure 5.5a–d, strongly depends on the unusual wave-vector path. The occupied valence bands are highly asymmetric to the unoccupied conduction bands about the Fermi level $E_F = 0$ (Figure 5.5a), directly reflecting the very complicated multi-orbital hybridizations in Li-O and Si-O bonds. The energy dispersions, which are shown along the high symmetry point, have strong anisotropic behaviors. For example, there exist the parabolic, oscillatory, and partially flat dispersion relations. Furthermore, the sub-band noncrossing, crossing and anti-crossing phenomena come to exist frequently. Most importantly, the highest occupied and the lowest unoccupied states, respectively, appear at the Z and Γ points (0, 0, 0.5) and (0.00, 0.00, 0.00). This indicates a very large indirect band gap of $E_g \approx 5.077$ eV. Such a value is only slightly lower than that $E_g \approx 5.5$ eV of diamonds.

Apparently, the optical threshold absorption frequency, which is measured from the reflection/absorption/transmission spectroscopies, should be higher than 5.1 eV [63]. That is to say, the various high-resolution optical measurements are very suitable in examining the obviously insulating behaviors [64]. In addition, the spin splitting is thoroughly in any energy bands, since the Li, Si, and O atoms cannot

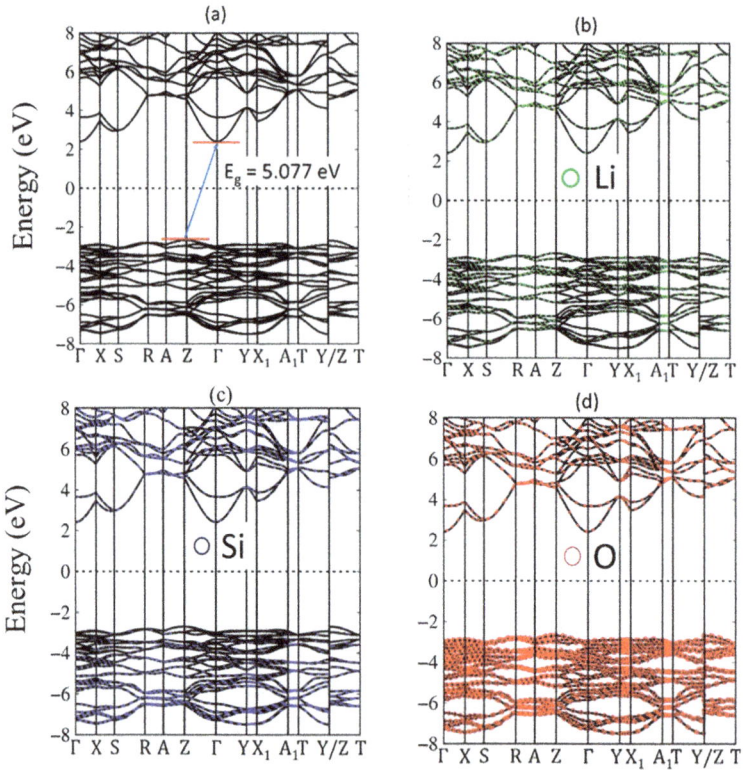

FIGURE 5.5 (a) Significant valence and conduction band of Li_2SiO_3 along the high symmetry points within the first Brillouin zone for the energy range $(-8.0\,eV, E^{cv}, 8\,eV)$ with the specific (b) lithium, (c) silicon, and (d) oxygen dominances (green, blue, and red balls), respectively.

create the spin-up or spin-down configurations. As a result, the magnetic moment is the vanishing and the spin density distribution becomes meaningless. In addition to the main features of band structure, the electronic Bloch wave functions for valence and conduction states could provide partial information on critical chemical bondings. Each state is regarded as the linear superposition of different orbitals; therefore, it can decompose into the distinct atomic contributions. The different atom dominances, being proportional to the ball radius, are available in understanding the important roles of chemical bonds on the rich electronic properties. The green, blue, and red balls, respectively, correspond to the Li, Si, and O contributions. In general, the effective valence and conduction states, which are closely related to the multi-orbital hybridizations of Li-O and Si-O bonds (discussed later in Figures 5.6 and 5.7), lie in the energy range of $-8.0\,eV$, E^{cv}, $8\,eV$. Any atoms have significant contributions in the whole band structure. This unusual phenomenon clearly indicates the wide-range modulations of the chemical bonding strengths in Li-O and Si-O bonds (strongly modulated hopping integrals). However, it might be difficult to observe the obvious Li-contributions (small green balls) as illustrated in Figure 5.5b because

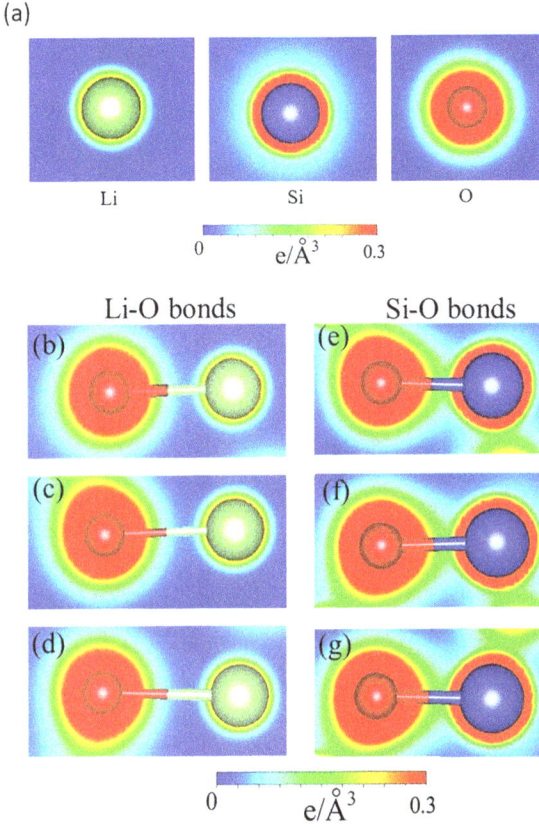

FIGURE 5.6 Spatial charge density distributions for (a) the isolated (Li, Si, and O) atoms; and the shortest/medium/longest (b)/(c)/(d) Li-O and (e)/(f)/(g) Si-O bonds

of the single 2s orbitals. On the contrary, the Si- and O-contributions are easy to observe in the electronic energy spectrum, e.g., the sufficiently large blue and red balls shown in Figure 5.5c and d. Specifically, the oxygen atoms dominate all the valence and conduction states, since they are associated with the whole chemical bonds. On the experimental side, the high-resolution ARPES is the only method in examining the wave-vector dependence of occupied electronic states, especially for the valence bands/the conduction bands below the Fermi level, with the diversified energy dispersion relations, in semiconductors/metals [65]. It is well known that their measurements are very successful for various condensed matter systems, such as the sp^2 based carbon materials [66], 3D LiXO ternary compounds [67], superconductors [40], and group-IV and group-V layered systems [68]. Very interestingly, a lot of ARPES experiments have verified the valence-band energy spectra of emergent graphene-related systems, being initiated from the K/K' prime or % Gamma valleys.

For example, there exist the linear Dirac-cone structures in monolayer/twisted bilayer graphenes [39] and monolayer graphene [40], the parabolic/parabolic and

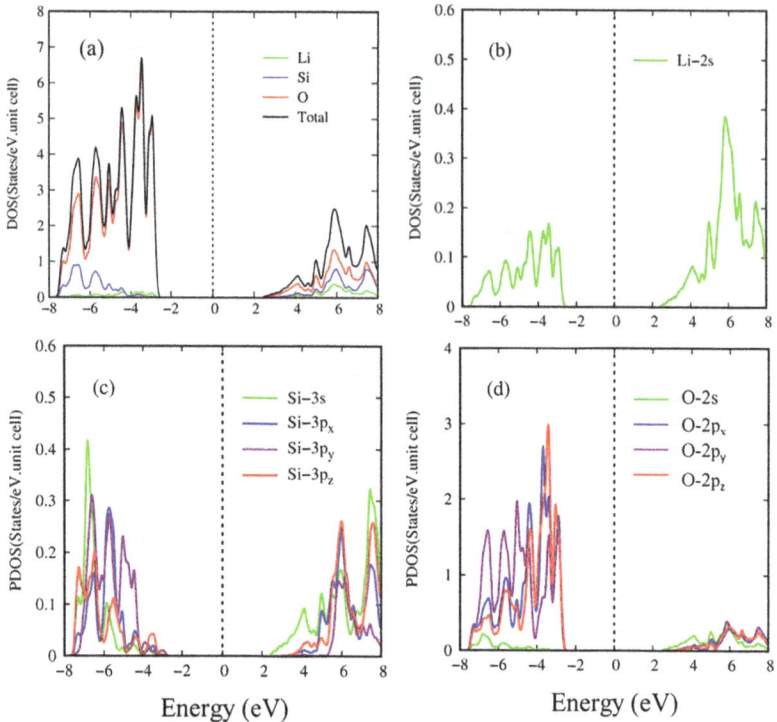

FIGURE 5.7 Atom- and orbital-projected DOS: those coming from (a) Li, Si, and O atoms (green, blue, and red curves), (b) Li-$2s$ orbitals (green curve), (c) Si-($3s$, $3p_x$, $3p_y$, $3p_z$) orbitals (green, blue, purple and red curves), and (d) O-($2s$, $2p_x$, $2p_y$, $2p_z$) orbitals (green, blue, purple and red curves)

linear bands in bilayer/trilayer AB stackings, the linear, partially flat, and Sombrero-shaped energy spectra in trilayer ABC stacking, the monolayer- and bilayer-like behaviors in AB-stacked graphite [39], and the energy gap and parabolic bands of 1D graphene nanoribbons [69]. Only monolayer graphene and 1D materials belong to zero- or finite-gap semiconductors [70]. Apparently, the unusual electronic properties directly reflect the pure and unique interlayer $2p_z$-$2p_z$ orbital hybridizations in the normal/enlarged/reduced honeycomb lattices [71]. The diverse behaviors are also confirmed by tight-binding model. The ARPES measurements could be utilized to detect the main features of band structure in 3D ternary Li_2SiO_3 material, in which they cover a very large indirect band gap, the highest occupied state at the Z point (0, 0, 0.5), the different energy dispersions along the high symmetry points, and the frequent noncrossing/crossing/anti-crossing behaviors. The experimental verifications could provide certain information in the multi-orbital hybridizations of chemical bonds. Very interestingly, the multi-orbital hybridizations in Li-O and Si-O chemical bonds could be roughly examined from the spatial charge distributions, as clearly illustrated in Figure 5.6a–g. Furthermore, the delicate chemical bondings are identified from the atom- and orbital-projected DOS (later discussions) depicted in

Figure 5.7a–d. It is well known that the carrier densities are very sensitive to the changes of bond lengths. Apparently, the 3D ternary Li$_2$SiO$_3$ has the highly modulated chemical bonds (Figure 5.2 and Table 5.2), being closely related to the available orbitals in different atoms. Concerning the shortest Li-O bonds 1.943 Å as shown in Figure 5.6b, each lithium atom only provides a single $2s$ orbital, and its effective distribution range is 0.54 Å, as measured from the deep-red region of the Li$^+$-ion core to the light green one of the outmost orbital. However, the two $1s$ orbitals do not take part in the critical orbital hybridizations with the oxygen atoms. Most importantly, an apparent overlap of distinct orbitals between Li and O atoms is revealed in the 1.20–1.60 Å range (a distance with the O-core). The important oxygen orbitals, which correspond to the light green and yellow regions, are approximately predicted to be associated with the $(2p_x, 2p_y, 2p_z)$ ones. Specifically, the O-$2s$ orbitals are relatively far away that of the Li-atom, so they might play a minor role in Li-O chemical bondings with the increase of Li-O length 2.117–2.291 Å as shown in Figure 5.6c and d, respectively, and the orbital overlap phenomenon of Li- and O-atoms decline quickly. As a result, the multi-orbital $2s$-$(2s, 2p_x, 2p_y, 2p_z)$ hybridizations of Li–O bonds exhibit the diverse hopping integrals. There are less Si-O chemical bonds, but quite strong bonding strengths as depicted in Figure 5.6e-g, compared with those of Li-O bonds. The effective distribution range of Si atom is much higher than that of Li one because of the large atomic number. Most importantly, both silicon and oxygen atoms could be classified into the heavy red and yellow-green regions, which, respectively, correspond to $3s$ and $2s$ and $(3p_x, 3p_y, 3p_z)$ and $(2p_x, 2p_y, 2p_z)$ orbital. Very interestingly, the chemical bondings of the former are clearly illustrated by the deformed spherical distributions between two atoms. That is to say, the Si-O chemical bonds consist of the four-orbital hybridizations due to $(3s, 3p_x, 3p_y, 3p_z)(2s, 2p_x, 2p_y, 2p_z)$. The chemical bonding strength declines in the increase in bond length, being obviously indicated by the reduced charge density between Si and O atoms. The extremely nonuniform chemical bondings even for the similar bonds will induce the high calculation barriers in researching the suitable hopping integrals for the phenomenological models, e.g., the reliable parameters of the tight-binding model. The DOS, which is defined as the number of states within a very small energy range of dE, is capable of fully understanding the valence and conduction energy spectra simultaneously. Generally, its special structures are created by the band-edge states with the vanishing group velocities. The critical points in the energy wave-vector space cover the local extreme points (minima and maxima), the saddle points, and all the partially flat energy dispersions. Apparently, the van Hove singularities are greatly diversified under the different dimensions. Of course, the discrete energy levels in 0D quantum dots [71] only present the delta-function-like peaks, as observed in the magnetically quantized Landau levels in emergent layered materials (e.g., few-layer graphene systems), and any-dimensional dispersionless energy bands [72]. The 1D parabolic and linear bands, respectively, the asymmetric divergent peaks in the square-root form and the plateau structures. As for the 2D condensed matter systems, the concave–convex-form dispersions, the Dirac-cone structure, the linear valence and conduction bands without energy spacing, the saddle points, and the constant-energy loops, respectively, create the broadening shoulders, V-shapes, the symmetric peaks related to the logarithmical divergence, and the square-root divergent peaks. Moreover, the higher dimension in 3D materials makes them only show symmetric/asymmetric peaks and the shoulder structures, mainly owing to the absence of quantum confinement. The latter come from the 3D parabolic energy dispersions. In the current work, the atom- and orbital-decomposed van Hove singularities are very useful in directly resolving the critical multi-orbital hybridizations in the Li-O and Si-O chemical bonds, as shown in Figure 5.7a–d. Three kinds of atoms have significant contributions within the whole energy spectrum $-8.0\,\text{eV}$, $E^{c\,v}$, $8\,\text{eV}$, clearly reflecting the large modulations of bond lengths (Table 5.2). For a wide-gap 3D Li_2SiO_3 compound, the DOS per unit cell is vanishing within a large energy range of $E_g \approx 5.077\,\text{eV}$ centered about the Fermi level (black curve) illustrated in Figure 5.7a. Maybe, an outstanding insulation property will lead to the experimental barrier in the further scanning tunneling spectroscopy (STS) measurements (later discussions) [73–77]. The value of DOS is dominated by the valence states ($E < 0$), but not the conduction ones ($E > 0$). That is, the hole and electron energy spectra are highly asymmetric about the Fermi level. There exist some prominent asymmetric/symmetric peaks and broadening shoulders in a 3D material, as discussed earlier. These van Hove singularities mainly arise from the valence and conduction band-edge states along the high symmetry-point paths, in which they belong to the local minimum, maximum, saddle, and dispersionless points in the energy wave-vector spaces. At least, ten and six special structures are, respectively, revealed in the valence and conduction energy spectra. As a result of the frequent crossings and anti-crossings, each one directly reflects the merged phenomenon of certain neighboring band-edge states under the finite broadening effects. This clearly indicates that the experimental measurements would be difficult to examine the number of energy sub-bands and their band-edge states. Apparently, the lithium and oxygen atoms (green and blue balls), respectively, make the weakest and strongest contributions to the total DOS. Such result is principally determined by the whole number of available atom orbitals in a primitive unit cell. Generally speaking, the important contributions, which arise from the various orbitals of different atoms, cover (I) Li-$2s$ orbitals (green curve) shown in Figure 5.7b, (II) Si-($3s$, $3p_x$, $3p_y$, $3p_z$) orbitals (green, blue, purple, and red curves) shown in Figure 5.7c, and (III) O-($2s$, $2p_x$, $2p_y$, $2p_z$) orbitals (green, blue, purple, and red curves) shown in Figure 5.7d. All the orbital contributions are merged together, especially for the number, energies, intensities, and forms of van Hove singularities. The main reason is that the distinct chemical bondings are associated with one another through the Li-O and Si-O bonds. Consequently, there exist the multi-orbital hybridizations of $2s$-($2s$, $2p_x$, $2p_y$, $2p_z$) and ($3s$, $3p_x$, $3p_y$, $3p_z$)($2s$, $2p_x$, $2p_y$, $2p_z$). The enlarged DOS for Li-$2s$ orbitals can provide the clear van Hove singularities and illustrates its weak, but significant feature as shown in Figure 5.7b. The observable contributions are obviously revealed in the half-occupied four orbitals of silicon atoms as shown in Figure 5.7c, in which their differences only reflect the strong anisotropy in a primitive unit cell (Figure 5.2). Most importantly, only three orbitals of O-($2p_x$, $2p_y$, $2p_z$) make comparable and dominating contributions as shown in Figure 5.7d, while the $2s$ ones are much smaller than the others in the opposite region. This unusual behavior might be consistent with the ($3s$, $3p_x$, $3p_y$, $3p_z$)–($2s$, $2p_x$, $2p_y$, $2p_z$) and ($2s$)($2p_x$, $2p_y$, $2p_z$) orbital hybridizations in the Si-O and Li-O bonds, respectively. It should be noticed that the O-$2s$ orbitals are almost fully forbidden in the semiconducting Li$_2$SiO$_3$ compound. The high-resolution STS is a very efficient technique in detecting the whole valence and conduction energy spectrum through the van Hove singularities, but not their wave-vector dependences (details in the previous books). Roughly speaking, a quantum tunneling current (a very small I) is initiated from a nanoscaled probe into a sample surface by a finite potential difference with substrate (V). Furthermore, its differential conductance, dI/dV, could be roughly regarded as the DOS [78]. Very interestingly, the up-to-date STS is successful in identifying the dimension-enriched van Hove singularities of graphene-related systems with the sp^2 chemical bondings. For example, there exist a lot of divergent peaks with the square root in 1D carbon nanotubes (discussed earlier) and graphene nanoribbons [40], a plateau structure across the Fermi level in metallic armchair tubules [79], symmetric V-shape vanishing at E_F for monolayer graphene (a zero-gap semiconductor) [39], the logarithmically symmetric peaks near E_F in twisted bilayer graphene systems (saddle point) [39], a gate-voltage-induced energy gap for bilayer AB and trilayer ABC stackings, a delta-function-like peak localized about E_F for ABC-stacked graphenes, a sharp dip structure close to E_F combined with a pair of square-root peaks under trilayer AAB stacking (a narrow-gap semiconductor with the constant-energy loops), and a finite DOS at E_F for the semimetallic Bernal graphite [80]. The similar STS experiments, being required for the 3D ternary Li$_2$SiO$_3$ materials, could verify a very large energy gap of $E_g \approx 5.077$ eV, six/three asymmetric/symmetric peaks and broadening shoulders in valence/conduction spectrum, and the high asymmetry about electron and hole states and their distinct distribution widths.

The experimental measurements, as well as the theoretical predictions on van Hove singularities, can identify the complicated band structure and thus multi-orbital hybridizations in Li-O and Si-O bonds. In the current LIBs [1–3], the 3D ternary Li$_2$SiO$_3$ material could serve as the solid-state electrolyte [6–8]. During the charging process, plenty of lithium ions are rapidly transferred from the cathode (e.g., Li/Fe/Co/NiO materials), into the electrolyte and then anode. The opposite is true for the discharging process. Apparently, the Li$^+$-flowing will generate the dramatic transformation of any materials. For example, the electrolyte compound should present a lot of intermediate configurations during the Li$^+$-transport processes that are characterized by the Moire superlattices with very large unit cells, as observed in the many-particle scattering phenomena. Obviously, too many atoms in a primitive unit cell induce numerical difficulties. This material also has certain metastable structures, such as Li$_x$Si$_y$O$_z$, Li$_x$Ge$_y$O$_z$, and Li$_x$Sn$_y$O$_z$. As a result, the structural transformation between them will come to exist, in which the optimal evolution paths are deduced to play a critical role in battery functionalities. How to resolve the dynamic phenomena of Li$^+$-transports becomes the most important issue in the near future. Whether the phenomenological method, the tight-binding model, could be built from the first-principles calculations on band structure and DOS is worthy of a closer investigation. As for the 3D ternary Li$_2$SiO$_3$ compound, various hopping integrals and Coulomb site energies in Hamiltonian (details in books) [81], corresponding to the multi-orbital hybridizations of Li-O and Si-O bonds, need to be considered simultaneously in the model evaluations. Furthermore, the extreme bonding strengths are clearly revealed in the largely modulated bond lengths (Table 5.2).

Apparently, it would be very difficult to establish the delicate and complicated intrinsic interactions. Another issue, which develops the theoretical framework for the mainstream electrolyte/cathode/anode/materials of Li^+-related batter materials [1–4], is very useful in thoroughly comprehending the diversified physical/chemical/material properties [82], e.g., the unusual quantization phenomena under a uniform perpendicular magnetic field in the 3D ternary LiXO-dependent compounds [83].

5.3.3 COMPARISONS, MEASUREMENTS, AND APPLICATIONS

The theoretical predictions on the essential properties of the 3D ternary Li_2SiO_3 compound could be examined by various experimental methods. The powder X-ray diffraction (PXRD) [84] is the most efficient technique in identifying the lattice symmetries of 3D materials. Apparently, it is very suitable to directly observe the 3D ternary Li_2SiO_3 compound. The lattice constant and phase information of Li_2SiO_3 have been successfully verified by the PXRD measurements. Most importantly, its bond lengths possess an observable range of 1.932 to 2.176 Å and 1.591 to 1.680 Å for Li-O and Si-O bonds, respectively [84], clearly revealing the evidence of multiorbital hybridizations. These results are well consistent with the current predictions. In addition to PXRD, the particle sizes and morphologies of samples are carried out by using the higher-resolution scanning electron microscopy (SEM) [85]. The side- and top-view nanoscale geometries are, respectively, examined by TEM [86] and scanning tunneling microscopy (STM) [87]. These measurements clearly illustrate the diversified geometric structures, such as layered graphene systems with different stacking configurations [88], coaxial carbon nanotubes in the presence of translation and rotation symmetries [89], and planar/curved/folded/scrolled graphene nanoribbons [90,91]. Both SEM and TEM are available to confirm the porous structures in Li_2SiO_3 [92]. This suggests the highly nonuniform chemical environment. However, the STM test might be absent up to now.

The details of chemical compositions for Li_2SiO_3 could be measured by X-ray photoelectron spectroscopy (XPS) [93], providing sufficient information about the type of atom/molecule as well as their concentration. Such measurements have been utilized to confirm the existence and ratio of oxygenated functional groups in graphene oxides (GOs) [93,94]. The similar investigations show that the surface elements in Li_2SiO_3-graphene composite (LSO-GE) exhibit O-1s and Si-2p, respectively, at 532 and 103.8 eV under the mole ratio of roughly 3.21:1 closely related to GOs [84]. Moreover, the pure phase of orthorhombic Li_2SiO_3 is confirmed to possess the ratio of 2:1 for lithium and silicon atoms [95].

One of the most important electronic properties, electronic energy spectra, is frequently examined by the optical spectroscopies [96], ARPES [97], and STS [98]. Very interestingly, a very wide energy gap ternary is 3D Li_2SiO_3, being due to the extremely strong covalent bondings, which is directly verified by using the optical absorption spectroscopy. The measured excitation frequency (~4.5 eV, in 91) is close to the VASP calculations (~5 eV). It is well known that the ARPES measurement is suitable for the direct examination of the occupied energy spectrum, with different dispersion relations along the high symmetry points in the first Brillouin zone. On the contrary, STS measurement, an extension of STM, can efficiently examine the energy-dependent DOS for valence and conduction bands in condensed matter systems. Up to now, both ARPES and STS measurements are very successful in verifying the diverse electronic properties in graphene nanoribbons [99] carbon nanotubes [100] few-layer graphene [101], adatom-adsorbed graphene systems [102], and graphites [103]. These measurements can provide sufficient information for the significant effects of chemical bondings in emergent materials. Nevertheless, the high-resolution ARPES and STS measurements are required for Li$_2$SiO$_3$ in near-future researches.

As a result of the rich and unique properties, the ternary 3D Li$_2$SiO$_3$ is expected to have potential applications in the next-generation energy storage, e.g., a good candidate as an electrolyte in LIBs. The chemical doping of graphene can enhance the conductivity of Li$_2$SiO$_3$ anode, being reported in the previous work [104]. The surface modifications, being made by simply mixing followed by sintering, might serve as a positive strategy to enhance the electrochemical performance of Li$_4$Ti$_5$O$_{12}$ anode [105]. In addition, the Li$_2$SiO$_3$ compound in the Li$_2$SiO$_3$@Li$_2$SnO$_3$/SnO$_2$ composite can decrease the particle size and enhance the electrochemistry performance of Li$_2$SnO$_3$ anode [106].

5.4 CONCLUDING REMARKS

The rich and unique fundamental properties of the emergent 3D Li$_2$SiO$_3$ compound are thoroughly investigated from the first-principles calculations. The critical multi-orbital hybridizations, which survive in Li-O and Si-O bonds, are accurately identified from the atom-dominated band structure, the charge density distributions in the greatly modulated chemical bonds, and the atom- and orbital-projected van Hove singularities. The theoretical framework could be further developed for the other electrolyte, anode, and cathode materials of LIBs [1–5], e.g., the important differences among the various LiXO-related compounds [83], and the diversified phenomena driven by various components. Very interestingly, the highly anisotropic and nonuniform environments need to be included in the phenomenological models. For example, the suitable tight-binding model is expected to have the position-dependent hopping integrals and the orbital-create on-site Coulomb potentials, in simulating the VASP energy bands and DOS. Whether the intrinsic atomic interactions of Hamiltonian could be expressed in the analytic form is the near-future studying focus; furthermore, they are very useful in exploring the other diverse phenomena, such as the magnetic quantization in a uniform perpendicular magnetic field. The solid-state anode material of Li$_2$SiO$_3$ with 24 atoms in a primitive unit cell is an orthorhombic structure. There are 32 Li-O and 16 Si-O chemical bonds, in which each atom has four neighboring other atoms. Their bond lengths Li-O and Si-O, respectively, present the large modulations over 13% and 5.6%. Most importantly, the very strong orbital hybridizations create a very wide indirect gap of $E_g \approx 5.077$ eV, being close to the largest one in diamond ≈ 5.5 eV. Its magnitude is even lower than the optical threshold absorption frequency. The high-resolution optical reflection/absorption/transmission spectroscopies [83,105] are available in detecting the obvious insulating property of the transparent material [107]. The occupied valence and the unoccupied conduction bands are highly asymmetric to each other about the Fermi level. Furthermore, there exist the wide energy range of (-8.0 eV, $E^{c\,v}$, 8.0 eV), the strong/various dispersion relations with the wave vector, the high anisotropy, and the frequently noncrossing/ crossing/anti-crossing behaviors. Moreover, the van Hove singularities, which arise from the band-edge states, appear as the six/three dominating/minor special structures in DOS of the valence/conduction energy spectrum ($E < 0$, $E > 0$).

They only belong to the broadening asymmetric/symmetric peaks and shoulders and play a critical role in examining the multi-orbital hybridizations of Li-O and Si-O bonds, $2s$-($2s$, $2p_x$, $2p_y$, $2p_z$) and ($3s$, $3p_x$, $3p_y$, $3p_z$)–($2s$, $2p_x$, $2p_y$, $2p_z$). The diverse covalent bondings are partially supported by the atom-dominated band structure and charge density distributions in modulated chemical bonds. The theoretical predictions on the optimal geometry, occupied wave-vector-dependent valence bands, and valence and conduction DOS, could be verified from X-ray elastic scatterings, LEED [45], and ARPES [46], respectively. The calculated results clearly illustrate that LiSiO-based compounds have certain metastable configurations, and even infinite intermediate ones during the charging and discharging processes for the Li⁺ transport in batteries. The similar structural transformations between two metastable structures [108] are expected to occur at any time. The optimal evolution paths, which might become an emergent issue, are under the current investigations.

REFERENCES

1. Nan Chen, Haiqin Zhang, Renjie Chen Li and Shaojun Guo Ionogel. Electrolytes for high-performance lithium batteries: A review. *Adv. Energy Mater.* **2018**, 8(12), 1702675.
2. Nancy J. Dudney and Bernd J. Neudecker. Solid state thin-film lithium battery systems. *Solid State Mater. Sci.* **1999**, 4, 479–482.
3. Rotem Marom, S. Francis Amalraj, Nicole Leifer, David Jacob and Doron Aurbach. A review of advanced and practical lithium battery materials. *J. Mater. Chem.* **2011**, 21, 9938–9954.
4. Arul Manuel Stephan. Review on gel polymer electrolytes for lithium batteries. *Eur. Polym. J.* **2006**, 42(1), 21–42.
5. Bruno Scrosati and Jürgen Garche. Lithium batteries: Status, prospects and future. *J. Power Sources* **2010**, 195, 2419–2430.
6. Reng Zhenga, Masashi Kotobukia, Shufeng Songa, Man on Lai and Li Lu. Review on solid electrolytes for all-solid-state lithium-ion batteries. *J. Power Sources* **2018**, 389, 198–213.
7. Jung-Joon Kim, Kyungho Yoon, Inchul Park and Kisuk Kang. Progress in the development of sodium-ion solid electrolytes. *Small Methods* **2017**, 1, 1700219.
8. Sheng Shui Zhang. A review on electrolyte additives for lithium-ion batteries. *J. Power Sources* **2006**, 162, 1379–1394.
9. Navaratnarajah Kuganathan, Apostolos Kordatos and Alexander Chroneos. Li₂SnO₃ as cathode material for lithium-ion batteries: Defects, lithium ion difusion and dopants. *Sci. Rep.* **2018**, 8, 12621.
10. Md Mokhlesur Rahman, Irin Sultana, Tianyu Yang, Zhiqiang Chen, Neeraj Sharma, Alexey M. Glushenkov and Ying Chen. Lithium germanate (Li₂GeO₃): A high-performance anode material for lithium-ion batteries. *Angew. Chem. Int. Ed.* **2016**, 5ah5, 16059–16063.
11. Nasrin Sulaiman, M. A. Hannan, Azah Mohamed, Edy Herianto Mjlan and Wan Ramil Wan Daud. A review on energy management system for fuel cell hybrid electric vehicle: Issues and challenges. *Renewable Sustainable Energy Rev.* **2015**, 52, 802–814.

12. Luis Oliveira, Maarten Messagie, Surendraprabu Rangaraju, Javier Sanfelix, Maria Hernandez Rivas and Joeri Van Mierlo. Key issues of lithium-ion batteries: From resource depletion to environmental performance indicators. *J. Cleaner Prod.* **2015**, 1, 354–362.

13. Jung-Ho Wee. A feasibility study on direct methanol fuel cells for laptop computers based on a cost comparison with lithium-ion batteries. *J. Power Sources* **2007**, 8(1), 424–436.

14. Michael Jones, Kevin Grogg, John Anschutz and Ruth Fierman. Sip-and-puff wireless remote control for the Apple iPod. *Off. J. RESNA* **2008**, 20(2), 107–110.

15. Tabbi Wilberforce, Zaki El-Hassan, F. N. Khatib, Ahmed Al Makky, Ahmad Baroutaji, James G. Carton, Abdul G. Olabi. Developments of electric cars and fuel cell hydrogen electric cars. *Int. J. Hydrogen Energy* **2017**, 40(5), 25695–25734.

16. Zhonghui Cui, Xiangxin Guo and Hong Li. High performance MnO thin-film anodes grown by radio-frequency sputtering for lithium ion batteries. *J. Power Sources* **2013**, 244(15), 731–735.

17. Candace K. Chan, Hailin Peng, Gao Liu, Kevin McIlwrath, Xiao Feng Zhang, Robert A. Huggins and Yi Cui. High-performance lithium battery anodes using silicon nanowires. *Nat. Nanotechnol.* **2008**, 3, 31–35.

18. Kyu-Jin Lee, Kandler Smith, Ahmad Pesaran and Gi-Heon Kim. Three dimensional thermal-, electrical-, and electronic. *J. Power Sources* **2013**, 241(1), 20–32.

19. Rotem Marom, S. Francis Amalraj, Nicole Leifer, David Jacob and Doron Aurbach. A review of advanced and practical lithium battery materials. *J. Mater. Chem.* **2011**, 21, 9938–9954.

20. Li Li, Jing Ge, Renjie Chen, Feng Wu and Shi Chen. Environmental friendly leaching reagent for cobalt and lithium recovery from spent lithium-ion batteries. *Waste Manage.* **2010**, 30(12), 2615–2621.

21. J. Hafner. Materials simulations using VASP: A quantum perspective to materials science. *Comput. Phys. Commun.* **2007**, 177(2), 6–13.

22. Zen-ichiro Takehara. Dissolution and precipitation reactions of lead sulfate in positive and negative electrodes in lead acid battery. *J. Power Sources* **2000**, 85(1), 29–37.

23. Nicholas A. Kotov. Charge transport Dilemma of solution-processed nanomaterials. *Chem. Mater.* **2014**, 26(1), 134–152.

24. Hongkang Wang, He Huang, Chunming Niu and Andrey L. Rogach. Ternary Sn-Ti-O based nanostructures as anodes for lithium ion batteries. *Small* **2014**, 11(12), 1364–1383.

25. Yuchen Ma. Simulation of interstitial diffusion in graphite. *Phys. Rev. B* **2007**, 76, 075419.

26. Kyoungmin Min, Seung-Woo Seo, Byungjin Choi, Kwangjin Park and Eunseog Cho. Computational screening for design of optimal coating materials to suppress gas evolution in Li-ion battery cathodes. *ACS Appl. Mater. Interfaces* **2017**, 9(21), 17822–17834.

27. Xue Bai, Tao Li, Zhiya Dang, Yong-Xin Qi, Ning Lun and Yu-Jun Bai. Ionic conductor of Li$_2$SiO$_3$ as an effective dual-functional modifier to optimize the electrochemical performance of Li$_4$Ti$_5$O$_{12}$ for high-performance Li-ion batteries. *ACS Appl. Mater. Interfaces* **2017**, 9(2), 1426–1436.

28. Minato Egashiraa, Hirotaka Todob, Nobuko Yoshimotoa, Masayuki Moritaa, Jun-Ichi Yamaki. Functionalized imidazolium ionic liquids as electrolyte components of lithium batteries. *J. Power Sources* **2007**, 174(2), 560–564.

29. Bruno Scrosati and Jürgen Garche. Lithium batteries: Status, prospects and future. *J. Power Sources* **2010**, 195(9), 2419–2430.

30. Jie Hu, Ao Ma and Aaron R. Dinner. Monte Carlo simulations of biomolecules: The MC module in CHARMM. *J. Comput. Chem.* **2006**, 27, 203–216.

31. Henk Eshuis, Jefferson E. Bates and Filipp Furche. Electron correlation methods based on the random phase approximation. *Theor. Chem. Acc.* **2012**, 131, 1084.

32. Shih-Yang Lin, Shen-Lin Chang, Feng-Lin Shyu, Jian-Ming Lu and Ming-Fa Lin. Feature-rich electronic properties in graphene ripples. *Carbon* **2015**, 86, 207–216.
33. Raffaele Resta. Macroscopic polarization in crystalline dielectrics: The geometric phase approach. *Rev. Mod. Phys.* **1994**, 66, 899.
34. W. Kohn, A. D. Becke and R. G. Parr. Density functional theory of electronic structure. *J. Phys. Chem.* **1996**, 100(31), 12974–12980.
35. Diandra L. Leslie-Pelecky and Reuben D. Rieke. Magnetic properties of nanostructured materials. *Chem. Mater.* **1996**, 8(8), 1770–1783.
36. Mark Fox and George F. Bertsch. Optical properties of solids. *Am. J. Phys.* **2002**, 70, 1269.
37. Maciej M. Maśka. Reentrant superconductivity in a strong applied field within the tight-binding model. *Phys. Rev. B* **2002**, 66, 054533.
38. Chiun-Yan Lin, Ching-Hong Ho, Jhao-Ying Wu, Thi-Nga Do, Po-Hsin Shih, Shih-Yang Lin and Ming-Fa Lin. Diverse Quantization Phenomena in Layered Materials. Boca Raton: CRC Press, **2020**, 1, 345.
39. Johan Nilsson, A. H. Castro Neto, F. Guinea and N. M. R. Peres. Electronic properties of graphene multilayers. *Phys. Rev. Lett.* **2006**, 97, 26680.
40. Yafei Li, Zhen Zhou, Panwen Shen and Zhongfang Chen. Structural and electronic properties of graphane nanoribbons. *J. Phys. Chem. C* **2009**, 113(33), 15043–1504.
41. Abdolali Alemi, Shahin Khademinia1, Sang Woo Joo, Mahboubeh Dolatyari and Akbar Bakhtiari. Lithium metasilicate and lithium disilicate nanomaterials: Optical properties and density functional theory calculations. *Int. Nano Lett.* **2013**, 3, 14.
42. Jürgen Hafner. Ab-initio simulations of materials using VASP: Density-functional theory and beyond. *Comput. Solid-State Chem.* **2008**, 29(13), 2039–2310.
43. Vatsal Dwivedi and Victor Chua, Physics of bulk and boundaries. Generalized transfer matrices for tight-binding models. *Phys. Rev. B* **2016**, 93, 134304.
44. Thi-Nga Do, Po-Hsin Shih, Godfrey Gumbs, Danhong Huang, Chih-Wei Chiu and Ming-Fa Lin. Diverse magnetic quantization in bilayer silicene. *Phys. Rev. B* **2018**, 97, 125416.
45. A. Guinier. X-ray diffraction in crystals, imperfect crystals, and amorphous bodies. *Chem. Educ.* **1964**, 41(5), 292.
46. Ari Deibert Palczewski. Angle-resolved photoemission spectroscopy (ARPES) studies of cuprate superconductors. *United States* **2010**, 10, 2172.
47. G. Binnig and H. Rohrer. Scanning tunneling microscopy. *Surf. Sci.* **1983**, 126(1), 3236–244.
48. Tomas Hansson, Chris Oostenbrink and Wilfred Fvan Gunsteren. Molecular dynamics simulations. *Curr. Opin. Struct. Biol.* **2002**, 12(2), 190–196.
49. Hans Persson, Ingvar Lindgren and Sten Salomonson. A new approach to the electron self-energy calculation. *Phys. Scr.* 1993, T46, 125–131.
50. Erich Runge and E. K. U. Gross. Density-functional theory for time-dependent systems. *Phys. Rev. Lett.* **1984**, 52, 997.
51. John P. Perdew, Kieron Burke and Matthias Ernzerhof. Generalized gradient approximation made simple. *Phys. Rev. Lett.* **1996**, 77, 3865–3868.
52. P. E. Blöchl. Projector augmented-wave method. *Phys. Rev. B* **1994**, 50, 17953.
53. H. J. Monkhorst and J. D. Pack. Special points for Brillouin-zone integrations. *Phys. Rev. B.* **1976**, 13, 5188.
54. Hamid Oughaddou, Hanna Enriquez, Mohammed Rachid Tchalala, Handan Yildirim, Andrew J. Mayne, Azzedine Bendounan, Gérald Dujardin, Mustapha Ait Ali, Abdelkader Kara. Silicene a promising new 2D material. *Prog. Surf. Sci.* **2015**, 90(1), 46–83.
55. Fields J. Zak and W. Zawadzki. Effective-mass approximation for electrons in crossed electric and magnetic. *Phys. Rev.* **1966**, 145, 536.

56. Thi-Nga Do, Po-Hsin Shih, Cheng-Peng Chang, Chiun-Yan Lin and Ming-Fa Lin. Rich magneto-absorption spectra of AAB-stacked trilayer graphene. *Phys. Chem. Chem. Phys.* **2016**, 18, 17597–17605.
57. Y. Lin, C. Lin, Y. Ho, T. Do and M. Lin, Magneto-optical properties of ABC-stacked trilayer graphene. Phys. Chem. Chem. Phys. **2015**, 17, 15921.
58. Fanyao Qu and Pawel Hawrylak. Magnetic exchange interactions in quantum dots containing electrons and magnetic ions. *Phys. Rev. Lett.* **2005**, 95, 217206.
59. Wahyu Setyawan and Stefano Curtarolo. High-throughput electronic band structure calculations: Challenges and tools. *Comput. Mater. Sci.* **2010**, 49, 299–312.
60. Jing Wang, Biao Lian, Xiao-Liang Qi and Shou-Cheng Zhang. Quantized topological magnetoelectric effect of the zero-plateau quantum anomalous Hall state. *Phys. Rev. B* **2015**, 92, 081107.
61. Xiaoyu Li and Huaming Yang. Morphology-controllable Li$_2$SiO$_3$ nanostructures. *Cryst. Eng. Comm.* 2014, 16, 4501–4507.
62. Ziying Wang and Ying Shirley Meng. Analytical electron microscopy: Study of all solid-state batteries. *Mater. and Energy* , **2015**, pp. 109–131.
63. Bertrand Philippe, Rémi Dedryvère, Joachim Allouche, Fredrik Lindgren, Mihaela Gorgoi, HåkanRensmo, DanielleGonbeau and Kristina Edström. Nanosilicon electrodes for lithium-ion batteries: Interfacial mechanisms studied by hard and soft X-ray photoelectron spectroscopy. *Chem. Mater.* **2012**, 24(6), 1107–1115.
64. Abhijit Prasad, Amitabha Basu and Mano J. Kumar Mahata. Impedance and conductivity analysis of Li$_2$SiO$_3$ ceramic. *Chalcogenide Lett.* **2011**, 8(8), 505–510.
65. R. J. Wilson, S. Chiang and D. D. Chambliss. Imaging semiconductors, metals and molecules with scanning tunneling microscopy. *Aust. J. Phys.* **1990**, 43(5), 393–400.
66. R. Unwin, Aleix G. Güell and Guohui Zhang. Nanoscale electrochemistry of sp^2 carbon materials: From graphite and graphene to carbon nanotubes. *Acc. Chem. Res.* **2016**, 49(9), 2041–2048.
67. Jincheng Du and L. Rene Corrales. Characterization of the structural and electronic properties of crystalline lithium silicates. *J. Phys. Chem.* **2006**, 110, 22346.
68. Fazel Shojaei, Hong Seok Kang. Electronic structures and Li-diffusion properties of group IV–V layered materials: Hexagonal germanium phosphide and germanium arsenide. *J. Phys. Chem. C* **2016**, 120(41), 23842–23850.
69. Wei Sun Leong, Muhammad Afiq Nurudin, Sohail Anwar, Mohammad Taghi Ahmadi and Razali Ismail. Effect of graphene nanoribbons layers on its band energy and the electrical properties. *J. Computat. Theor. Nanosci.* **2012**, 9(12), 2082–2085.
70. Eduardo V. Castro, Nuno Peres, Joao M. B. Lopes dos Santos, Antonio H. Castro Neto and Francisco Guinea. Localized states at zigzag edges of Bilayer graphene. *Phys. Rev. Lett.* **2008**, 100, 026802.
71. Hanjun Sun, Li Wu, Weili Wei and Xiaogang Qu. Recent advances in graphene quantum dots for sensing. *Mater. Today* **2013**, 16(11), 433–442.
72. P. E. Kornilovitch and A. S. Alexandrov. Isotope effect on the electron band structure of doped insulators. *Phys. Rev. B* **2004**, 70, 224511.
73. Enyue Zhao, Xiangfeng Liu, Hu Zhao, Xiaoling Xiao and Zhongbo Hu. Ion conducting Li$_2$SiO$_3$-coated lithium-rich layered oxide exhibiting high rate capability and low polarization. *Chem. Commun.* **2015**, 51, 9093–9096.
74. Anton Du Plessis, Igor Yadroitsev, Ina Yadroitsava and Stephan G. Le Roux. X-ray microcomputed tomography in additive manufacturing: A review of the current technology and applications. *3D Print. Addit. Manuf.* **2018**, 5(3), 227–247.
75. A. A. Kordyuk, V. B. Zabolotnyy, D. V. Evtushinsky, A. N. Yaresko, B. Büchner and S. V. Borisenko, Electronic band structure of Ferro-Pnictide superconductors from ARPES experiment. *J. Supercond. Novel Magn.* **2013**, 26(9), 2837–2841.

76. P. Richet, B. O. Mysen and D. Andrault. Melting and premelting of silicates, Raman spectroscopy and X-ray diffraction of Li_2SiO_3 and Na_2SiO_3. *Phys. Chem. Miner.* **1996**, 23, 157–172.

77. Abhijit Prasad and Amitabha Basu. Structural and dielectric studies of Li_2SiO_3 ceramic. *Mater. Lett.* **2012**, 66, 1–3.

78. WuZhaoyin Xiangwei, Wen Xiaogang, Xu Xiuyan and Wang Jiu Li. Synthesis and characterization of Li_4SiO_4 nano-powders by a water-based sol–gel process. *J. Nucl. Mater.* **2009**, 392(3), 471–475.

79. Zhiyong Zhou, Michael Steigerwald, Mark Hybertsen, Louis Brusm and Richard A. Friesner. Electronic structure of tubular aromatic molecules derived from the metallic (5,5) armchair single wall carbon nanotube. *J. Am. Chem. Soc.* **2004**, 126(11), 3597–3607.

80. Ching-Hong Ho, Yen-Hung Ho, Ying-Yen Liao, Yu-Huang Chiu, Cheng-Peng Chang and Ming-Fa Lin. Diagonalization of landau level spectra in rhombohedral graphite. *J. Phys. Soc. Jpn.* **2012**, 81, 024701.

81. Christian Müller, Egbert Zienicke, Stefan Adams, Junko Habasaki and Philipp Maass. Comparison of ion sites and diffusion paths in glasses obtained by molecular dynamics simulations and bond valence analysis. *Phys. Rev. B* **2007**, 75, 014203.

82. C. Fernandes, M.J. Gaspar, J. Pires, A. Alves, R. Simões, J.C. Rodrigues, M.E. Silva, A. Carvalho, J.E. Brito, J.L. Lousada. Physical, chemical and mechanical properties of Pinus sylvestris wood at five sites in Portugal. *iForest – Biogeosciences and Forestry,* **2017**, 10, 669–679.

83. Balázs Dóra, Janik Kailasvuori and R. Moessner. Lattice generalization of the Dirac equation to general spin and the role of the flat band. *Phys. Rev. B* **2011**, 84, 195422.

84. Shuai Yang, Qiufen Wang, Juan Miao, Jingyang Zhang, Dafeng Zhang, Yumei Chen and Hong Yang. Synthesis of graphene supported Li_2SiO_3 as high performance anode material for lithium-ion batteries. *Appl. Surf. Sci.* **2018**, 444, 522–529.

85. Fabian Pease and W.C. Nixon. High resolution scanning electron microscopy. *J. Sci. Strum* **1965**, 42, 81.

86. J. Garnaes. Transmission electron microscopy of scanning tunneling tips. *J. Vac. Sci. Technol. A* **1990**, 8, 441.

87. Paul K. Hansma. Scanning tunneling microscopy. *J. Appl. Phys.* **1987**, 61(2), R1–R24.

88. Mikito Koshino and Edward McCann. Multilayer graphenes with mixed stacking structure: Interplay of Bernal and rhombohedral stacking. *Phys. Rev. B* **2013**, 87, 045420.

89. Dmitry V. Kosynkin, Amanda L. Higginbotham, Alexander Sinitskii and Jay R. Lomeda, A. Dimiev, B. K. Price and J. M. Tour. Longitudinal unzipping of carbon nanotubes to form graphene nanoribbons. *Nature* **2009**, 458, 872–876.

90. Zheng Liu, Kazu Suenaga, Peter J. Harris and Sumio Iijima. Open and closed edges of graphene layers. *Phys. Rev. Lett.* **2009**, 102, 015501.

91. Lisa M. Viculis, Julia J. Mack and Richard B. Kaner. A chemical route to carbon nanoscrolls. *Science* **2003**, 299, 1361.

92. Georgina Mondragon-Gutierrez, Daniel Cruz, Heriberto Pfeiffer and Silvia Bulbulian. Low temperature synthesis of Li_2SiO_3: Effect on its morphological and textural properties. *Res. Lett. Mater. Sci.* **2008**, 2008, 1–4.

93. Y. Lu, Y. Jiang, W. Wei, H. Wu, M. Liu, L. Niu and W. Chen. Novel blue light emitting graphene oxide nanosheets fabricated by surface functionalization. *J. Mater. Chem* **2012**, 22, 2929–2934.

94. Yang-Fan Xu, Mu-Zi Yang, Bai-Xue Chen, Xu-Dong Wang, Hong-Yan Chen, Dai-Bin Kuang and Cheng-Yong Su. Graphene oxide composite for photocatalytic CO_2. *Reduct. J. Am. Chem. Soc.* **2017**, 139, 5660–5663.

95. Abdolali Alemi, Shahin Khademinia, Sang Woo Joo, Mahboubeh Dolatyari and Akbar Bakhtiari. Lithium metasilicate and lithium disilicate nanomaterials: Optical properties and density functional theory calculations. *Int. Nano Lett.* **2013**, 3, 14.

96. Duckhwan Lee and Andreas C. Albrecht. On the interaction operator in the optical spectroscopies. *J. Chem. Phys.* **1983**, 78, 3382.

97. H. Wadati, T. Yoshida, A. Chikamatsu, H. Kumigashira, M. Oshima, H. Eisaki, Z.-X. Shen, T. Mizokawa and A. Fujimori. Angle-resolved photoemission spectroscopy of perovskite-type transition-metal oxides and their analyses using tight-binding band structure. *Phase Transitions* **2006**, 79(8), 617–635.

98. T. Cuk, D. Lu, X. Zhou, Z.-X. Shen, T. Devereaux and N. Nagaosa. A review of electron-phonon coupling seen in the high-T$_c$ superconductors by angle-resolved photoemission studies (ARPES). *Phys. Status Solidi* **2005**, 242, 11–29.

99. Han Huang, Dacheng Wei, Jiatao Sun, Swee Liang Wong, Yuan Ping Feng, A. H. Castro Neto and Andrew Thye Shen Wee. Spatially resolved electronic structures of atomically precise armchair graphene nanoribbons. *Sci. Rep.* **2012**, 2, 983.

100. G. Li, A. Luican, J. L. Dos Santos, A. C. Neto, A. Reina, J. Kong and E. Andrei. Observation of Van Hove singularities in twisted graphene layers. *Nat. Phys.* **2010**, 6, 109–113.

101. V. Cherkez, G. T. de Laissardiere, P. Mallet and J.-Y. Veuillen. Van Hove singularities in doped twisted graphene bilayers studied by scanning tunneling spectroscopy. *Phys. Rev. B* **2015**, 91, 155428.

102. G. Li, A. Luican and E. Y. Andrei. Scanning tunneling spectroscopy of graphene on graphite. *Phys. Rev. Lett.* **2009**, 102, 176804.

103. W. Y. Ching, Y. P. Li, B. W. Veal and D. J. Lam. Electronic structures of lithium meta-silicate and lithium disilicate. *Phys. Rev. B* **1985**, 32, 1203.

104. Xue Bai, Tao Li, Zhiya Dang, Yong-Xin Qi, Ning Lun and Yu-Jun Bai. Ionic conductor of Li$_2$SiO$_3$ as an effective dual-functional modifier to optimize the electrochemical performance of Li$_4$Ti$_5$O$_{12}$ for high-performance Li-ion batteries. *ACS Appl. Mater. Interfaces* **2017**, 9(2), 1426–1436.

105. A. C. T. North, D. C. Phillips and F. S. Mathews. A semi-empirical method of absorption correction. *Acta Cryst.* **1968**, 24, 351–359.

106. Shuai Yang, Jingyang Zhang, Qiufen Wang, Juan Miao, Chenli Zhang, Lin Zhao and Yanlei Zhang. Li$_2$SiO$_3$ and Li$_2$SnO$_3$/SnO$_2$ as a high performance lithium-ion battery. *Mater. Lett.* **2019**, 234, 375–378.

107. Long Xia, Yanan Yang, Xinyu Zhang, Jian Zhang, Bo Zhong, Tao Zhang and Huatao Wang. Crystal structure and wave-transparent properties of lithium aluminum silicate glass-ceramics. *Ceram. Int.* **2018**, 44(12), 14896–14900.

108. N. Kuganathan, L. H. Tsoukalas and A. Chroneos. Defects, dopants and Li ion diffusion in Li$_2$SiO$_3$. *Solid State Ionics* **2019**, 335, 61–66.

109. H. Völlenkle. Verfeinerung der Kristallstrukturen von Li$_2$SiO$_3$ und Li$_2$GeO$_3$. *Zeitschrift fur Kristallographie* **1981**, 154, 77–81.

110. Tao Tang, Piheng Chen, Wenhua Luo, Deli Luo and Yu Wang. Crystalline and electronic structures of lithium silicates: A density functional theory study. *J. Nucl. Mater.* **2012**, 420, 31–38.

111. Shenggui Ma, Shichang Li, Tao Gao, Yanhong Shenm, Xiaojun Chen, Chengjian Xiao and Tiecheng Lu. Molecular dynamics simulations of structural and melting properties of Li$_2$SiO$_3$. *Ceram. Int.* **2018**, 44(3), 3381–3387.

6 Electrolytes for High-Voltage Lithium-Ion Battery

A New Approach with Machine Learning

Ming-Hsiu Wu, Chih-Ao Liao,
Ngoc Thanh Thuy Tran, and Wen-Dung Hsu
National Cheng Kung University

CONTENTS

6.1 INTRODUCTION

In recent years, with a high demand on an electric vehicle, developing safe and efficient high-voltage lithium-ion battery (LIB) becomes more and more urgent [1]. However, the decomposition of traditional organic electrolyte compounds during charging and discharging significantly impedes the development. Therefore, it is desired to design new electrolyte molecules that can withstand in high voltage to replace the old ones. There are usually two directions to design a new molecule, including functional group modification and machine learning method. In this chapter, we will discuss the latter method. This method provides a whole new road to approach material design.

6.2 METRICS FOR MOLECULAR SELECTION

According to the frontier molecular orbital theory, K. Fukui et al. [2] have pointed out that electrochemical stable window of the electrolytes refers to the lowest unoccupied molecular orbital (LUMO), the highest occupied molecular orbital (HOMO), and the interval in between LUMO and HOMO. The reactivity can be predicted by comparing the electrochemical potential of the electrode (the electron Fermi level) with the LUMO and HOMO of the electrolyte molecule. The LUMO and HOMO are usually referred to as the electron orbitals of a molecule in a neutral state. In redox reaction, however, ability of electron accepting, or withdrawing, is the direct property to determine if the reaction would occur or not. Therefore, instead of LUMO and HOMO, electron affinity (EA) and ionization energy (IE) are considered in most researches. The energy levels of EA and IE can be defined as the redox standard and also can take the structural changes into account. In other words, the value of IE/EA can be important properties (characteristics) for electrolyte molecular selection.

For an ideal molecule for electrolyte, its IE value should be higher than the work function of cathode materials to prevent oxidation by cathode materials. For high-voltage batteries, the work function of cathode materials is lower than conventional ones; thus, a high-IE electrolyte molecule is needed as shown in Figure 6.1. Similarly, at the anode side in order to prevent reduction by anode materials, the EA value for the electrolyte molecule should be lower than the work function of anode materials.

To do the molecular selection for developing new electrolyte, other properties, such as structural stability during reduction or oxidation, dipole moment, and polarizability, are also supposed to be taken into consideration. However, because of limited molecular properties included in current databases, such as materials project [3], QM9 [4], and ZINC [5], full property selection is not possible, IE and EA, the most important characteristics, were used to do the first step selection. Additionally, viscosity of electrolyte is also an important property to consider. The viscosity of

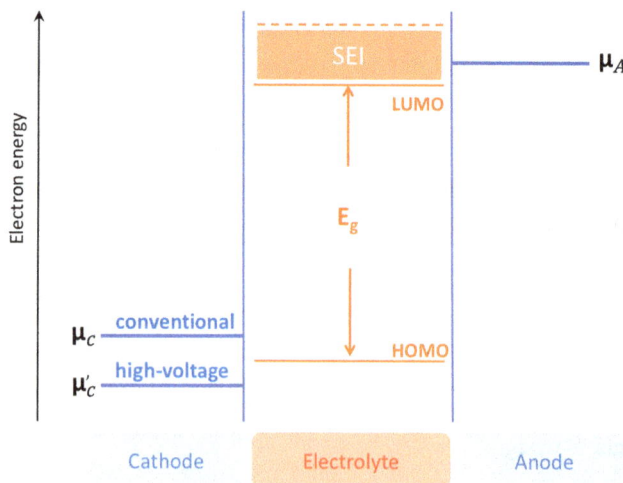

FIGURE 6.1 Chemical potential of LIBs.

molecule is highly related to its dipole moment and polarizability, so one can take these properties into account in the next-step selection in the future. Furthermore, stability of molecular structure can play a dominant role in molecular selection, since it indirectly but closely affects the stability and durability of battery. When molecules are oxidized or reduced by electrode, molecular bonds possibly break and then decompose into flammable small molecules, which lead to devastating result.

6.3 EXPERIMENTS, FIRST-PRINCIPLES CALCULATION, AND MACHINE LEARNING

Conventional molecule design is a time-consuming, laborious, as well as expert-based mission. The most common method is trial-and-error. Countless material development experiments are often evitable, and whole process is often high investment involved, no matter in time or human power. Historically, the timescale for the development of new material, from laboratory to commercial application, is about 15–20 years [6]. The number of possible small molecular structures is estimated to be about on the order of 10^{60} [7], constituting a well-known chemical space. It is almost impossible to explore this space via the conventional trial-and-error experimental method. First-principles calculation can provide material developer's another approach that circumvents the arduousness of experimentation. By quantum mechanics, physical and chemical properties of these molecules can be exactly obtained through solving Schrodinger equation. The first-principles calculation strategy significantly reduces the cost of developing new molecules, though vast chemical space still cannot be efficiently exploited, and virtually only little part of molecules in the world has been investigated until now.

The conception of the machine learning has existed for a long time since Arthur Samuel first came up with the phrase "machine learning" in 1952. However, the lack of material data significantly impeded the development of the machine learning in the material science field. During the past 10 years, a large amount of data including material structures and its corresponding properties have been accumulated, especially from first-principles calculation. Hence, the power of the machine learning has drawn people's attention away from conventional molecules. A large number of studies are dedicated to implementing machine learning to learn the relationship between material structures between their properties from existing data in recent years. Gradually, the machine learning becomes a powerful method for investigating materials at a large scale in the initial stage of material design.

6.4 MACHINE LEARNING REGRESSION MODEL AND PROPERTY PREDICTOR

In recent years, the machine learning has seen a burgeoning method for the application to molecular design. The machine learning model has been used as property predictor aiming to predict molecular properties from its structure by learning implicit relation between them, as shown in Figure 6.3a. The model could accurately predict thousands of molecules within minutes. Previous works have successfully integrated this technology into molecule screening pipelines [8,9]. In machine learning framework, molecular structures are transformed into a digital representation that serves as input for the machine learning model. Two ideal attributes for molecular representation are uniqueness and invertibility. Uniqueness means that a specific molecule can only be represented by a unique molecular representation. Invertibility means molecular representation can be transformed back to a specific single molecule. There are mainly two types of representation, including 3D geometry and 2D molecular graphs. The latter can be further subdivided into four categories: string-based, image-based, tensor, and others. In the following, we will briefly describe the extended-connectivity fingerprint (ECFP) [10], simplified molecular-input line-entry system (SMILES) [11], and Coulombic matrix (CM) [8] methods.

The ECFP is a topological fingerprint that is originally used for the analysis of molecular characterization, substructure, and similarity. Nowadays, it is also adopted to perform the machine learning and statistical analysis of molecules. It basically has four steps: (a) assign each atom with an identifier, (b) atom identifiers are augmented with information from the atom neighborhood, (c) delete the duplicated substructure, and (d) hash list of identifiers into a fixed-length bit vector. It does not have the attribute of invertibility. Divided molecular substructure cannot be recombined together into the original molecule exactly. Furthermore, due to the fixed-length bit vector, each bit may represent multiple substructures, which often leads to the difficulty of the analysis. It is worth mentioning that initially each binary bit represents whether the substructure exists in a molecule or not, but here in this study, the bit is replaced with the number of the substructure in the molecule, as a so-called ECFPNUM. The ECFPNUM is further used for molecular representation since it is more suitable for property prediction than ECFP.

The SMILES is a molecular representation in the form of a line notation for describing the molecular structure by using ASCII characters. For example, benzene is denoted in the form of SMILES as c1ccccc1. It is the most popular representation in the field of machine learning since it follows particular grammar syntax and can be directly applied to natural language processing (NLP) models. In practice, because many machine learning algorithms cannot process characters (strings) as input directly, it has to be converted into numeric form. The way to standardize converting SMILES into numeric form is first setting every different character as an atom type. Then, for a given molecule every character of its SMILES is converted into bit vectors formed by the atom types. The bit vectors then combined follow the sequence of characters appearing in its SMILES to form a binary matrix that can directly operate as machine learning input. This scheme is known as one-hot encoding. Invertibility is a main advantage of SMILES, since one-hot encoding representation can be converted back to original molecules directly. SMILES, however, also suffers drawback at the same time. One molecule can have multiple SMILES representations. The nonuniqueness SMILES stem from the arbitrary starting atom in a molecule can be used to construct its SMILES. Some cheminformatic packages, such as RDKit [12], have the function to canonize the SMILES. However, Bjerrum et al. argue that the latent space created from canonical SMILES may have problems, since only specific grammar syntax has been learned, instead of the general underlying rule of molecule structure [13].

The CM is a molecular representation to describe the electrostatic interaction between atoms in a molecule. It is calculated by Equation (6.1).

$$
M_{ij}^{Coulomb} = \begin{cases} 0.5Z_i^{2.4} & \text{for } i = j \\[2mm] \dfrac{Z_i Z_j}{R_{ij}} & \text{for } i \neq j \end{cases} \tag{6.1}
$$

The diagonal element is the polynomial fit of the potential energy of the atom itself (self-energy), while the off-diagonal elements correspond to the energy of Coulombic interaction between pairs of different atoms in the molecules. The CM can be obtained by performing quantum mechanical calculations. Through the machine learning correlation between molecular structure and its CM could be revealed. Thus, the learned model has exhibited strong an ability to predict molecular electronic properties. The CM also suffers from a number of issues. A different number of atoms in a molecule leads to different sizes of CMs. The solution is to pad by the vacancy of matrix of little molecule. Another issue is that different atom labeling schemes lead to different CMs. A simple solution to this is to simply sort the matrices in the order of specific atomic property.

6.5 PROPERTY PREDICTOR

Figure 6.2 shows the prediction performance of three different property predictor models. The data were collected from Material Project (MP) [3] that contains 21,165 molecular structures and their corresponding values of IE and EA in the solvent state calculated through *ab* inito calculation based Q-chem software. In this study, 80% of the database was used for training, 10% of them were used for validation, and the remaining 10% was used for testing. The first model used ECFPNUM as molecular representation and adopted fully connected NN to do machine learning. The second model used SMILES as molecular representation, which is converted into a binary matrix through one-hot encoding and adopted RNN to do machine learning. We

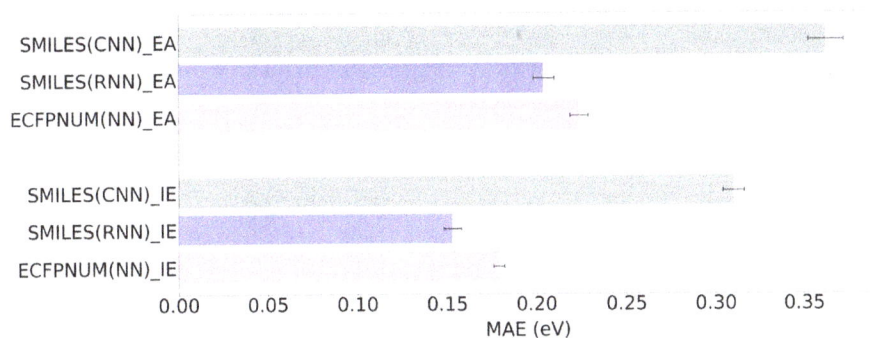

FIGURE 6.2 Prediction performance of three different property predictors.

call this model SMILES (RNN). The last model also used SMILES as a molecular representation with one-hot encoding to convert them to digit matrix and adopted convolution neural network (CNN) jointing with a fully connected NN to perform machine learning. We call this model SMILES (CNN). The property predictor performance is evaluated by mean absolute error (MAE) on the test set. The results show that using SMILES representation with RNN model can outperform ECFPNUM representation with NN model and the SMILES representation with CNN jointing NN model. Although these models can predict IE/EA from molecular representation swiftly, the optimal MAE is still about 0.2 eV for EA and 0.15 eV for IE. These errors are still large if one wants to do molecular selection that needs the errors smaller than 0.1 eV or less. Thus, the models just provide us a tool to do the first screening of molecules from large-scale candidates. The more exact values should rely on more expensive verification methods, such as first-principles calculations or experimental measurements.

6.6 INVERSE DESIGN AND DEEP GENERATIVE MACHINE LEARNING MODEL

In this work, several aforementioned strategies for molecule selection have been utilized, which indeed significantly accelerate the pace of material development. In conventional material design, the properties of a specific material or molecular structure could be obtained through experiments or simulations, but the process is not quietly efficient as mentioned above since exploring chemical space at a large scale is a daunting task. Another strategy has been proposed – starting with desired molecular properties and inversely searching for ideal molecular structures [14,15]. The process is known as inverse design. Inverse design can be approached in many different methods. One of these methods is high-throughput virtual screening (HTVS) [16,17]. HTVS allows the developers or researchers rapidly screen millions of molecules and enable them to identify promising molecules. Candidate molecules will be further evaluated by more expensive methodologies. HTVS methodology has been successful at designing high-performing molecules. For example, in organic photovoltaics, molecules have been screened for their HOMO/LUMO energies and photovoltaic conversion efficiency [18,19]. For organic light-emitting diodes, molecules have been screened for their singlet–triplet gap and photoluminescent emission [20].

Another approach is harnessing the power of a deep generative model. The goal of the generative model is to learn how to generate data whose distribution is like the original data set. Generative model can be successfully applied in many different domains, such as image regeneration, audio regeneration, and even drug discovery and molecule regeneration. There have been three main generative models mainly used in molecular generation: Variational autoencoder (VAE) [21,22], generative adversarial network (GAN) [23,24], and reinforcement learning (RL) [25,26]. Here, we will not describe the details of the underlying mechanism of each model, but just briefly introduce them. VAE is derived from autoencoder (AE) shown as Figure 6.3b. The AE consists of an encoder and a decoder. The encoder is responsible for compressing molecular representation into a vector in low-dimensional space, known as latent space and the decoder aims to transform a vector back to original molecular

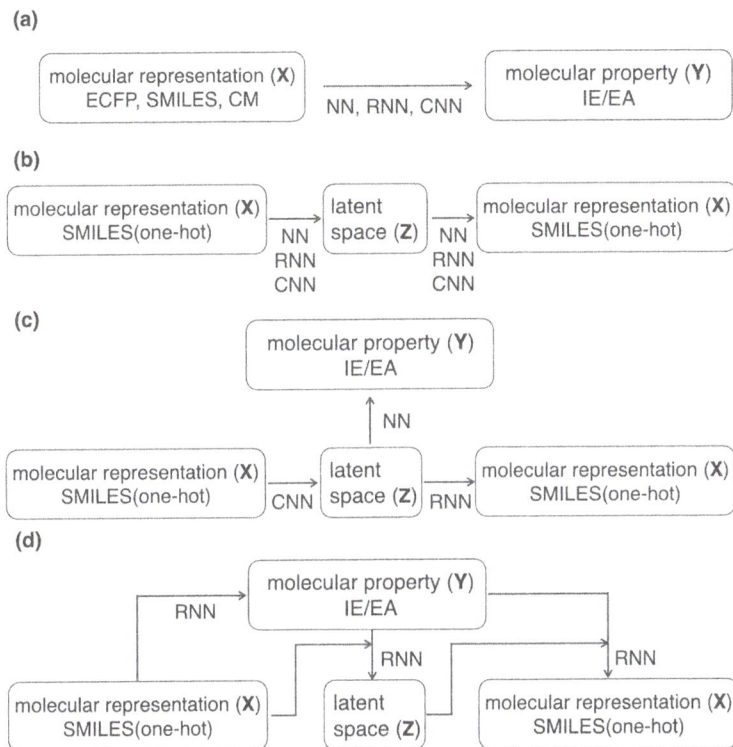

FIGURE 6.3   Graphical diagram of the machine learning application to molecular design: (a) property prediction, (b) autoencoder (AE), (c) variational autoencoder (VAE) proposed by Bombarelli et al. [21], and (d) conditional variational autoencoder (CVAE) proposed by Kang et al. [27]. (The figure was adapted from Kang et al. [27].).

representation. Nevertheless, because the sparse of latent space created by AE makes it difficult to exploit the space, a probability distribution has been introduced by VAE in order to fill uncovered latent space. Therefore, a molecule does not map as a fixed point in the latent space, but as a probability distribution, usually Gaussian distribution. The continuous representation of molecules in latent space allows us to generate a molecule easily by simple operation, such as decoding random sampling vector, perturbing known chemical structure, or interpolating between molecules. Another generative model is GAN. In this framework, the generator competes against discriminator. The generator attempts to generate artificial data from sampling noise space. The discriminator is responsible for determining whether the data are fake or real. The goal of this model is that generator has the ability to generate data that discriminator cannot distinguish as fake. As a result, we can generate new molecules from the generator. The last one is RL, which has been seen as the most promising model in molecular design recently. The mechanism of RL is that the generator learns how to take action in order to maximize the reward (properties) within an environment (SMILES generation).

6.7   DATA

There are only a few online accessible databases containing EA and IE values, such as MP [3] and National Institute of Standards and Technology (NIST). MP has about 35,336 molecular structures and its corresponding IE and EA values (not every molecule labeled both EA and IE) that are acquired through quantum mechanical calculation (continuously updated). Besides, the MP provides the user a convenient application programming interface (API), so users can easily access the molecular data according to their needs. NIST also stores molecular EA and IE values that are acquired by experiments. As NIST does not provide any API, it is hard for users to collect a large number of data, and thus, web crawling must be involved in automatic collection workflow. Additionally, QM9 [4,8] includes 133,885 molecular SMILES and its corresponding HOMO and LUMO. The HOMO and LUMO can be roughly mapped to EA and IE, since there is a high correlation between IE/HOMO and EA/LUMO. In terms of dipole moment and stability of molecule structure, as we know, there is a dipole moment database in NIST. At the same time, we are looking for chemical properties that can quantitatively describe the stability of the molecule. Furthermore, in order to train a deep generative model, collecting a large number of molecules is necessary. In our work, 6 million molecular SMILES data from ZINC [5] are collected to train the deep generative model. ZINC is a free database of commercially available compounds and contains over 750 million molecules that can be downloaded efficiently.

6.8   OUR ADAPTED MODEL AND EXPERIENCE

Since VAE has been introduced into the molecular design in 2016 by Bombarelli et al., over 40 works have been proposed by using different deep generative models and some variations of them. In this work, to design ideal molecules for electrolyte, two proposed models have been adapted to our generative model for designing electrolyte molecules. The first one is chemical VAE (chemVAE), proposed by Bombarelli et al. [21] (Figure 6.3c), and the other one, named conditional-molecular-design-ssvae (SSVAE), is proposed by Kang et al. [27] (Figure 6.3d). In the frame of chemVAE (Figure 6.4), SMILES is transformed into a one-hot representation ($\mathbf{x}$), and CNN is used as an encoder to map $\mathbf{x}$ to latent space, $\mathbf{z}$, with Gaussian distribution. The RNN decoder transforms $\mathbf{z}$ back to one-hot representation, $\mathbf{x}$. Additionally, another NN, named property predictor, mapped $\mathbf{z}$ to molecular properties shows the structure of the model. The continuous representation of molecules in latent space allows us to generate molecules easily by simple operation, such as decoding random sampling vector, perturbing known chemical structure, or interpolating between molecules. This work also introduces the Gaussian process model to guide the search of ideal molecules efficiently in latent space. In Bombarelli's model, three components (encoder, decoder, and property predictor) were trained simultaneously. In this study, a small portion of the data used in this work are obtained from MP with IE/EA label and many other unlabeled molecules whose SMILES is from ZINC; hence, the VAE and property predictor cannot be trained simultaneously. A new approach is that VAE model can be trained by all the molecular SMILES from MP

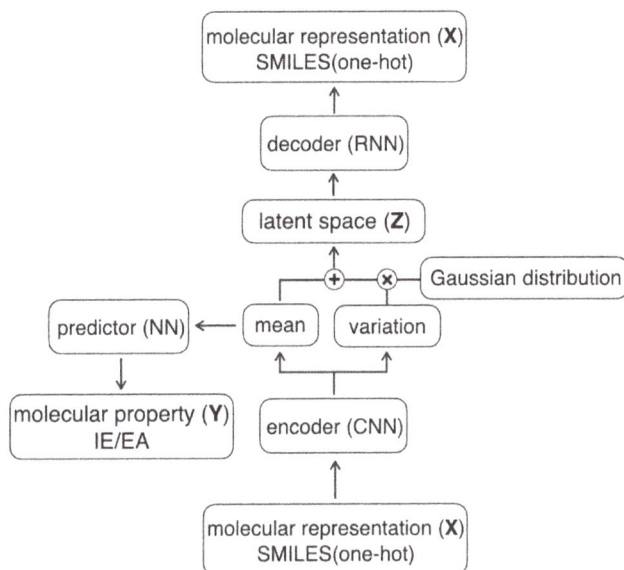

FIGURE 6.4   Structure of chemVAE

and ZINC first. After finishing VAE training, the encoders from VAE are extracted and combined with property predictor NN for the next step of the training of MP data containing molecular SMILES and its corresponding properties. In order to preserve VAE system, in the process of training this property predictor, all the parameters in the encoder need to be frozen and only optimize the parameters of the property predictor. However, the prediction performance of this model is not good enough to compare with our previous trained property predictors, since the parameters of the encoder cannot be optimized accordingly. In Bombarelli's original work, all the training data are labeled, but in our case, the labeled data only count a small portion of training data. The paper shows that latent space created by AEs jointly trained with the property prediction task shows a gradient by property values; molecules with high specific property values are located in one region, and molecules with the low values are in another. The AEs that were trained without the property prediction task do not show a discernible pattern with respect to property values. It seems like encoder, decoder, and property predictor if trained together molecules mapped in the latent space show property-related distribution. Evidently, the chemVAE model is not suitable for our case, since the molecules labeled with IE/EA only count a small portion of all data in our case.

In reality, a small portion of molecules were labeled with their properties is a very common situation in most of the database because the cost of acquiring properties is very expensive. Kang et al. try to solve this problem and proposed SSVAE [27]. The model combines semisupervised learning and conditional AE together and allows us to directly generate new molecules whose properties are close to a predetermined target condition without any extra optimization procedure. In this framework, the labeled data are used for building a property predictor that introduces the Gaussian

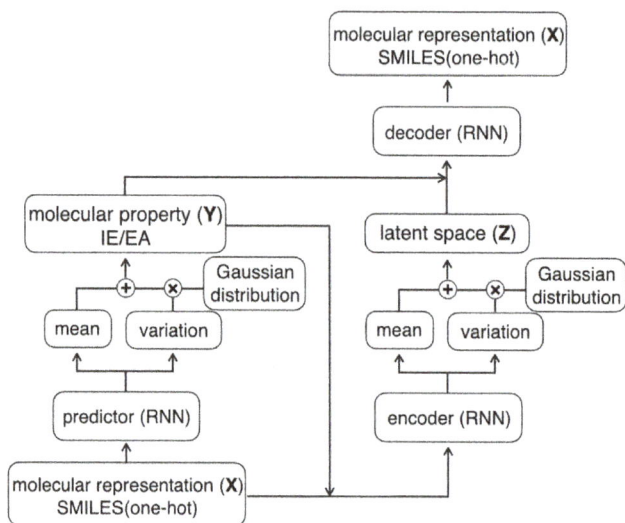

FIGURE 6.5 Structure of SSVAE

distribution in order to address the intractability of **y**. The missing **y** of unlabeled data is predicted by this property predictor. Next, the RNN encoder maps **x** and **y** to the latent space **z,** similarly introducing the Gaussian distribution. Lastly, RNN decoder maps **y** and **z** to original **x**. The structure of SSVAE is shown in Figure 6.5. In this work, MP data with IE/EA label and a large number of molecular SMILES from ZINC are used to train the SSVAE model.

The performance of SSVAE is evaluated through our previously trained SMILES(RNN) property predictor model. Based on setting a predetermined IE value, 1,000 molecules from the decoder of SSVAE mode are generated and then screened by our previously mentioned SMILES(RNN) property predictor (Figure 6.6).

Candidate molecules can be synthesized and characterized by various techniques for electrochemical testing. It is worthwhile noting that the energy level of IE can be measured by ultraviolet photoelectron spectroscopy and inverse photoemission spectroscopy [28]. The main principle of photoelectron spectroscopy is that the source of radiation can eject molecular electrons from core orbitals or valence orbitals. The IE is obtained by subtracting the energy of the source radiation from the measured kinetic energy of the ejected electron. For the energy level of EA, it can be also measured by photoelectron spectroscopy and laser photodetachment [29]. Both methods directly measure threshold energy for the removal of an electron from the molecular anion. In addition, by using the voltammetric technique, the redox potentials (reduction/ oxidation potentials that are proportional to IE/EA values, respectively) can be verified by cyclically sweeping a potential between a working electrode and a reference electrode in the solution, while measuring the current response [30]. The experimental results can be compared with the density-functional theory calculation results.

Comparing conventional electrolyte molecules – EC, DMC, and VC, their IE values are obtained from experiments (NIST database) and first-principles calculation (MP database). The experimental values of IE are 10.4, 11.0, and 10.08 eV for

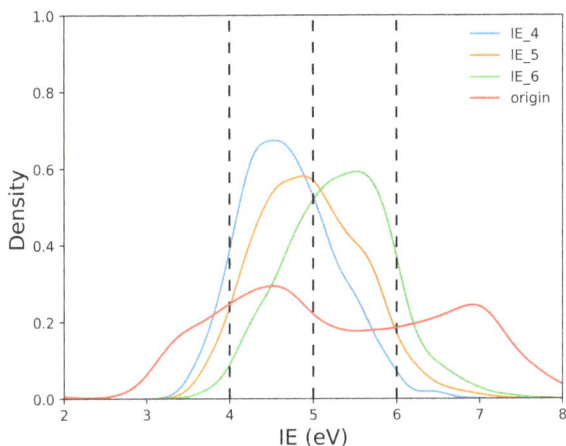

FIGURE 6.6 Value of IE distribution of all the original data and the generated molecules whose IE values were predicted by property predictor. Although the result is not as good as the paper reported [27], the trend of the distribution is consistent with the increase in IE values. In this study, IE value is set to 4, 5, and 6 eV. The tolerance for the screening is 0.2 eV. The success rates are 15.3%, 22.9%, and 15.2% for 4, 5, and 6 eV, respectively.

EC, DMC, and VC, respectively [31,32]. And, the first-principles calculation values of IE are 8.44, 8.3, and 7.04 eV for EC, DMC, and VC, respectively [3]. The order of experimental IE value of EC and DMC is not consistent with the order of first-principles calculation value. The discrepancy in order can be probably attributed to several of the factors. First, the experiments measure the value of IE in the gas phase, but for the first-principles calculation in MP, the solvent environment has been taken into consideration. Furthermore, first-principles calculation solves the Schrodinger equation through the series of approximations and simplifications. The final result could vary by the selection of different basis sets and DFT functionals. Nevertheless, in both methods, the IE values of DMC and EC are higher than that of VC, which means VC could be oxidized more easily than DMC and EC. The result suggests that VC can be a suitable electrolyte additive molecule, and EC and DEC can be suitable for electrolyte molecules.

6.9 CONCLUSIONS

In this work, the power of advanced machine learning technology has been harnessed to design brand-new electrolyte molecules that are suitable for high-voltage LIB electrolyte. The energy level of IE/EA of molecules is regarded as a metric to gauge the potential of a candidate molecule. Three property predictors that have an ability to map molecular structure to its chemical properties (IE/EA) rapidly have been built. Among them, SMILES (one-hot encoding) with RNN has better prediction performance than the others. Furthermore, in order to carry out the inverse design, the proposed generative model is adapted to our application. So far, new molecules with predetermined property values can be automatically generated. Nevertheless, unlike the other fields such as in particular speech recognition and computer vision, the scarcity of labeled data in material and molecular databases significantly impedes the development of the machine learning in this field. In addition, even though the machine learning exhibits phenomenal prediction ability, the difficulty of interpreting machine learning models makes people hard to unveil the real latent relation between the molecule and its corresponding properties. Undoubtedly, the machine learning in molecules and material design deserves more material or molecular knowledge involved. All in all, this work is expected to pave the way for the resolutions to the aforementioned problems and electrolyte molecular design.

REFERENCES

1. Sun, Y., N. Liu, and Y. Cui, Promises and challenges of nanomaterials for lithium-based rechargeable batteries. *Nature Energy*, 2016, 1(7): p. 16071.
2. Fukui, K., T. Yonezawa, and H. Shingu, A molecular orbital theory of reactivity in aromatic hydrocarbons. *Journal of Chemical Physics*, 2004, 20: pp. 722–725.
3. Qu, X., et al., The electrolyte genome project: A big data approach in battery materials discovery. *Computational Materials Science*, 2015, 103: pp. 56–67.
4. Blum, L.C. and J.-L. Reymond, 970 million druglike small molecules for virtual screening in the chemical universe database GDB-13. *Journal of the American Chemical Society*, 2009, 131(25): pp. 8732–8733.
5. Irwin, J.J. and B.K. Shoichet, ZINC: A free database of commercially available compounds for virtual screening. *Journal of Chemical Information and Modeling*, 2004, 45(1): pp. 177–182.
6. Maine, E. and E. Garnsey, Commercializing generic technology: The case of advanced materials ventures. *Research Policy*, 2006, 35(3): pp. 375–393.
7. Virshup, A.M., et al., Stochastic voyages into uncharted chemical space produce a representative library of all possible drug-like compounds. *Journal of the American Chemical Society*, 2013, 135(19): pp. 7296–7303.
8. Rupp, M., et al., Fast and accurate modeling of molecular atomization energies with machine learning. *Physical Review Letters*, 2012, 108(5): p. 058301.
9. Hansen, K., et al., Machine learning predictions of molecular properties: Accurate many-body potentials and nonlocality in chemical space. *Journal of Physical Chemistry Letters*, 2015, 6(12): pp. 2326–2331.
10. Rogers, D. and M. Hahn, Extended-connectivity fingerprints. *Journal of Chemical Information and Computer Sciences*, 2010, 50(5), pp. 742–754.
11. Weininger, D., SMILES, a chemical language and information system. 1. Introduction to methodology and encoding rules. *Journal of Chemical Information and Computer Sciences*, 2002, 28(1): pp. 31–36.
12. Landrum, G., RDKit: Open-source cheminformatics.
13. Sattarov, E.J.B.A.B., Improving chemical autoencoder latent space and molecular de novo generation diversity with heteroencoders. arXiv e-prints:1806.09300, June 2018.
14. Kuhn, C. and D.N. Beratan, Inverse strategies for molecular design. *Journal of Physical Chemistry*, 1996, 100(25): pp. 10595–10599.
15. Zunger, A., Inverse design in search of materials with target functionalities. *Nature Reviews Chemistry*, 2018, 2(4): pp. 1–16.
16. Broach, J. and J. Thorner, High-throughput screening for drug discovery. *Nature*, 1996, 384(6604): pp. 14–16.
17. Hoelder, S., P.A. Clarke, and P. Workman, Discovery of small molecule cancer drugs: Successes, challenges and opportunities. *Molecular Oncology*, 2012, 6(2): pp. 155–176.

18. Xiao, D., et al., Inverse design and synthesis of acac-coumarin anchors for robust TiO$_2$ sensitization. *Journal of the American Chemical Society*, 2011, 133(23): pp. 9014–9022.

19. Lopez, S. A., B.S.-L.J. de Goes Soares, and A. Aspuru-Guzik, Design principles and top non-fullerene acceptor candidates for organic photovoltaics. *Joule*, 2017, 1(4): pp. 857–870.

20. Gómez-Bombarelli, R., et al., Design of efficient molecular organic light-emitting diodes by a high-throughput virtual screening and experimental approach. *Nature Materials*, 2016, 15(10): pp. 1120–1127.

21. Gómez-Bombarelli, R., et al., Automatic chemical design using a data-driven continuous representation of molecules. *ACS Central Science*, 2018, 4(2): pp. 268–276.

22. Kingma, D.P. and M. Welling, Auto-encoding variational bayes, *ICLR 2014*, Banff, Canada, 2013.

23. Goodfellow, I.J., et al., Generative adversarial networks, 2014.

24. Kadurin, A., et al., druGAN: An advanced generative adversarial autoencoder model for de novo generation of new molecules with desired molecular properties in silico. *Molecular Pharmaceutic*, 2017, 14(9): pp. 3098–3104.

25. Olivecrona, M., et al., Molecular de-novo design through deep reinforcement learning. *Journal of Cheminformatics*, 2017, 9(1): pp. 1–14.

26. Popova, M., O. Isayev, and A. Tropsha, Deep reinforcement learning for de novo drug design. *Science Advances*, 2018, 4(7): p. 7885.

27. Kang, S. and K. Cho, Conditional molecular design with deep generative models. *Journal of Chemical Information and Modeling*, 2018, 59(1): pp. 43–52.

28. Akaike, K., et al., Ultraviolet photoelectron spectroscopy and inverse photoemission spectroscopy of [6, 6]-phenyl-C 61-butyric acid methyl ester in gas and solid phases. *Journal of Applied Physics*, 2008, 104(2): p. 023710.

29. Kebarle, P. and S. Chowdhury, Electron affinities and electron-transfer reactions. *Chemical Reviews*, 1987, 87(3): pp. 513–534.

30. Brushett, F.R., J.T. Vaughey, and A.N. Jansen, An all-organic non-aqueous lithium-ion redox flow battery. *Advanced Energy Materials*, 2012, 2(11): pp. 1390–1396.

31. McGlynn, S. and J. Meeks, Photoelectron spectra of carbonyls, carbonates, oxalates and esterification effects. *Journal of Electron Spectroscopy and Related Phenomena*, 1976, 8(2): pp. 85–93.

32. Wittel, K., et al., Photoelectron spectra and molecular properties, Xlviii carbonates and thiocarbonates. *Zeitschrift für Naturforschung B*, 1975, 30(11–12): pp. 862–874.

7 Geometric and Electronic Properties of Li+-Based Battery Cathode *Li$_x$Co/NiO$_2$ Compounds*

Shih-Yang Lin, Hsin-Yi Liu,
Sing-Jyun Tsai, and Ming-Fa Lin
National Chung Cheng University

CONTENTS

7.1   INTRODUCTION

Li+-based battery cathode: Li$_x$Co/NiO$_2$ compounds have attracted a lot of experimental theoretical studies. The up-to-date experimental measurements on crystal symmetries clearly show that the high-resolution X-ray elastic scatterings [1] and low-energy electron diffractions [LEED in Ref. 2] are available for the accurate identifications for the 3D ternary Li$_x$Co/NiO$_2$ cathode materials. Furthermore, the delicate tunneling microscopy [STM in Ref. 3] and tunneling electron microscopy [TEM in Ref. 4]. Specifically, the latter are very successful in resolving the unusual morphologies of surfaces and nanostructures. They can delicately identify the complex relations among the honeycomb lattice, the finite-width quantum confinement, the flexible geometry, and the distinct chemical bondings of carbon atoms, such as the chiral and achiral hexagonal arrangements of 1D planar graphene nanoribbons and carbon nanotubes, and their curved, folded, scrolled, and stacked configurations [5–10]. On the other hand, the former could be utilized to fully investigate the vacancy-dependent optimal geometries of the 3D ternary Li$_x$Co/NiO$_2$ compounds, being helpful in examining the theoretical predictions and thus promoting the further understanding of electronic and magnetic properties.

(a)

(b)

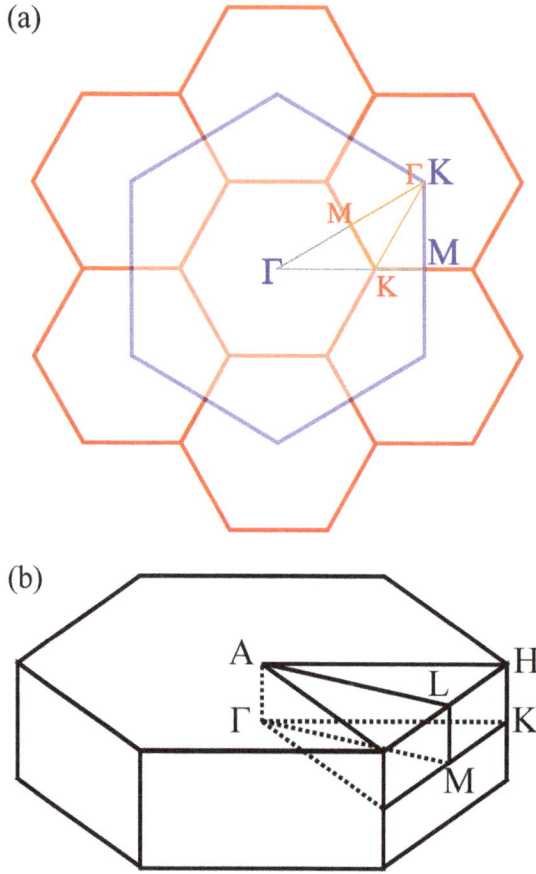

FIGURE 7.4 The first Brillouin zone with the trigonal symmetry: (a) the larger/smaller projection on the (k_x, k_y) plane associated with the smaller/larger atom number in $LiCoO_2$-related compounds, and (b) the 3D structure and high-symmetry points.

7.3 UNUSUAL CRYSTAL STRUCTURES OF 3D TERNARY LI$_x$CO/NIO$_2$ MATERIALS

Based on the delicate first-principles calculations for the optimal geometric structures, the 3D ternary LiCo/NiO$_2$ cathode materials present the unusual crystal lattice symmetries. Certain metastable configurations are chosen to clearly illustrate the complex physical and chemical environments: (a) $Li_{12}Co_{12}/Ni_{12}O_{24}$, (b) $Li_9Co_{12}/Ni_{12}O_{24}$, (c) $Li_6Co_{12}/Ni_{12}O_{24}$, and (d) $Li_3Co_{12}/Ni_{12}O_{24}$ (Figure 7.3a–d, respectively). All the cathode systems possess the trigonal symmetry, but have some vacancies under the unsaturated cases of $x = 0.75$, 0.50, and 0.25. The metastable configurations are expected to be very important in initiating/achieving the outstanding charging and discharging processes, since they are closely related to a plenty of intermediate states during the battery operation. Apparently, the total ground-state energy [E_t] becomes

Energy is revealed in the diverse forms through the everyday living, e.g., the commercialized chemical batteries [11], solar electromagnetic fields [12], hydrogen gases [13], flowing waters [14], blowing winds [15], nuclear powers [16], oil mines [17], oil gases [18], and coal mines [19]. In order to largely reduce the greenhouse effects [20], a lot of theoretical [21,22] and experimental [22,23] researches are conducted for developing the emergent green energy materials. Up to now, the highly potential applications include the battery-operated cell phones [24] and electric vehicles [25], the solar-cell companies [12], the hydrogen-driven buses [26], the methane gas [27], the water-created electric power [28], and the wind turbines [29]. It is well known that a chemical battery strongly depends on the ion transport. The whole experimental progress clearly shows that the Li^+-, Al^+-, and Fe^+-based batteries present the strongly competitive relations in the near-future applications [30–32]. In general, the outstanding performances are based on the low cost, lightweight, more safety, long lifetime, rapid charging/discharging, high temperature, and wide voltage range [30]. Moreover, the various theoretical models are developed to fully explore their diverse physical, chemical, and material properties, such as the first-principles calculations under the local charge density approximation [33], the molecular dynamics simulations [34], and the neutral network methods [35]. The first ones are reliable and available in thoroughly exploring the fundamental properties in various compounds, as done in this book chapter.

Specifically, the Li^+-based batteries [LIBs; Figure 7.1; Ref. 36], which belong to one of the mainstream materials in the basic science, engineering, and application, will be chosen for a model study in this work [37–39]. Apparently, LIBs are mainly made up of the electrolyte [40], cathode [41], anode [42], and electrolyte subsystems [43], in which all the subsystems might have similar behaviors. The theoretical [44] and experimental [45] studies clearly show each component presents a unique crystal structure within a rather large unit cell [46], directly illustrating the various chemical bonding strengths of oxide compounds. These large modulations of bond lengths are the most important common characteristics; furthermore, they would become the critical factors in solving the optimal matches between the core components and determining the efficient Li^+-ion transports [46]. Obviously, how to get the best LIBs, with the highest performance, should be achieved within a unified engineering viewpoint. Very interesting, three kinds of LIB components [all the subsystems] could belong to the 3D lithium oxide compounds, e.g., the cathode/anode/electrolyte materials corresponding to LiCoO/LiTiO/LiSiO systems [46–48], respectively. The up-to-date LIBs are available in many electronic devices, such as radio, watches, clocks, cell phones, laptops, iPods, and cars/buses [49–52]. This work only explores the geometric and electronic properties of the 3D ternary LiCo/NiO cathode materials in LIBs using the first-principles calculations.

Most important, the Li^+ transports (Figure 7.1), with the lightest ion mass, come to exist at any time during the charging and discharging processes according to the paths of cathode $\rightarrow$ electrolyte $\rightarrow$ anode & anode $\rightarrow$ electrolyte $\rightarrow$ cathode [53]. Furthermore, a separator membrane [54] is added to avoid the internal short circuit and only allows the smallest Li^+ to freely move between the positive and negative electrodes [54]. The two electrodes are externally connected to an electric supply after the initial charging process, where electron carriers quickly escape from

Charge Process

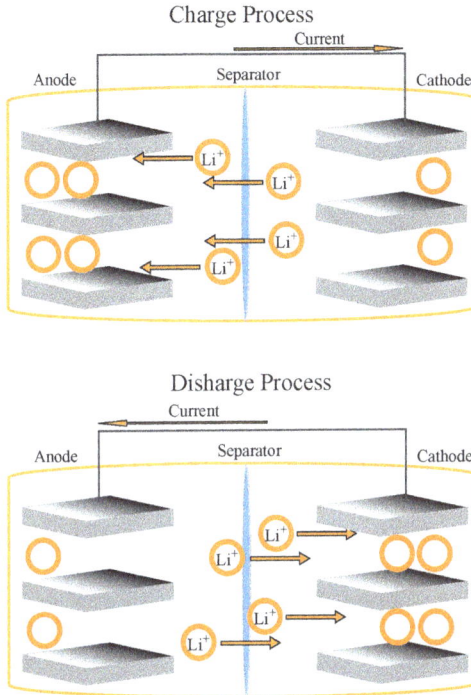

Disharge Process

FIGURE 7.1   Charging and discharging processes in Li$^+$-based batteries.

cathode and transport using the external lead to anode, resulting in the charge current. In order to remain the electric neutrality, the Li$^+$ ions rapidly transport through the parallel direction internally from cathode to anode through the solid-/liquid-state electrolyte [55]. With this efficient process, the external energy from the electrical supply is stored in LIBs under the chemical energy. Another chemical process, in which electrons move from anode to cathode through the external lead and Li$^+$ ion transport back to cathode via the specific electrolyte, is able to provide the electric power and thus do works on electronic devices [49]. Apparently, the drastic changes of geometric structures are revealed in the cathode/electrolyte/anode materials. The structural transformations between two metastable configurations will be rather complicated or very difficult to solve such new issues [56].

On the theoretical side, the first-principles calculations, but not the phenomenological models, are frequently utilized to comprehend the geometric, electronic, and magnetic properties of the cathode/anode/electrolyte materials of LIBs. Apparently, the extremely non-uniform chemical bonds in a large unit cell are responsible for the creation of very complicated atomic interactions [multi-orbital hybridizations] and thus almost impossible to express the analytic Hamiltonian. According to the previous theoretical predictions [57–65], three critical components of LIBs, which are built from 3D ternary LiXO compounds, belong to the large-gap insulators, such as $E_g \sim 1.50$, 2.60, and 5.077 eV for the LiCo/NiO [66], Li$_4$Ti$_5$O$_{12}$ [67], and Li$_2$SiO$_3$ systems [cathode, anode, and electrolyte in LIBs]. Roughly speaking, such studies only provide the atom-dependent density of states and cannot solve the most important multi-orbital hybridization in chemical bonds. The diversified fundamental properties, which are close to the high-potential functionalities of LIBs, will be thoroughly clarified by exploring the spatial charge density distributions and orbital-decomposed van Hove singularities.

In this book chapter, the LIB cathode, the 3D ternary Li_xCo/NiO_2 compound [68,69], is thoroughly studied for its crystal symmetries and electronic properties through the first-principles calculations. Furthermore, $x = 1.00, 0.75, 0.50$, and 0.25 are available for the component-enriched physical/chemical/material phenomena. The delicate analyses, which are conducted on the various chemical bonds in a large unit cell, the atom-created valence and conduction bands, the spatial charge/spin density distributions, and the atom- and orbital-projected van Hove singularities, are capable of providing the critical multi-orbital hybridizations. Furthermore, the spin-induced behaviors, the spin-split electronic energy spectrum, the finite magnetic moment, and the spatial spin arrangement will be completely tested whether they could survive in the transition-metal-related materials. That is, the spin-dependent Coulomb interactions will be included in the current VASP calculations. A detailed comparison is also made among the Li_xCoO_2 and Li_xNiO_2 compounds [or for the cathode, anode, and electrolyte materials] in terms of the main features of essential properties. Also, a close and complicated relation of the numerical VASP calculations with the tight-binding model is discussed in detail. The theoretical predictions on the optimal lattice, the electronic energy spectrum and band gap, and the rich special structures in valence and conduction density of states could be verified from the high-resolution measurements of X-ray diffraction/low-energy electron diffraction [LEED; Ref. 70], angle-resolved photoemission spectroscopy [ARPES; Ref. 71], and scanning tunneling spectroscopy [STS; Ref. 72], respectively.

7.2 DELICATELY NUMERICAL VASP CALCULATIONS

The theoretical researches cover the numerical simulations and the phenomenological models, in which the former and the latter are frequently utilized for the complicated and simple crystal structures, respectively. For example, the emergent few-layer graphene systems, which only have the interlayer atomic interactions due to the $C-2p_z$ single-orbital hybridizations, are well described by the first-principles calculations and the tight-binding model simultaneously in terms of the essential properties. Specifically, the generalized tight-binding model, being developed for external electric and magnetic fields, is successful in predicting the rich and unique quantization phenomena, e.g., the diversified Landau-level energy spectra in monolayer graphene [73], bilayer AA- and AB-stacked stackings [74], and trilayer AAA-, ABA-, ABC-, and AAB-stacked ones [75]. In addition, the numerical method is impossible to deal with the magnetic-field issues. The close relations between them are very interesting topics under the current investigations. Apparently, the first-principles calculations are available for fully exploring the geometric and electronic properties in the 3D ternary Li_xCo/NiO_2 cathode materials.

The Vienna ab initio simulation package, VASP [76], which is built from the density functional theory [DFT; Ref. 77], is utilized to achieve the optimal geometric symmetries, band structure, charge density distributions in different chemical bonds, atom- and orbital-projected van Hove singularities, and spin-induced properties. Two very complicated intrinsic interactions, which are created by the electron quasiparticles, need to be solved in the numerical calculations. The many-particle exchange and correlation energies, the electron–electron Coulomb interactions beyond the classical electrodynamics, are based on the Perdew–Burke–Ernzerhof functional [PBE; Ref. 78] under the generalized gradient approximation. Furthermore, the frequent electron-ion scatterings are consistent with the projector-augmented wave [PAW; Ref. 79] pseudopotentials. In general, the plane waves, with the kinetic energy cutoff of 500 eV, are very reliable in serving as a complete set. That is, their linear superposition is rather suitable in characterizing Bloch wave functions and band structures. For the 3D ternary Li$_x$Co/NiO$_2$ compounds, the first Brillouin zone is sampled by $9 \times 9 \times 9$ and $100 \times 100 \times 100$ k-point meshes within the Monkhorst-Pack scheme [80] for the geometric optimizations and electronic energy spectra, respectively. Most important, the convergence condition of the ground-state energy is ~10^{-5} eV between two consecutive evaluation steps, in which the maximum Hellmann–Feynman force acting on each ion is less than 0.01 eV Å^{-1} during the atom relaxations.

The delicate VASP calculations, as clearly illustrated in the flowing chart of Figure 7.2, are finished through the self-consistent processes. One first chooses the initial local charge density [$\rho(\mathbf{r})$], calculates the electron-ion potential energies and the classical and many-body exchange-correlation electron–electron interactions [the effective Coulomb interactions], solves the Kohn–Sham equations, evaluates the intermediate electron density ground-state energy, tests the convergent condition about the Feynman force, and then determines whether the outputs are reliable.

The calculated physical quantities, which cover the atom-created valence and conduction states, the spatial charge densities after/before the creation of chemical bond, the atom- and orbital-projected density of states, the atom-related spin density distributions, the spin-split or degenerate states across the Fermi level, and the finite or zero magnetic moments, are able to determine the critical multi-/single-orbital hybridizations of oxidized bondings and the atom-induced spin configurations in the current cathode materials. The above-mentioned theoretical framework, which is built from the first-principles results, has been successfully conducted on the systematic investigations of 2D layered graphene systems [book in Ref. 81] and 1D graphene nanoribbons [book in Ref. 82], such as the diversified phenomena of geometric, electronic, and magnetic properties due to the different dimensionalities, planar or buckled honeycomb lattices, layer numbers, stacking configurations, adatom chemisorptions, and guest-atom substitutions.

Most important, how to achieve the combinations of the numerical simulations with phenomenological models becomes one of the emergent issues in the near-future studies [83], in which the latter can provide more physical/chemical pictures on the critical mechanisms and thus explore the other essential properties [e.g., magneto-electronic [84], optical [85], and transport properties [86]]. The VASP

Initial Guess
$\rho(\mathbf{r})$

Calculate Effective Potential
$V_{eff}(\mathbf{r})=V_{eff}(\mathbf{r})+\int\frac{\rho(r\prime)}{|r-r\prime|}dr\prime+V_{xc}[(\rho(\mathbf{r})]$

Solve Kohn-Sham Equation
$[-\frac{\hbar^2}{2m_e}\nabla_i^2+V_{eff}]\psi_i=\varepsilon_i\psi_i$

Calculate Electron density & Total Energy
$\rho(\mathbf{r})=\sum_i|\psi_i(r)|^2\rightarrow E_{tot}[\rho(\mathbf{r})]=...$

No   Converged?

Yes

Exit and output results
$\rho_0(r),\ E_i[\rho_0(r)],\ Forces,\ Eigenvalue,\ ...$

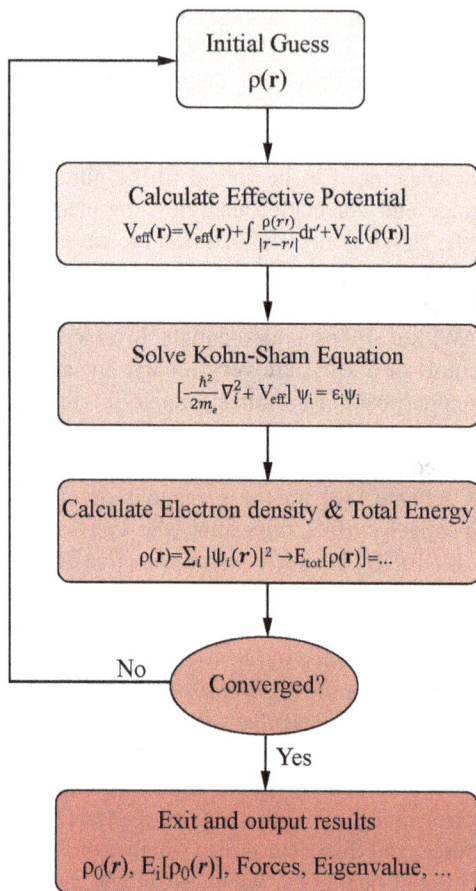

FIGURE 7.2   Flowing chart during the VASP calculations.

low-lying valence and conduction bands, being relatively close to the Fermi level, would be well simulated through a good fitting of the parameterized tight-binding model/the effective-mass approximation [87], while the electron and hole energy spectra only possess the simple dispersion relations, such as those of AA- and AB-stacked few-layer graphene systems. That is, this point of view is successful in studying the diversified phenomena of layered graphenes, because only the single-orbital hybridizations of $2p_z$–$2p_z$ serve as the weak, but important, interlayer van der Waals interactions. Furthermore, the generalized tight-binding model is further built for graphene systems under a uniform perpendicular magnetic field, and the parameterized Hamiltonian matrix, being characterized by a lot of magnetic hopping integrals, is a very large Hermitian one. By its diagonalization, the rich magnetic quantization phenomena are clearly revealed in layered group IV and group V systems, e.g., the diversified Landau-level energy spectra of graphene [73], silicene [88,89], germanene [89], tinene [90], bismuthene [91], and phosphorene [92]. This chapter will show that the 3D ternary Li_xCo/NiO_2 materials have very extremely

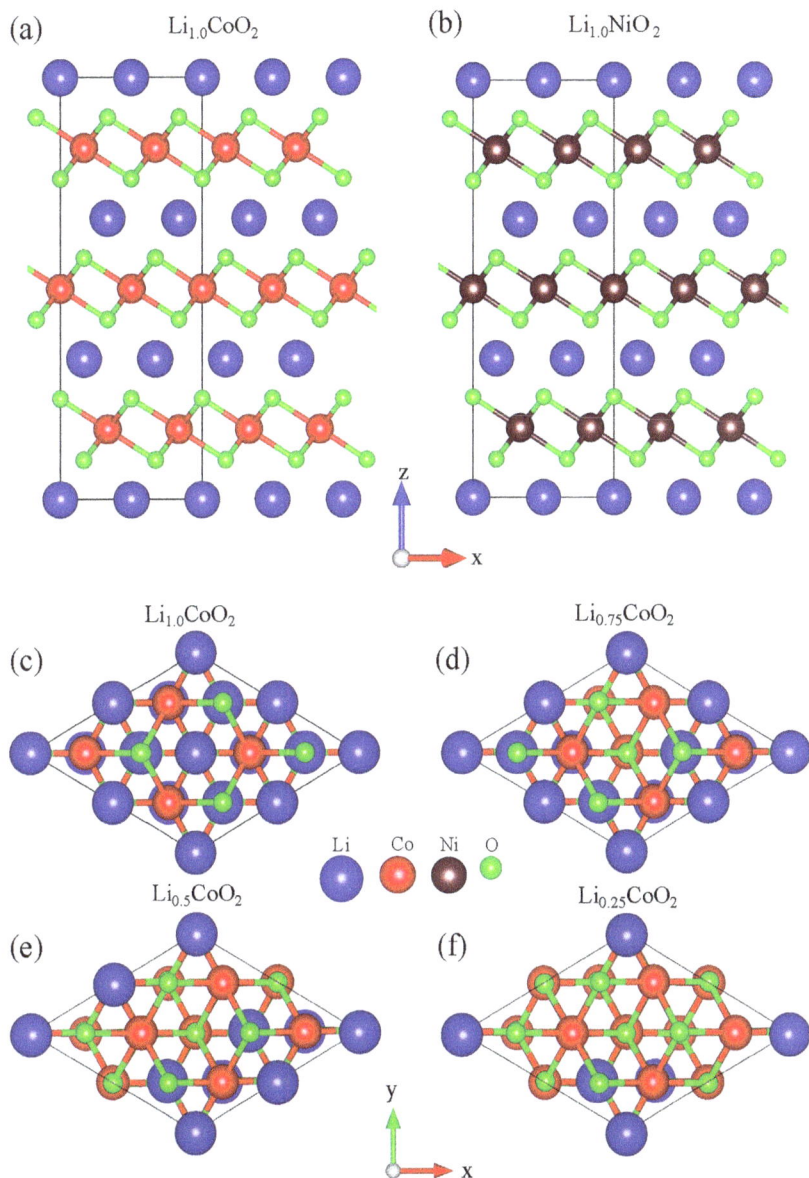

FIGURE 7.3 According to the delicate VASP calculations on the optimal crystal structures of the cathode materials: (a) Li$_{12}$Co$_{12}$/Ni$_{12}$O$_{24}$, (b) Li$_9$Co$_{12}$/Ni$_{12}$O$_{24}$, (c) Li$_6$Co$_{12}$/Ni$_{12}$O$_{24}$, and (d) Li$_3$Co$_{12}$/Ni$_{12}$O$_{24}$ under the trigonal symmetry unit cell.

non-uniform chemical environment in a primitive unit cell (see Figure 7.3) and thus complicated band structures (Figure 7.4). As a result, the tight-binding models, with the various hopping integrals due to the different chemical bonding strengths, would be rather difficult in simulating the first-principles electronic structures.

TABLE 7.1

Optimal Geometric Parameters for the Li$^+$-Based Battery Cathodes: Li$_x$CoO$_2$ and Li$_x$NiO$_2$ in the presence of x = 1.00, 0.75, 0.50, and 0.25

	Li–O (Å)	(Co–O)/ (Ni–O) (Å)	# of Li–O	# of (Co–O)/ (Ni–O)	Magnetic Moment (μB)	Ground-State Energy (eV)
Li$_{1.00}$CoO$_2$	2.090	1.923	72	72	NM	−296.911
Li$_{0.75}$CoO$_2$	2.085–2.147	1.888–1.937	54	72	−2.940	−282.907
Li$_{0.50}$CoO$_2$	2.087–2.143	1.881–1.933	36	72	−0.340	−267.706
Li$_{0.25}$CoO$_2$	2.080–2.155	1.878–1.943	18	72	−2.900	−251.399
Li$_{1.00}$NiO$_2$	2.121	1.973	72	72	−11.320	−268.144
Li$_{0.75}$NiO$_2$	2.098–2.192	1.898–1.987	54	72	6.830	−254.546
Li$_{0.50}$NiO$_2$	2.096–2.190	1.891–1.985	36	72	−1.610	−239.772
Li$_{0.25}$NiO$_2$	2.089–2.196	1.886–1.982	18	72	0.850	−224.711

higher in the decrease in lithium concentration, e.g., the E_t change of −296.991 eV → −251.399 eV/−268.114 eV → −224.711 eV during the variation of x = 1.0 → x = 0.25 for the Li$_x$Co/NiO$_2$ compounds (Table 7.1). As a result, Li$_x$CoO$_2$ is relatively stable [easily produced in experimental syntheses], compared with Li$_x$NiO$_2$.

The modulated chemical bonds, which are sensitive to the changes of lithium concentrations, will appear within a similar unit cell (Figure 7.3a and b). There exist 72/72, 54/72, 36/72, and 18/72 Li–O/Co–O and Ni–O Chemical bonds, respectively, corresponding to x = 1.00, 0.75, 0.50, and 0.25. In the fully saturated configuration, as indicated in Table 7.1 of x = 1.00, this case only presents the single Li–O and Co–O and Ni–O bond lengths [2.09 and 1.92, and 1.92 Ås]. However, with the reduced x-value, the Li-related vacancies will create a weak, but significant, distortion of the specific crystal structure. This further induces the minor modulations of the chemical bond lengths. For example, the Li$_{0.75}$CoO$_2$ compound displays the modulated lengths in the ranges of ~2.09–2.15 and ~1.89–1.94 Å, respectively, for the Li–O and Co–O bonds, according to the nine independent ones. Apparently, the modulation percentage is enhanced by the increase in lithium vacancies. The x-dependent charge density distributions might have the obvious changes, and thus, the multi-orbital hybridizations in chemical bonds would become more non-uniform. That is to say, the different nearest-neighbor hopping integrals are required to be included in the phenomenological model, e.g., the various atomic interactions in the Hamiltonian of the tight-binding model [80,81]. In addition, most of Li$_x$Co/NiO$_2$ systems possess the net magnetic moments, clearly illustrating the ferromagnetic spin configurations. How to introduce the spin-induced many-body Coulomb interactions [the Hubbard-like on-site electron–electron interactions; Refs. 80,81] is an emergent issue, such as its role during the battery operations.

7.4 RICH AND UNIQUE ELECTRONIC PROPERTIES

It is very interesting that the 3D ternary Li$_x$Co/NiO$_2$ compounds, the Li$^+$-based battery cathode materials, possess the unusual crystal structures and thus display the diversified electronic and magnetic properties. As a result of the trigonal symmetry in each Li-related system, many valence and conduction bands, as clearly illustrated in Figure 7.5a–h, present the various energy dispersions along the 3D high-symmetry points (Figure 7.4b). In addition, this first Brillouin zone is similar to that of the layer graphite. Apparently, the occupied energy spectra are highly asymmetric to the unoccupied ones about the Fermi level $E_F = 0$. This behavior might be closely related to the complex orbital hybridizations in Li–O and Co–O/Ni–O bonds. Furthermore, the former are the dominating ones; that is, they have more energy subbands. Specially, the conduction-band states present a giant energy spacing of $E^{c, v} > 2.0\,eV$ within a specific spectral range, being never observed in the other condensed-matter systems. Such feature survives in any cathode compounds, in which the critical mechanism is worthy of further investigations. Generally speaking, electronic structures are very complicated under the oscillatory dispersion relations, the highly anisotropic behaviors, and the frequent crossings/anti-crossings. It would be very difficult to characterize the width of each energy subband, or it is almost impossible to distinguish/identify/examine the various valley structures for the different subbands. A lot of band-edge states might come to the high-symmetry points or the other wave vectors. These critical points in the energy-wave-vector space would induce the unique van Hove singularities [93] and thus create the strong absorption structures in optical properties [94]. The predicted unusual energy spectra will become the next-step challenges in the high-resolution ARPES measurements [95].

The non-magnetic semiconductors or ferromagnetic metals are mainly determined by the valence and conduction energy subbands near to the Fermi level, being sensitive to strongly depending on the kind of transition metal and lithium-atom concentration. According the delicate VASP calculations, the $LiCoO_2$ compound belongs to a direct-gap semiconductor (Figure 7.5a), in which band gap of $E_g \sim 1.05\,eV$ arises from the energy spacing of the highest occupied and lowest unoccupied states in between the $M\Gamma$ direction. Any electronic states are doubly degenerate under spin degree of freedom. This system might be a non-magnetic or anti-ferromagnetic semiconductor, where the magnetic configurations depend on the spatial spin density distributions. As for the normal lithium concentration of $x = 0$, the Fermi level is just situated at the center of band gap. Furthermore, there are three observable groups of energy subbands: (I) the deeper valence subbands, (II) the shallower ones, and (III) the conduction subbands, respectively, corresponding to energy ranges of -8.00 eV $< E^v < -3.00\,eV$, -2.50 eV $< E^v < -1.00\,eV$, and 0.55 eV $< E^c < 1.50\,eV$. The first, second, and third groups are, respectively, dominated by the oxygen, cobalt, and cobalt atoms [the blue, red, and green open circles in Figure 7.6a–c, being consistent with the orbital-projected density of states in Figure 7.9]. The lithium contributions to all the energy subbands are rather weak, but very important. The energy-dependent atom dominances might be associated with the orbital ionization energies. On the other hand, the decreased variation will create the semiconductor–metal and magnetic configuration transitions, such as the ferromagnetic metals in Figure 7.5b–d. These are obviously revealed in the redshift Fermi level, as well as the spin splitting. The concentration-reduced effects lie in the generation of partially unoccupied valence subbands and the diminishing energy spacing between (I) and (II) valence-state groups. There exist more free carriers in the increase in x, and the first and second groups mix together under $x \geq 0.5$ (Figure 7.5c and d).

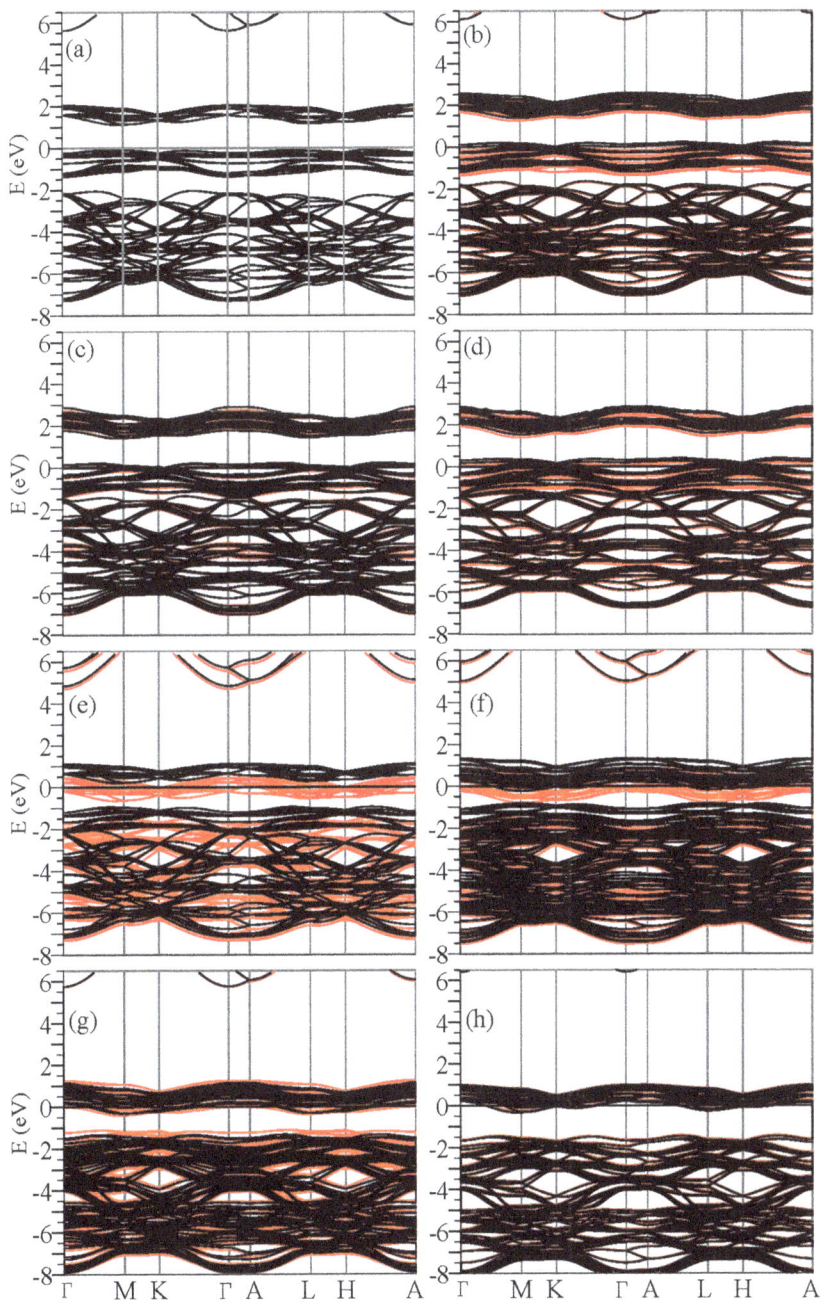

FIGURE 7.5 The rich and unique band structures with the spin-degenerate or spin-split valence and conduction bands along the high-symmetry points of the first Brillouin zone within the energy range of $-8.0\,\text{eV} \leq E^{c,\,v} \leq 6.5\,\text{eV}$ for the cathode compounds: (a)/(e) Li$_{12}$Co$_{12}$/Ni$_{12}$O$_{24}$, (b)/(f) Li$_9$Co$_{12}$/Ni$_{12}$O$_{24}$, (c)/(g) Li$_6$Co$_{12}$/Ni$_{12}$O$_{24}$, and (d)/(h) Li$_3$Co$_{12}$/Ni$_{12}$O$_{24}$. The black and red curves, respectively, correspond to the spin-up and spin-down electronic state.

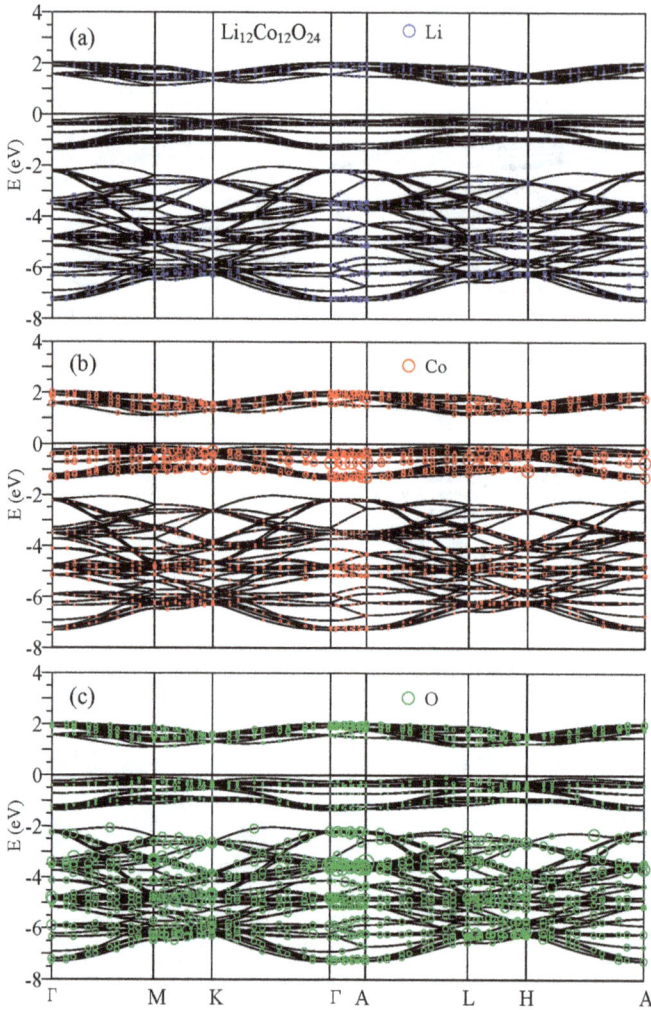

FIGURE 7.6 (a) Lithium, (b) cobalt, and (c) oxygen dominances of the ternary 3D $Li_{12}Co_{12}O_{24}$ compound by the blue, red, and green open circles, respectively.

There exist certain important differences between 3D ternary Li_xCoO_2 and Li_xNiO_2 compounds in the featured band structures, as clearly indicated in Figure 7.5a–h. First, the latter belong to the ferromagnetic metals even under the high lithium concentrations, e.g., electronic energy spectrum for Li_xNiO_2 of $x = 1$ and 0.75 in Figure 7.5e and f, respectively. Second, the observable energy spacing in the first and second groups of valence subbands, which could survive in the latter [Figure 7.5a–d], is thoroughly absent in the Ni-related materials under any Li-concentration cases. That is, the latter only present the continuous valence-state spectra. Third, the total width of the effective energy spectrum is wider in the Co-related case, compared with the Ni-dependent one. This directly reflects the strength of Co–O or Ni–O bond.

In terms of the Fermi level, band gap, free carrier density, number of energy subbands, energy dispersions, crossing/anti-crossing behaviors, spin-split electronic states, and effective ranges of electronic energy spectra, the first-principles electronic structures might be too complicated to be simulated by the phenomenological models. The extremely non-uniform chemical and physical environments, which survive in the unit cells of the 3D ternary Li$_x$Co/NiO$_2$ compounds, are responsible for the featured electronic structures. Such critical factors cover a lot of multi-orbital hybridizations in Li–O and Co–O/Ni–O bonds and the [Co, O]–/[Ni, O]-dependent spin configurations. The orbital-induced various hopping integrals, site energies, the spin-created many-body electron–electron Coulomb interactions [the Hubbard-like spin-dominated on-site ones], and lithium-concentration-created vacancies need to be included in the significant Hamiltonian simultaneously. Apparently, a plenty of reliable parameters are required in the further calculations. It would be very difficult to make a good Hamiltonian diagonalization, thus being almost impossible to achieve the concise physical pictures in the full understanding of the featured electronic energy spectra.

As for the high-resolution experimental measurements, ARPES is the only technique for the accurate identifications of the wave-vector-dependent energy spectra below the Fermi level, especially for the occupied electronic states [95]. The ARPES experiments, being conducted on the various condensed-matter systems, are very successful in observing the diversified band structures, such as the diverse energy dispersions in the sp^2-bonding carbon systems, 3D transition metal oxide compounds [96], superconductors [97], and group IV and group V layered materials [98,99]. For example, many measured results have verified the rich and unique valence band structures for the emergent graphene-related materials, in which they are initiated from the stable K/K′ or Γ, such as the Dirac-cone band structures in monolayer/twisted bilayer graphene systems and monolayer graphene, the parabolic/linear and parabolic energy dispersions in bilayer/trilayer AB stackings, the linear, partially flat, and Sombrero-shaped energy bands in trilayer ABC stacking, and the monolayer- and bilayer-like properties in Bernal graphite. Only monolayer graphene belongs to a zero-gap semiconductor, and the others are semimetals with the weak valence and conduction band overlaps. This is deduced to arise from the layer- and stacking-enriched interlayer atomic interactions of the $2p_z$–$2p_z$ orbital hybridizations in the normal or enlarged honeycomb lattices. Specifically, the spin-polarized/non-polarized ARPES measurements are available in detecting the main features of spin-split/spin-degenerate occupied energy subbands in 3D ternary Li$_x$Co/NiO$_2$ material. The experimental examinations cover the non-magnetic semiconducting behavior in LiCoO$_2$ or the ferromagnetic metals of the others, the strong dependences of one or two valence-subband groups on the transition metal atoms and their concentrations, the highly isotropic and oscillatory energy dispersions, the frequent subband crossings/anti-crossings, and the significant valence energy width of ~7–8 eV. Such verifications are helpful for providing certain first-step evidences in the multi-orbital hybridizations of chemical bonds.

Very interesting, the charge density distributions, as obviously indicated in Figure 7.7a–l, are capable of providing the diversified chemical environments. It should be noticed that more delicate chemical bondings are achieved by linking them

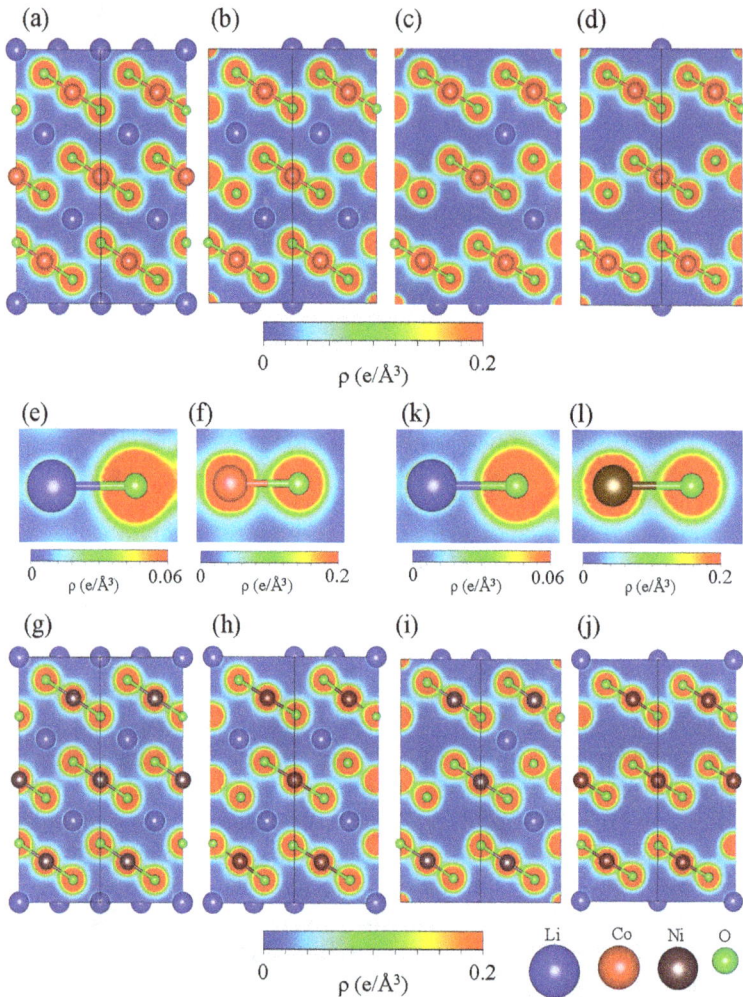

FIGURE 7.7 Spatial charge density distributions for (a) $Li_{12}Co_{12}O_{24}$, (b) $Li_9Co_{12}O_{24}$, (c) $Li_6Co_{12}O_{24}$, (d) $Li_3Co_{12}O_{24}$, a certain (e) Li–O/(f) Co–O bond, (g) $Li_{12}Ni_{12}O_{24}$, (h) $Li_9Ni_{12}O_{24}$, (i) $Li_6Ni_{12}O_{24}$, and (j) $Li_3Ni_{12}O_{24}$, and a certain (k) Li–O/(l) Ni–O bond.

with the atom- and orbital-projected density of states (see Figures 7.9–7.11), and the atom-dominated band structures (Figures 7.5 and 7.6). Apparently, the carrier densities strongly depend on the lithium concentration and the kind of transition atom. As for the ternary Li_xCoO_2 materials, they present similar charge distributions under the different xs except for those near the Li vacancies; i.e., the minor variations of carrier densities associated with the reduced lithium atoms are difficult to directly observe from Figure 7.7a–d. The strong orbital overlaps are easy to identify in the Co–O bonds, while the opposite is true for the Li–O bonds. That the ration of the former versus the latter is more than five times could be understood from the different scales in Figure 7.7e and f. As a result, the strength of Li–O bond is much weaker than that

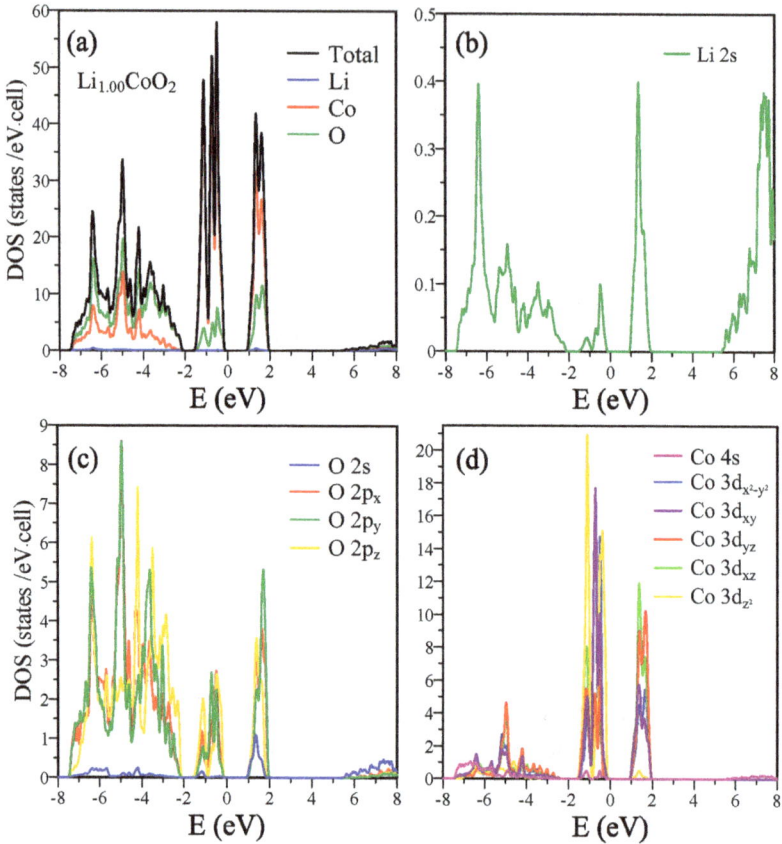

FIGURE 7.9 Atom- and orbital-decomposed density of states in the wide energy range of $-8.0\,\text{eV} \leq E^{c,\,v} \leq 8.0\,\text{eV}$ for the 3D ternary $Li_{12}Co_{12}/Ni_{12}O_{24}$ compound: those coming from (a) Li, Co, and O atoms [blue, red, and green curves], (b) Li-2s orbitals [green curve], (c) O–($2s$, $2p_x$, $2p_y$, $2p_z$) orbitals [blue, red, green, and yellow curves], and (d) Co–($4s$, $3d_{x^2-y^2}$, $3d_{xy}$, $3d_{yz}$, $3d_{xz}$, $3d_{z^2}$) [pink, blue, purple, red, green, and yellow curves].

is almost vanishing in the conduction energy spectrum. The above-mentioned rich behaviors are revealed in the $LiNiO_2$ compound except for the metallic property [a finite density at the Fermi level], the non-separated energy spectra of the first and second groups, and the spin-split characteristics (Figure 7.10a–d).

The density of states, as clearly shown in Figure 7.11a–f, is very sensitive to the variations in the Li concentrations. The drastic changes cover the position of the Fermi level, the spin-dependent free carrier density, the number, intensity, energy, and form of van Hove singularities, the separation or non-separation of the first and second valence groups, the atom and orbital dominances, and the effective whole width of energy spectra These should be closely related to the chemical bonding strengths in a primitive unit cell (Figure 7.3a–d). It should also be noticed that the important differences between the spin-up and spin-down density of states near the Fermi level are able to determine the net magnetic moment (Table 7.1). The of Co–O bond. This can account for the slight distortions of geometric structures as the lithium concentration declines. Such response is very important for the release and recovery of lithium ions from the cathode material during the charging and discharging processes. That is to say, the creation of Li vacancy is supported by the strong Co–O bondings and then drives the ion transport. A similar phenomenon is clearly revealed in Li–O and Li–Ni bonds (Figure 7.7g–l) (very important comments).

Most important, the concisely multi-orbital hybridizations of Li–O and Co–O/Ni–O bonds could be directly identified from the spatial charge density distributions around them, as clearly illustrated in Figure 7.7e and f & k–l. As for the Li–O bonds [~2.10 Å in Figure 7.7e and k], each lithium atom presents only a single 2s orbital, and its effective distribution width is ~0.54 Å, as observed from the deep blue region of the Li$^+$-ion core to the light green one of the outmost orbital. Consequently, the two 1s orbitals do not participate in the critical orbital hybridizations with the oxygen atoms. A weak, but significant, overlap of distinct orbitals between Li and O atoms is revealed within the ~1.20–1.60 Å range [a distance measured from the O core]. The important oxygen orbitals, which correspond to the light green and yellow regions, are approximately predicted to be associated with the [$2p_x$, $2p_y$, $2p_z$] ones. Specifically, the O-2s orbitals are relatively far away from those of the Li atom; therefore, they are almost negligible for the Li–O chemical bondings. There are more Co–O/Ni–O bonds and quite strong bonding strengths (Figure 7.7f–l), compared to those of Li–O bonds. The effective distribution range for transition metal atom is comparable to that of the O one. Furthermore, it could be further classified into heavy red, light red, and yellow-green regions, which, respectively, correspond to [$3s$, $3p_x$, $3p_y$, $3p_z$], [$4s$], and [$3d_{x^2-y^2}$, $3d_{xy}$, $3d_{yz}$, $3d_{xz}$, $3d_{z^2}$]. Obviously, the first ones are independent of the multi-orbital hybridizations in the Co–O/Ni–O bonds. The deformed carrier distributions, which come to exist between Co/Ni and O atoms, also cover the four orbital of the latter. According to the above-mentioned features, the Li–O and Co–O/Ni–O bonds are predicted to exhibit 2s–[$2p_x$, $2p_y$, $2p_z$] and [$2s$, $2p_x$, $2p_y$, $2p_z$]–[$4s$, $3d_{x^2-y^2}$, $3d_{xy}$, $3d_{yz}$, $3d_{xz}$, $3d_{z^2}$] orbital hybridizations, respectively. The multi-orbital chemical bondings are also supported from the atom- and orbital-decomposed van Hove singularities (Figures 7.9–7.11). In addition, the bonding strength in Li$_x$CoO$_2$ is slightly stronger than that of Li$_x$NiO$_2$. The above-mentioned highly non-uniform chemical environments might generate many difficulties in the phenomenological methods [e.g., the generalized tight-binding model].

The ferromagnetic configurations in all the metallic cathode compounds are worthy of further analyses from the net magnetic moments and the spatial spin density distributions, as displayed in Table 7.2a–g and Figure 7.8a–h, respectively. Only Li$_{12}$Co$_{12}$O$_{24}$ in Figure 7.8a is a non-magnetic semiconductor (Figure 7.8a). Generally speaking, the five Co–3d/Ni–3d orbitals play the critical roles in determining the magnetic properties, according to their most important contributions in the atom- and orbital-decomposed magnetic moments of Table 7.2a–g. Their values lie in the great modulation ranges: ~0.165–0.601 and ~0.165–0.601 μ, respectively, Co–3d and Ni–3d orbitals. Furthermore, the Li–2s and O–[$2s$, $2p_x$, $2p_y$, $2p_z$] orbitals only present minor contributions, being lower than the former under the one or two orders. As a result, the net magnetic moments are very sensitive to the changes in the kinds of transition metal atoms, separate orbitals, and their concentrations. These results

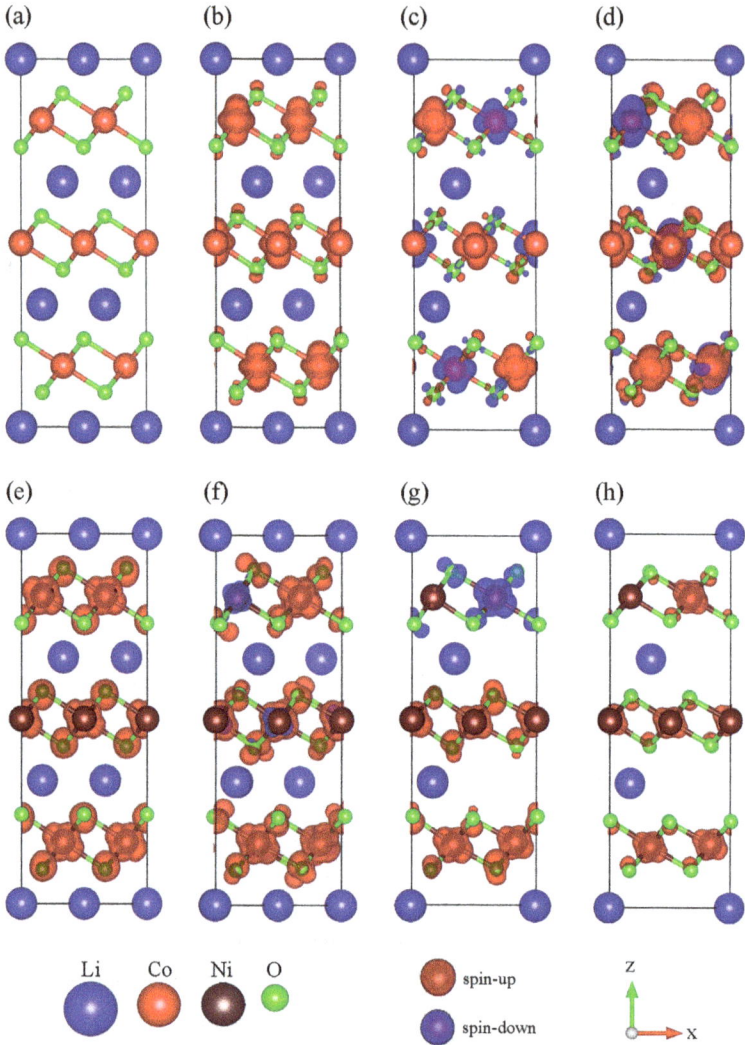

FIGURE 7.8 Spin density distributions due to the 3D ternary cathode compounds: (a) $Li_{12}Co_{12}O_{24}$ without magnetism, (b) $Li_9Co_{12}O_{24}$, (c) $Li_6Co_{12}O_{24}$, (d) $Li_3Co_{12}O_{24}$, (e) $Li_{12}Ni_{12}O_{24}$, (f) $Li_9Ni_{12}O_{24}$, (g) $Li_6Ni_{12}O_{24}$, and (h) $Li_3Ni_{12}O_{24}$.

clearly indicate that the rather high anisotropy of the physical environments comes to exist, especially for those during the variation of lithium concentration.

Most important, the spatial spin density distributions (Figure 7.8b–h) are able to provide the atom- and position-dependent magnetic configuration. There exist (I) the full spin-up/spin-down arrangement, (II) the mixed, but dominated by the former, and (III) the comparable mixings, in which they, respectively, correspond to $Li_9Co_{12}O_{24}/Li_{12}Ni_{12}O_{24}$ (Figure 7.8b and e; Table 7.2a and d) and $Li_3Co_{12}O_{24}/Li_9Ni_{12}O_{24}/Li_6Ni_{12}O_{24}/Li_3Co_{12}O_{24}$ (Figure 7.8d,f,g–h; Table 7.2c,e–g). Three kinds of

TABLE 7.2

Atom- and Orbital-Dependent Magnetic Moments: (a)/(b)/(c) Li$_x$CoO$_2$ with x = 0.75/0.50/0.25 and (d)/(e)/(f)/(g) Li$_x$NiO$_2$ with x = 1.00/0.75/0.50/0.25

(a) Li$_9$Co$_{12}$O$_{24}$					(b) Li$_6$Co$_{12}$O$_{24}$					(c) i$_3$Co$_{12}$O$_{24}$				
Atom	s	p	d	Total	Atom	s	p	d	Total	Atom	s	p	d	Total
Li$_1$	0	0	0	0	Li$_1$	−0.001	−0.001	0	−0.001	Li$_1$	0.001	0	0	0.001
Li$_2$	−0.001	0	0	0	Li$_2$	−0.001	−0.001	0	−0.001	Li$_2$	0.001	0	0	0.001
Li$_3$	0	0	0	0	Li$_3$	0.001	0	0	0.001	Li$_3$	−0.001	−0.001	0	−0.001
Li$_4$	−0.001	0	0	0	Li$_4$	0.001	0	0	0.001					
Li$_5$	0	0	0	0	Li$_5$	−0.001	−0.001	0	−0.001					
Li$_6$	0	0	0	0	Li$_6$	−0.001	−0.001	0	−0.001					
Li$_7$	0	0	0	0										
Li$_8$	0	0	0	0										
Li$_9$	0	0	0	0										
Co$_1$	−0.003	−0.002	−0.189	−0.194	Co$_1$	−0.001	−0.003	0.307	0.303	Co$_1$	−0.001	0	−0.597	−0.598
Co$_2$	−0.004	−0.004	−0.232	−0.239	Co$_2$	0.001	0.002	−0.327	−0.324	Co$_2$	−0.001	−0.001	−0.497	−0.498
Co$_3$	−0.004	−0.004	−0.229	−0.236	Co$_3$	−0.001	−0.003	0.307	0.303	Co$_3$	−0.003	−0.009	−0.271	−0.284
Co$_4$	−0.003	−0.002	−0.188	−0.193	Co$_4$	0.001	0.002	−0.327	−0.324	Co$_4$	0	−0.011	0.606	0.595
Co$_5$	−0.004	−0.004	−0.233	−0.24	Co$_5$	0	0.001	−0.379	−0.378	Co$_5$	−0.001	0	−0.597	−0.598
Co$_6$	−0.003	−0.002	−0.189	−0.194	Co$_6$	−0.001	−0.003	0.262	0.258	Co$_6$	−0.001	−0.001	−0.493	−0.494
Co$_7$	−0.004	−0.004	−0.229	−0.236	Co$_7$	0	0.001	−0.379	−0.378	Co$_7$	0	−0.011	0.606	0.595
Co$_8$	−0.003	−0.002	−0.188	−0.193	Co$_8$	−0.001	−0.003	0.262	0.258	Co$_8$	−0.003	−0.009	−0.275	−0.287
Co$_9$	−0.004	−0.003	−0.252	−0.258	Co$_9$	−0.001	−0.003	0.307	0.303	Co$_9$	−0.002	−0.008	−0.395	−0.406
Co$_{10}$	−0.004	−0.003	−0.252	−0.258	Co$_{10}$	0.001	0.002	−0.327	−0.324	Co$_{10}$	0	−0.01	0.595	0.585
Co$_{11}$	−0.003	−0.003	−0.165	−0.171	Co$_{11}$	−0.001	−0.003	0.307	0.303	Co$_{11}$	−0.001	−0.001	−0.486	−0.488
Co$_{12}$	−0.003	−0.003	−0.168	−0.174	Co$_{12}$	0.001	0.002	−0.327	−0.324	Co$_{12}$	−0.001	−0.001	−0.482	−0.485
O$_1$	0	−0.017	0	−0.017	O$_1$	0	0.008	0	0.008	O$_1$	0	−0.018	0	−0.018
O$_2$	0	−0.017	0	−0.017	O$_2$	0	−0.023	0	−0.023	O$_2$	0	−0.029	0	−0.029
O$_3$	0	−0.011	0	−0.011	O$_3$	0	0.008	0	0.008	O$_3$	0	−0.031	0	−0.031
O$_4$	0	−0.012	0	−0.012	O$_4$	0	−0.023	0	−0.023	O$_4$	−0.001	−0.013	0	−0.014
O$_5$	−0.001	−0.017	0	−0.017	O$_5$	0	0.005	0	0.005	O$_5$	0	−0.014	0	−0.014
O$_6$	0	−0.008	0	−0.009	O$_6$	0	−0.007	0	−0.007	O$_6$	0	−0.016	0	−0.017
O$_7$	0	−0.02	0	−0.021	O$_7$	0	0.005	0	0.005	O$_7$	−0.001	−0.039	0	−0.04
O$_8$	0	−0.011	0	−0.011	O$_8$	0	−0.007	0	−0.007	O$_8$	−0.001	−0.02	0	−0.021
O$_9$	0	−0.008	0	−0.009	O$_9$	0	−0.009	0	−0.009	O$_9$	0	−0.014	0	−0.014
O$_{10}$	−0.001	−0.016	0	−0.017	O$_{10}$	0	0.021	0	0.021	O$_{10}$	0	−0.016	0	−0.017
O$_{11}$	0	−0.02	0	−0.021	O$_{11}$	0	−0.009	0	−0.009	O$_{11}$	−0.001	−0.02	0	−0.022
O$_{12}$	0	−0.011	0	−0.011	O$_{12}$	0	0.021	0	0.021	O$_{12}$	−0.001	−0.04	0	−0.04
O$_{13}$	0	−0.017	0	−0.017	O$_{13}$	0	−0.009	0	−0.009	O$_{13}$	0	−0.018	0	−0.018
O$_{14}$	0	−0.017	0	−0.018	O$_{14}$	0	0.021	0	0.021	O$_{14}$	0	−0.029	0	−0.029
O$_{15}$	0	−0.011	0	−0.011	O$_{15}$	0	−0.009	0	−0.009	O$_{15}$	−0.001	−0.013	0	−0.014
O$_{16}$	0	−0.012	0	−0.012	O$_{16}$	0	0.021	0	0.021	O$_{16}$	0	−0.031	0	−0.031
O$_{17}$	−0.001	−0.016	0	−0.017	O$_{17}$	0	0.005	0	0.005	O$_{17}$	−0.001	−0.028	0	−0.029
O$_{18}$	0	−0.008	0	−0.009	O$_{18}$	0	−0.007	0	−0.007	O$_{18}$	−0.001	−0.015	0	−0.016
O$_{19}$	0	−0.02	0	−0.021	O$_{19}$	0	0.005	0	0.005	O$_{19}$	0	−0.025	0	−0.025

(Continued)

TABLE 7.2 (*Continued*)

Atom- and Orbital-Dependent Magnetic Moments: (a)/(b)/(c) Li_xCoO_2 with $x = 0.75/0.50/0.25$ and (d)/(e)/(f)/(g) Li_xNiO_2 with $x = 1.00/0.75/0.50/0.25$

(a)	$Li_9Co_{12}O_{24}$				(b)	$Li_6Co_{12}O_{24}$				(c)	$i_3Co_{12}O_{24}$			
Atom	s	p	d	Total	Atom	s	p	d	Total	Atom	s	p	d	Total
O_{20}	0	−0.011	0	−0.011	O_{20}	0	−0.007	0	−0.007	O_{20}	0	−0.015	0	−0.016
O_{21}	0	−0.008	0	−0.009	O_{21}	0	0.008	0	0.008	O_{21}	−0.001	−0.029	0	−0.029
O_{22}	−0.001	−0.016	0	−0.017	O_{22}	0	−0.023	0	−0.023	O_{22}	−0.001	−0.015	0	−0.016
O_{23}	0	−0.021	0	−0.021	O_{23}	0	0.008	0	0.008	O_{23}	0	−0.015	0	−0.016
O_{24}	0	−0.011	0	−0.011	O_{24}	0	−0.023	0	−0.023	O_{24}	0	−0.025	0	−0.025
Total	−0.05	−0.37	−2.51	−2.94	Total	0	−0.03	−0.31	−0.34	tot	−0.03	−0.59	−2.29	−2.9

(d)	$Li_{12}Ni_{12}O_{24}$				(e)	$Li_9Ni_{12}O_{24}$			
Atom	s	p	d	Total	Atom	s	p	d	Total
Li_1	−0.004	−0.008	0	−0.012	Li_1	0.004	0.01	0.001	0.015
Li_2	−0.004	−0.008	0	−0.012	Li_2	0.001	0.003	0	0.004
Li_3	−0.004	−0.008	0	−0.012	Li_3	0.001	0.003	0	0.004
Li_4	−0.004	−0.008	0	−0.012	Li_4	0.001	0.003	0	0.004
Li_5	−0.004	−0.008	0	−0.012	Li_5	0.004	0.01	0.001	0.015
Li_6	−0.004	−0.008	0	−0.012	Li_6	0.001	0.003	0	0.004
Li_7	−0.004	−0.008	0	−0.012	Li_7	0.001	0.002	0	0.003
Li_8	−0.004	−0.008	0	−0.012	Li_8	0.001	0.002	0	0.003
Li_9	−0.004	−0.008	0	−0.012	Li_9	0.004	0.01	0.001	0.015
Li_{10}	−0.004	−0.008	0	−0.012					
Li_{11}	−0.004	−0.008	0	−0.012					
Li_{12}	−0.004	−0.008	0	−0.012					
Ni_1	0.001	0	−0.754	−0.753	Ni_1	0.001	−0.001	1.186	1.185
Ni_2	0.001	0	−0.754	−0.753	Ni_2	−0.001	−0.001	−0.075	−0.077
Ni_3	0.001	0	−0.754	−0.753	Ni_3	−0.001	−0.001	−0.104	−0.106
Ni_4	0.001	0	−0.754	−0.753	Ni_4	0.004	−0.001	0.724	0.728
Ni_5	0.001	0	−0.754	−0.753	Ni_5	−0.001	−0.001	−0.075	−0.077
Ni_6	0.001	0	−0.754	−0.753	Ni_6	0.001	−0.001	1.186	1.185
Ni_7	0.001	0	−0.754	−0.753	Ni_7	−0.001	−0.001	−0.104	−0.106
Ni_8	0.001	0	−0.754	−0.753	Ni_8	0.004	−0.001	0.724	0.728
Ni_9	0.001	0	−0.754	−0.753	Ni_9	−0.001	0	−0.1	−0.101
Ni_{10}	0.001	0	−0.754	−0.753	Ni_{10}	−0.001	0	−0.099	−0.101
Ni_{11}	0.001	0	−0.754	−0.753	Ni_{11}	0.001	−0.002	1.211	1.21
Ni_{12}	0.001	0	−0.754	−0.753	Ni_{12}	0.004	−0.002	0.769	0.771
O_1	−0.012	−0.077	−0.001	−0.089	O_1	0.005	0.085	0	0.089
O_2	−0.012	−0.077	−0.001	−0.089	O_2	0.004	0.057	0	0.061
O_3	−0.012	−0.077	−0.001	−0.089	O_3	0.006	0.05	0	0.056
O_4	−0.012	−0.077	−0.001	−0.089	O_4	0.007	0.052	0	0.059
O_5	−0.012	−0.077	−0.001	−0.089	O_5	0.005	0.072	0	0.077
O_6	−0.012	−0.077	−0.001	−0.089	O_6	0.006	0.055	0	0.061

(*Continued*)

TABLE 7.2 (Continued)

Atom- and Orbital-Dependent Magnetic Moments: (a)/(b)/(c) Li$_x$CoO$_2$ with $x = 0.75/0.50/0.25$ and (d)/(e)/(f)/(g) Li$_x$NiO$_2$ with $x = 1.00/0.75/0.50/0.25$

(d)	Li$_{12}$Ni$_{12}$O$_{24}$				(e)	Li$_9$Ni$_{12}$O$_{24}$			
Atom	s	p	d	Total	Atom	s	p	d	Total
O$_7$	−0.012	−0.077	−0.001	−0.089	O$_7$	0.004	0.029	0	0.033
O$_8$	−0.012	−0.077	−0.001	−0.089	O$_8$	0.007	0.068	0	0.075
O$_9$	−0.012	−0.077	−0.001	−0.089	O$_9$	0.006	0.055	0	0.061
O$_{10}$	−0.012	−0.077	−0.001	−0.089	O$_{10}$	0.005	0.072	0	0.077
O$_{11}$	−0.012	−0.077	−0.001	−0.089	O$_{11}$	0.004	0.029	0	0.033
O$_{12}$	−0.012	−0.077	−0.001	−0.089	O$_{12}$	0.007	0.068	0	0.075
O$_{13}$	−0.012	−0.077	−0.001	−0.089	O$_{13}$	0.004	0.057	0	0.061
O$_{14}$	−0.012	−0.077	−0.001	−0.089	O$_{14}$	0.005	0.084	0	0.089
O$_{15}$	−0.012	−0.077	−0.001	−0.089	O$_{15}$	0.006	0.05	0	0.056
O$_{16}$	−0.012	−0.077	−0.001	−0.089	O$_{16}$	0.007	0.052	0	0.059
O$_{17}$	−0.012	−0.077	−0.001	−0.089	O$_{17}$	0.004	0.025	0	0.03
O$_{18}$	−0.012	−0.077	−0.001	−0.089	O$_{18}$	0.007	0.063	0	0.07
O$_{19}$	−0.012	−0.077	−0.001	−0.089	O$_{19}$	0.004	0.076	0	0.08
O$_{20}$	−0.012	−0.077	−0.001	−0.089	O$_{20}$	0.006	0.064	0	0.071
O$_{21}$	−0.012	−0.077	−0.001	−0.089	O$_{21}$	0.007	0.063	0	0.07
O$_{22}$	−0.012	−0.077	−0.001	−0.089	O$_{22}$	0.004	0.025	0	0.03
O$_{23}$	−0.012	−0.077	−0.001	−0.089	O$_{23}$	0.004	0.076	0	0.08
O$_{24}$	−0.012	−0.077	−0.001	−0.089	O$_{24}$	0.006	0.064	0	0.07
Total	−0.32	−1.93	−9.07	−11.32	Total	0.15	1.43	5.25	6.83

(f)	Li$_6$Ni$_{12}$O$_{24}$				(g)	Li$_3$Ni$_{12}$O$_{24}$			
Atom	s	p	d	Total	Atom	s	p	d	Total
Li$_1$	0.001	0.001	0	0.002	Li$_1$	0.001	0.001	0	0.002
Li$_2$	0.001	0.001	0	0.002	Li$_2$	0.001	0.001	0	0.002
Li$_3$	−0.002	−0.005	0	−0.007	Li$_3$	0.001	0.001	0	0.002
Li$_4$	−0.002	−0.005	0	−0.007					
Li$_5$	0.001	0.001	0	0.002					
Li$_6$	0.001	0.001	0	0.002					
Ni$_1$	0.001	0.001	−0.39	−0.388	Ni$_1$	0	0	0.013	0.013
Ni$_2$	0.001	0.001	−0.258	−0.256	Ni$_2$	0	0	0.105	0.104
Ni$_3$	0.001	0.001	−0.39	−0.388	Ni$_3$	0	0	0.105	0.104
Ni$_4$	0.001	0.001	−0.258	−0.256	Ni$_4$	0	0	−0.006	−0.006
Ni$_5$	−0.001	−0.001	−0.034	−0.035	Ni$_5$	0	0	0.013	0.013
Ni$_6$	0	−0.001	0.739	0.738	Ni$_6$	0	0	0.104	0.103
Ni$_7$	−0.001	−0.001	−0.034	−0.035	Ni$_7$	0	0	−0.006	−0.006
Ni$_8$	0	−0.001	0.739	0.738	Ni$_8$	0	0	0.105	0.105
Ni$_9$	0.001	0.001	−0.39	−0.388	Ni$_9$	0	0	0.013	0.013
Ni$_{10}$	0.001	0.001	−0.258	−0.256	Ni$_{10}$	0	0	−0.009	−0.009
Ni$_{11}$	0.001	0.001	−0.39	−0.388	Ni$_{11}$	0	0	0.098	0.097

(Continued)

TABLE 7.2 (Continued)

Atom- and Orbital-Dependent Magnetic Moments: (a)/(b)/(c) Li_xCoO_2 with $x = 0.75/0.50/0.25$ and (d)/(e)/(f)/(g) Li_xNiO_2 with $x = 1.00/0.75/0.50/0.25$

(f)	$Li_6Ni_{12}O_{24}$				(g)	$Li_3Ni_{12}O_{24}$			
Atom	s	p	d	Total	Atom	s	p	d	Total
Ni_{12}	0.001	0.001	−0.258	−0.256	Ni_{12}	0	0	0.097	0.096
O_1	−0.007	−0.052	0	−0.059	O_1	0.001	0.005	0	0.006
O_2	−0.006	−0.032	0	−0.039	O_2	0.001	0.009	0	0.009
O_3	−0.007	−0.052	0	−0.059	O_3	0.001	0.005	0	0.006
O_4	−0.006	−0.032	0	−0.039	O_4	0.002	0.01	0	0.012
O_5	0.004	0.049	0	0.054	O_5	0.001	0.002	0	0.002
O_6	0.006	0.038	0	0.044	O_6	0.001	0.014	0	0.015
O_7	0.004	0.049	0	0.054	O_7	0.001	0.002	0	0.003
O_8	0.006	0.038	0	0.044	O_8	0.002	0.015	0	0.018
O_9	−0.004	−0.037	0	−0.042	O_9	0.001	0.002	0	0.002
O_{10}	−0.007	−0.062	0	−0.07	O_{10}	0.001	0.014	0	0.015
O_{11}	−0.004	−0.037	0	−0.042	O_{11}	0.002	0.015	0	0.018
O_{12}	−0.007	−0.062	0	−0.07	O_{12}	0.001	0.002	0	0.003
O_{13}	−0.004	−0.037	0	−0.041	O_{13}	0.001	0.005	0	0.006
O_{14}	−0.007	−0.062	0	−0.07	O_{14}	0.001	0.009	0	0.009
O_{15}	−0.004	−0.037	0	−0.041	O_{15}	0.002	0.01	0	0.012
O_{16}	−0.007	−0.062	0	−0.07	O_{16}	0.001	0.005	0	0.006
O_{17}	0.004	0.049	0	0.054	O_{17}	0.001	0	0	0.001
O_{18}	0.006	0.038	0	0.044	O_{18}	0.002	0.016	0	0.018
O_{19}	0.004	0.049	0	0.054	O_{19}	0.001	0.004	0	0.005
O_{20}	0.006	0.038	0	0.044	O_{20}	0.002	0.013	0	0.015
O_{21}	−0.007	−0.052	0	−0.059	O_{21}	0.001	0	0	0.001
O_{22}	−0.006	−0.032	0	−0.039	O_{22}	0.002	0.016	0	0.018
O_{23}	−0.007	−0.052	0	−0.059	O_{23}	0.002	0.013	0	0.015
O_{24}	−0.006	−0.032	0	−0.039	O_{24}	0.001	0.004	0	0.005
Total	−0.05	−0.38	−1.18	−1.61	Total	0.03	0.19	0.63	0.85

ferromagnetic configurations are greatly diversified by the transition metal atoms and lithium concentrations. For example, the local spin densities/magnetic moments strongly rely on the Co/Ni positions. Furthermore, the partial minor magnetic contributions are due to the lithium and oxygen atoms, especially for the latter being closely related to the Li vacancies. The unusual spin arrangements are revealed under the strong competitions/cooperations with the multi-orbital interactions in chemical bonds, according to the spin- and charge-induced Hamiltonian in the Hubbard tight-binding model [100]. All the intrinsic interactions should cover the orbital-created hopping integrals and ionization energies and the prominent on-site electron–electron Coulomb interactions [not the spin-orbital couplings]. How to achieve the ground states for the various doping cases is a very interesting topic in the near-future researches. Based on the previous theoretical and experimental works on the stationary currents, it is very difficult to directly link the electronic spin states with them. The spin configuration becomes an important issue due to the strong effect on the ion transport of batteries.

The density of states [DOS], being characterized as the number of electronic states within a rather small energy range of dE, is able to reveal the main features valence and conduction energy spectra simultaneously, as clearly illustrated in Figures 7.9–7.11. Its special structures, van Hove singularities, mainly originate from the band-edge states with zero group velocities (Figures 7.5 and 7.6). Such critical points in the energy-wave-vector space include the local extreme points [minima and maxima], the saddle points, and all the dispersionless energy subbands. Very interesting, the different dimensions can create the diversified van Hove singularities, e.g., the delta-function-like prominent peaks for the 0D discrete energy levels, the square-root asymmetric peaks/the plateau structures of the 1D parabolic/linear energy subbands, and the shoulder structures/V shapes/logarithmic peaks/square-root asymmetric peaks in the 2D parabolic/Dirac-cone/saddle-point/constant-energy-loop band structures. The 3D ternary Li$_x$CoO$_2$ compounds have a lot of oscillatory energy subbands, with the frequent crossings/anti-crossings; therefore, van Hove singularities are displayed as many asymmetric/symmetric peaks and normal/irregular shoulders. Each special structure might be due to some merged band-edge state with very close energies lower than the broadening energy.

In this chapter, the atom- and orbital-projected van Hove singularities, as indicated in Figures 7.9a–d and 7.10a–d, respectively, for LiCoO$_2$ and LiNiO$_2$ compounds, could be directly utilized to resolve the important multi-orbital hybridizations in the Li–O and Co–O/Ni–O bonds. Three kinds of atoms [the blue, red, and green curves] have the significant contributions and similar van Hove singularities in the whole energy range. This strongly suggests that there exist the close relations among any neighboring chemical bonds, or the mixed effects of different chemical bonds cannot be distinguished from one another. As to the LiCoO$_2$ system (Figure 7.9a–d), the whole energy spectrum could be classified into three groups: (I) the deeper valence region [-8.00 eV $\leq E^v \leq -3.00$ eV], (II) the shallower one [-2.50 eV $\leq E^v \leq -1.00$ eV], and (III) the conduction region [0.55 eV $\leq E^v \leq -1.50$ eV]. An energy gap of ~1.09 eV, with a vanishing density of states, covers the Fermi level and separates the very high asymmetric valence and conduction energy spectra. Most of electronic states are displayed as the occupied spectrum, but not the unoccupied one. According to the delicate VASP calculations, the lithium contributions [the blue curve in Figure 7.9b] are lowest; furthermore, the oxygen and cobalt atoms (Figure 7.9c and d), respectively, play a dominating role in the deeper valence-state region [e.g., $E < -2.50$ eV] and the opposite region. Only the single Li–2s orbital takes part in the chemical bondings with the three O–[$2p_x$, $2p_y$, $2p_z$] orbitals, being strongly supported by the spatial charge density distribution in Li–O bonds (Figure 7.7). Very interesting, the four O–($2s$, $2p_x$, $2p_y$, $2p_z$) [the blue, red, green, and yellow curves in Figure 7.9c] orbitals strongly hybridize with the six Co–($4s$, $3d_{x^2-y^2}$, $3d_{xy}$, $3d_{yz}$, $3d_{xz}$, $3d_z^2$) ones [the pink, blue, purple, red, green, and yellow curves in Figure 7.9d], since their van Hove singularities come to exist together. Among their available orbitals, the O–2s orbitals only make minor contributions at any energies, and the Co–4s contribution

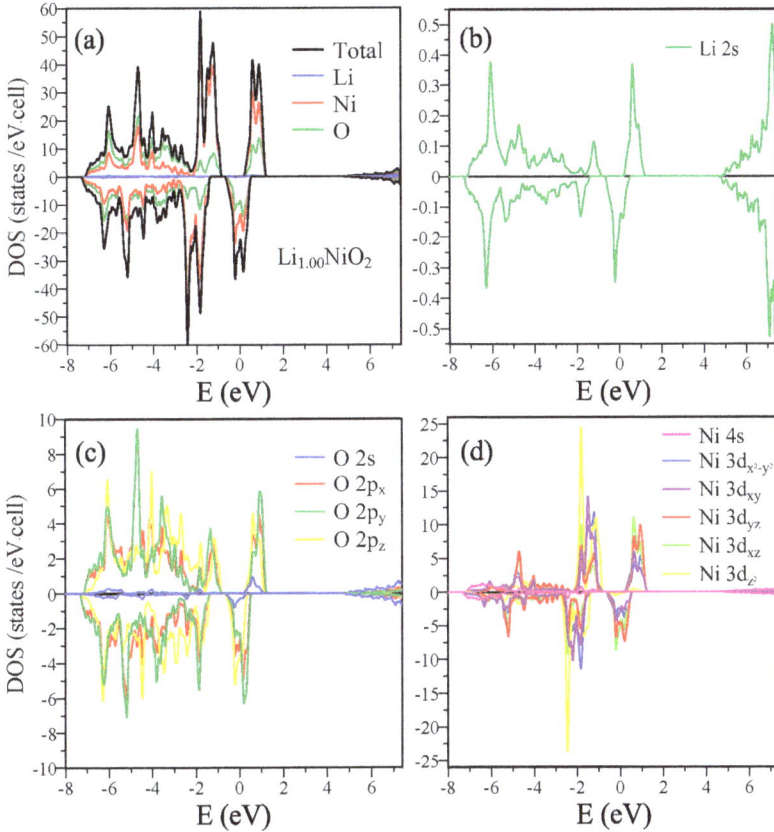

FIGURE 7.10 A similar plot as Figure 7.9a–d, but shown for the 3D ternary Li$_{12}$Ni$_{12}$O$_{24}$ compound. The spin-split density of states shows the metallic ferromagnetism.

lithium-concentration-related effects only directly reflect the highly non-uniform chemical and physical environments [details in the spatial charge/spin density distributions indicated by Figures 7.7 and 7.8]. Similar results are revealed in other cathode compounds, e.g., the aluminum- and iron-ion battery cathodes. Such unusual phenomena might be the critical mechanisms for the release and recovery of the Li$^+$ ions from the Li$_x$Co/NiO cathode materials. That is to say, whether there exist some lithium vacancies/defects would be very important in initiating the charging process and performing the high-speed and long-term cycle operation.

The high-resolution STS and spin-polarized STS [experimental details in Ref. 101] are rather efficient techniques in thoroughly examining the whole valence and conduction van Hove singularities, being totally different from ARPES and SP-ARPES for the wave-vector-dependent occupied energy subbands. Their functionalities come from a quantum tunneling current [a very small I], being initiated from a nano-scaled probe with a gate voltage [V]. The measured differential conductance, dI/dV, is characterized as the density of states. Specifically, the previous experimental measurements clearly show that STS is very successful in verifying the

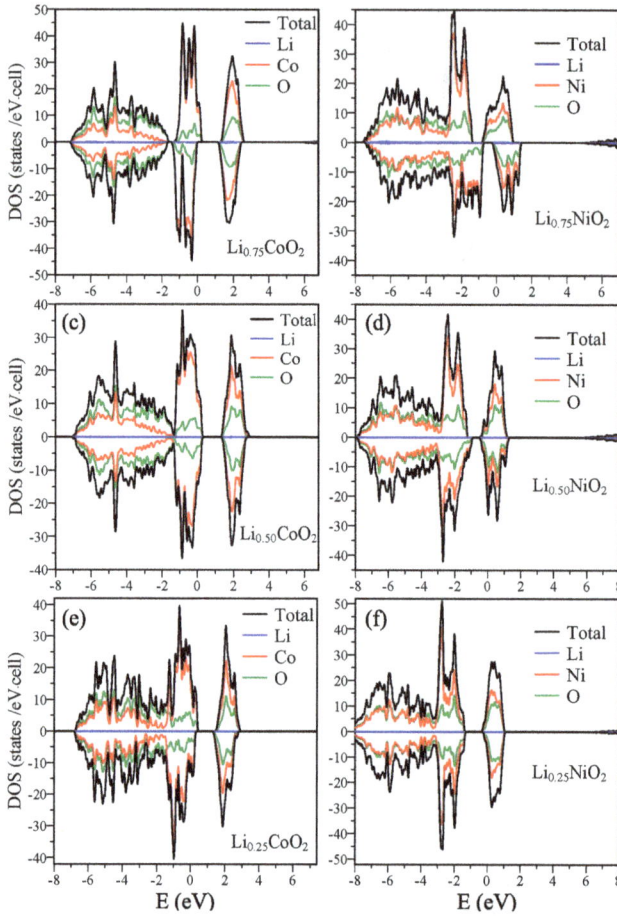

FIGURE 7.11 Strong lithium-concentration dependences of the atom-projected densities of states for (a)/(b) $Li_9Co_{12}/Ni_{12}O_{24}$, (c)/(d) $Li_6Co_{12}/Ni_{12}O_{24}$, and (e)/(f) $Li_3Co_{12}/Ni_{12}O_{24}$.

dimension-diversified van Hove singularities of graphene-related systems with the sp^2 honeycomb lattices, e.g., many divergent peaks under the square-root form from 1D planar graphene nanoribbons and cylindrical carbon nanotubes and graphene nanoribbons, a plateau structure across the Fermi level in metallic armchair nanotubes, a symmetric V shape vanishing at E_F for monolayer graphene [a zero-gap semiconductor], the logarithmically symmetric peaks near E_F in twisted bilayer graphenes [the saddle points], an electric-field-created band gap for bilayer AB and trilayer ABC stackings, a delta-function-like peak centered about E_F for ABC-stacked graphenes [surface states], a sharp dip structure close to E_F combined with a pair of square-root peaks from trilayer AAB stacking [a narrow-gap semiconductor with the constant-energy loops], and a finite density of states at E_F for the semimetallic AB-stacked bulk graphite. Similar STS and SP-STS experiments, which are conducted on the 3D ternary Li_xCo/NiO_2 cathode materials of Li^+-ion batteries, are able to fully examine the position of the Fermi level, the magnitude of band gap, many van Hove singularities, the wide/narrow valence/conduction energy spectra and their total widths, the separated or non-separated valence ones, and the spin-degenerate or spin-split behaviors. Such measurements can provide very useful information about the strong dependences on the kind of transition metal atom and the lithium concentration, but not the orbital-decomposed contribution.

In addition to LiCo/NiO$_2$ compounds, both LiFeO$_2$ and LiFePO$_4$ materials are also frequently utilized as the cathodes of the Li$^+$-based batteries. The delicate VASP calculations and detailed analyses, as done for the former (Figures 7.5–7.11), should be very useful in resolving the diverse phenomena associated with the latter, such as the Fe- and P-dependent multi-orbital hybridizations and magnetic configurations. The geometric, electronic, magnetic, transport, and optical properties are expected to be greatly diversified by the coupled effects. The critical mechanisms cover many atoms/chemical bonds in a large unit cell, the atom-induced or atom-deleted vacancies [e.g., the reduced/recovered lithium concentration in this work], the significant impurities [e.g., the substitution of Li by Na/K/Rb/Cs], the doping of extra components [four or five components], and the atom-dominated spin distributions [especially for transition metal atoms]. How to link the above-mentioned factors and the stationary ion currents will become an emergent issue. As for the cathode materials in aluminum- and iron-ion-based batteries, the AB-stacked graphite [the Bernal system] and V$_2$O$_5$, respectively, serve outstanding roles for the ion transports. Apparently, the chemical intercalations and de-intercalations of multi-element anions will be responsible for the rich and unique essential properties. For example, the simultaneous optimal distribution of AlCl$_4^-$ and Al$_2$Cl$_7^-$ anions, which come to exist between two neighboring graphitic layers, determines a primitive unit cell and thus the crystal symmetry. On the other side, the graphite alkali-intercalation compounds belong to the excellent Li$^+$-ion battery anode. These two systems should be totally different from each other in terms of the fundamental physical/chemical/material properties.

7.5   CONCLUDING REMARKS

The first-principles calculations clearly show that the 3D ternary Li$_x$Co/NiO$_2$ compounds, the cathode materials of the Li$^+$-ion batteries, exhibit the unusual geometric, electronic, and magnetic properties; furthermore, the lithium concentration plays an important role in determining the featured phenomena [the doping can greatly diversify the fundamental properties]. This study is able to provide certain meaningful information about the critical mechanisms in the ion transport. There are many Li–O and Co–O/Ni–O bonds in a primitive unit cell with the trigonal symmetry, and the significant modulations of bond lengths come to exist under the non-saturated cases of $x \neq 1$. The extremely non-uniform chemical [charge density distributions] and physical [spin density distributions] environments further diversify band structures and density of states. The main features of electronic energy spectra cover the x-dependent position of the Fermi level [the strong dependence of free carrier density on x], a lot of highly asymmetric valence and conduction subbands, the irregular energy dispersions without the stable valley structures, the frequent crossings and anti-crossings, the separated or non-separated valence energy spectra, the distinct atom dominances within the specific energy ranges, and the spin-split and spin-degenerate states near E_F [the ferromagnetic and non-magnetic characteristics]. According to the delicate analyses on the atom-dominated band structures, spatial charge densities, and atom- and orbital-decomposed van Hove singularities, the critical multi-orbital hybridizations are deduced to be [Li; $2s$]–[O; $2p_x$, $2p_y$, $2p_z$] and [O; $2s$, $2p_x$, $2p_y$, $2p_z$]–[Co/Ni; $4s$, $3d_{x^2-y^2}$, $3d_{xy}$, $3d_{yz}$, $3d_{xz}$, $3d_{z^2}$]. The unique metallic ferromagnetism, which principally comes from five Co-/Ni-3D orbitals, is predicted to survive in most of the materials, but absent in the saturated $LiCoO_2$ system. Lithium and oxygen atoms only make minor, but significant, contributions. There exist three kinds of ferromagnetic configurations according to the spin arrangements, being strongly competed/cooperated with the chemical bondings. Both of them should be the intrinsic interactions in the Hamiltonian. That is to say, the orbital-dependent hopping integrals and ionization energies and the spin-created on-site electron–electron Coulomb interactions are included in the Hubbard-like tight-binding model. The calculated optimal geometries, band structures, density of states, and net magnetic moments could be verified by the X-ray diffraction, ARPES, STS, and SQUID, respectively. The delicate VASP calculations and analyses are expected to be very suitable for the outstanding cathode materials, $LiFeO_2$ and $LiFePO_4$. That the 3D three and four components are critical factors in creating the diverse phenomena is under the current investigations.

REFERENCES

1. Zhu, C., Tuchband, M. R., Young, A., Shuai, M., Scarbrough, A., Walba, D. M., et al. 2016 Resonant carbon K-edge soft X-ray scattering from lattice-free heliconical molecular ordering: soft dilative elasticity of the twist-bend liquid crystal phase. *Physical Review Letters*, 116, 147803.
2. Dai, Z., Jin, W., Grady, M., Sadowski, J. T., Dadap, J. I., Osgood Jr, R. M., & Pohl, K. 2017 Surface structure of bulk 2H-MoS₂ (0001) and exfoliated suspended monolayer MoS_2: a selected area low energy electron diffraction study. *Surface Science*, 660, 16–21.
3. Carstens, T., Ispas, A., Borisenko, N., Atkin, R., Bund, A., & Endres, F. 2016 In situ scanning tunneling microscopy (STM), atomic force microscopy (AFM) and quartz crystal microbalance (EQCM) studies of the electrochemical deposition of tantalum in two different ionic liquids with the 1-butyl-1-methylpyrrolidinium cation. *Electrochimica Acta*, 197, 374–387.
4. Feist, A., Bach, N., da Silva, N. R., Danz, T., Möller, M., Priebe, K. E., et al. 2017 Ultrafast transmission electron microscopy using a laser-driven field emitter: femtosecond resolution with a high coherence electron beam. *Ultramicroscopy*, 176, 63–73.
5. Van der Lit, J., Jacobse, P. H., Vanmaekelbergh, D., & Swart, I. 2015 Bending and buckling of narrow armchair graphene nanoribbons via STM manipulation. *New Journal of Physics*, 17, 053013.
6. Maitra, U., Matte, H. S. S., Kumar, P., & Rao, C. N. R. 2012 Strategies for the synthesis of graphene, graphene nanoribbons, nanoscrolls and related materials. *CHIMIA International Journal for Chemistry*, 66, 941–948.
7. Zhang, J., Xiao, J., Meng, X., Monroe, C., Huang, Y., & Zuo, J. M. 2010 Free folding of suspended graphene sheets by random mechanical stimulation. *Physical Review Letters*, 104, 166805.
8. Kelly, K. F., Chiang, I. W., Mickelson, E. T., Hauge, R. H., Margrave, J. L., Wang, X., et al. 1999 Insight into the mechanism of sidewall functionalization of single-walled nanotubes: an STM study. *Chemical Physics Letters*, 313, 445–450.

9. Biedermann, L. B., Bolen, M. L., Capano, M. A., Zemlyanov, D., & Reifenberger, R. G. 2009 Insights into few-layer epitaxial graphene growth on 4*H*-SiC (0001) substrates from STM studies. *Physical Review B*, 79, 125411.

10. Klusek, Z., Kozlowski, W., Waqar, Z., Datta, S., Burnell-Gray, J. S., Makarenko, I. V., et al. 2005 Local electronic edge states of graphene layer deposited on Ir (1 1 1) surface studied by STM/CITS. *Applied Surface Science*, 252, 1221–1227.

11. Nishide, H., & Oyaizu, K. 2008 Toward flexible batteries. *Science*, 319, 737–738.

12. Orilall, M. C., & Wiesner, U. 2011 Block copolymer based composition and morphology control in nanostructured hybrid materials for energy conversion and storage: solar cells, batteries, and fuel cells. *Chemical Society Reviews*, 40, 520–535.

13. Cook, B. 2002 Introduction to fuel cells and hydrogen technology. *Engineering Science & Education Journal*, 11, 205–216.

14. Boelens, R. 2014 Cultural politics and the hydrosocial cycle: water, power and identity in the Andean highlands. *Geoforum*, 57, 234–247.

15. Staffell, I., & Pfenninger, S. 2016 Using bias-corrected reanalysis to simulate current and future wind power output. *Energy*, 114, 1224–1239.

16. Newmark, N. M., Blume, J. A., & Kapur, K. K. 1973 *Seismic Design Spectra for Nuclear Power Plants*. Consulting Engineering Services, Urbana, IL.

17. Jiang, S., Wang, W., Xue, X., Cao, C., & Zhang, Y. 2016 Fungal diversity in major oil-shale mines in China. *Journal of Environmental Sciences*, 41, 81–89.

18. Luo, S., & Feng, Y. 2017 The production of fuel oil and combustible gas by catalytic pyrolysis of waste tire using waste heat of blast-furnace slag. *Energy Conversion and Management*, 136, 27–35.

19. Mark, C., & Gauna, M. 2016 Evaluating the risk of coal bursts in underground coal mines. *International Journal of Mining Science and Technology*, 26, 47–52.

20. Wang, W. C., Yung, Y. L., Lacis, A. A., Mo, T. A., & Hansen, J. E. 1976 Greenhouse effects due to man-made perturbations of trace gases. *Science*, 194(4266), 685–690.

21. GhaffarianHoseini, A., Dahlan, N. D., Berardi, U., GhaffarianHoseini, A., Makaremi, N., & GhaffarianHoseini, M. 2013 Sustainable energy performances of green buildings: a review of current theories, implementations and challenges. *Renewable and Sustainable Energy Reviews*, 25, 1–17.

22. Greeley, J., & Markovic, N. M. 2012 The road from animal electricity to green energy: combining experiment and theory in electrocatalysis. *Energy & Environmental Science*, 5(11), 9246–9256.

23. Zhang, Q., Sun, Y., Xu, W., & Zhu, D. 2014 Organic thermoelectric materials: emerging green energy materials converting heat to electricity directly and efficiently. *Advanced Materials*, 26(40), 6829–6851.

24. Shih, E., Bahl, P., & Sinclair, M. J. 2002 September Wake on wireless: an event driven energy saving strategy for battery operated devices. In Proceedings of the 8th Annual International Conference on Mobile Computing and Networking, 160–171, ACM, New York.

25. Dong, J., Liu, C., & Lin, Z. 2014 Charging infrastructure planning for promoting battery electric vehicles: an activity-based approach using multiday travel data. *Transportation Research Part C: Emerging Technologies*, 38, 44–55.

26. Frey, H. C., Rouphail, N. M., Zhai, H., Farias, T. L., & Gonçalves, G. A. 2007 Comparing real-world fuel consumption for diesel-and hydrogen-fueled transit buses and implication for emissions. *Transportation Research Part D: Transport and Environment*, 12, 281–291.

27. Liu, J., & Barnett, S. A. 2003 Operation of anode-supported solid oxide fuel cells on methane and natural gas. *Solid State Ionics*, 158, 11–16.

28. Prasad, N. R., Ranade, S. J., Hoang, H. T., & Phuc, N. H. 2012 Hydropower energy recovery (Hyper) from water-flow systems in Vietnam. In 10th International Power & Energy Conference (IPEC), 92–97. IEEE, Ho Chi Minh City.

29. Morren, J., De Haan, S. W., Kling, W. L., & Ferreira, J. A. 2006 Wind turbines emulating inertia and supporting primary frequency control. *IEEE Transactions on Power Systems*, 21, 433–434.

30. Yu, Y., Gu, L., Zhu, C., Van Aken, P. A., & Maier, J. 2009 Tin nanoparticles encapsulated in porous multichannel carbon microtubes: preparation by single-nozzle electrospinning and application as anode material for high-performance Li-based batteries. *Journal of the American Chemical Society*, 131, 15984–15985.

31. Ambroz, F., Macdonald, T. J., & Nann, T. 2017 Trends in aluminium-based intercalation batteries. *Advanced Energy Materials*, 7, 1602093.

32. Christiansen, A. S., Johnsen, R. E., Norby, P., Frandsen, C., Mørup, S., Jensen, S. H., et al. 2015 In Situ Studies of Fe^{4+} Stability in β-Li_3Fe_2 $(PO_4)_3$ cathodes for Li ion batteries. *Journal of the Electrochemical Society*, 162, A531–A537.

33. Langreth, D. C., & Mehl, M. J. 1983 Beyond the local-density approximation in calculations of ground-state electronic properties. *Physical Review B*, 28, 1809.

34. Kresse, G., & Hafner, J. 1993 Ab initio molecular dynamics for liquid metals. *Physical Review B*, 47, 558.

35. Lee, T. H., White, H., & Granger, C. W. 1993 Testing for neglected nonlinearity in time series models: a comparison of neural network methods and alternative tests. *Journal of Econometrics*, 56, 269–290.

36. Gabano, J. P. 1983 *Lithium Batteries*. Academic Press, London and New York, 467.

37. Wang, Z. L., Xu, D., Huang, Y., Wu, Z., Wang, L. M., & Zhang, X. B. 2012 Facile, mild and fast thermal-decomposition reduction of graphene oxide in air and its application in high-performance lithium batteries. *Chemical Communications*, 48, 976–978.

38. Scott, I. D., Jung, Y. S., Cavanagh, A. S., Yan, Y., Dillon, A. C., George, S. M., & Lee, S. H. 2010 Ultrathin coatings on nano-$LiCoO_2$ for Li-ion vehicular applications. *Nano Letters*, 11, 414–418.

39. Huang, B., Jang, Y. I., Chiang, Y. M., & Sadoway, D. R. 1998 Electrochemical evaluation of $LiCoO_2$ synthesized by decomposition and intercalation of hydroxides for lithium-ion battery applications. *Journal of Applied Electrochemistry*, 28, 1365–1369.

40. Zhang, S. S. 2006 A review on electrolyte additives for lithium-ion batteries. *Journal of Power Sources*, 162, 1379–1394.

41. Whittingham, M. S. 2004 Lithium batteries and cathode materials. *Chemical Reviews*, 104, 4271–4302.

42. Stevens, D. A., & Dahn, J. R. 2000 High capacity anode materials for rechargeable sodium-ion batteries. *Journal of the Electrochemical Society*, 147, 1271–1273.

43. Moura, S. J., Argomedo, F. B., Klein, R., Mirtabatabaei, A., & Krstic, M. 2016 Battery state estimation for a single particle model with electrolyte dynamics. *IEEE Transactions on Control Systems Technology*, 25, 453–468.

44. Zhang, Y., Pelliccione, C. J., Brady, A. B., Guo, H., Smith, P. F., Liu, P., et al. 2017 Probing the Li insertion mechanism of $ZnFe_2O_4$ in Li-ion batteries: a combined X-ray diffraction, extended X-ray absorption fine structure, and density functional theory study. *Chemistry of Materials*, 29, 4282–4292.

45. Xiang, H. F., Li, Z. D., Xie, K., Jiang, J. Z., Chen, J. J., Lian, P. C., et al. 2012 Graphene sheets as anode materials for Li-ion batteries: preparation, structure, electrochemical properties and mechanism for lithium storage. *RSC Advances*, 2, 6792–6799.

46. Ning, F., Li, S., Xu, B., & Ouyang, C. 2014 Strain tuned Li diffusion in $LiCoO_2$ material for Li ion batteries: a first principles study. *Solid State Ionics*, 263, 46–48.

47. Wang, Z., Huang, S., Chen, B., Wu, H., & Zhang, Y. 2014 Infiltrative coating of $LiNi_{0.5}Co_{0.2}Mn_{0.3}O_2$ microspheres with layer-structured $LiTiO_2$: towards superior cycling performances for Li-ion batteries. *Journal of Materials Chemistry A*, 2, 19983–19987.

48. Rahaman, O., Mortazavi, B., & Rabczuk, T. 2016 A first-principles study on the effect of oxygen content on the structural and electronic properties of silicon suboxide as anode material for lithium ion batteries. *Journal of Power Sources*, 307, 657–664.

49. Goodenough, J. B., & Park, K. S. 2013 The Li-ion rechargeable battery: a perspective. *Journal of the American Chemical Society*, 135, 1167–1176.

50. Zheng, G., Cui, Y., Karabulut, E., Wågberg, L., Zhu, H., & Hu, L. 2013 Nanostructured paper for flexible energy and electronic devices. *MRS Bulletin*, 38, 320–325.

51. Andreasen, S. J., Ashworth, L., Remon, I. N. M., & Kær, S. K. 2008 Directly connected series coupled HTPEM fuel cell stacks to a Li-ion battery DC bus for a fuel cell electrical vehicle. *International Journal of Hydrogen Energy*, 33, 7137–7145.

52. Wee, J. H. 2007 A feasibility study on direct methanol fuel cells for laptop computers based on a cost comparison with lithium-ion batteries. *Journal of Power Sources*, 173, 424–436.

53. Yamakawa, S., Yamasaki, H., Koyama, T., & Asahi, R. 2013 Numerical study of Li diffusion in polycrystalline LiCoO$_2$. *Journal of Power Sources*, 223, 199–205.

54. Hwang, Y. J., Nahm, K. S., Kumar, T. P., & Stephan, A. M. 2008 Poly (vinylidene fluoride-hexafluoropropylene)-based membranes for lithium batteries. *Journal of Membrane Science*, 310, 349–355.

55. Sagane, F., Abe, T., Iriyama, Y., & Ogumi, Z. 2005 Li$^+$ and Na$^+$ transfer through interfaces between inorganic solid electrolytes and polymer or liquid electrolytes. *Journal of Power Sources*, 146, 749–752.

56. Ziebarth, B., Klinsmann, M., Eckl, T., & Elsässer, C. 2014 Lithium diffusion in the spinel phase Li$_4$Ti$_5$O$_{12}$ and in the rocksalt phase Li$_7$Ti$_5$O$_{12}$ of lithium titanate from first principles. *Physical Review B*, 89, 174301.

57. Ning, F., Li, S., Xu, B., & Ouyang, C. 2014 Strain tuned Li diffusion in LiCoO2 material for Li ion batteries: a first principles study. *Solid State Ionics*, 263, 46–48.

58. Petibon, R., Chevrier, V. L., Aiken, C. P., Hall, D. S., Hyatt, S. R., Shunmugasundaram, R., & Dahn, J. R. 2016 Studies of the capacity fade mechanisms of LiCoO$_2$/Si-alloy: graphite cells. *Journal of the Electrochemical Society*, 163, A1146–A1156.

59. Kramer, D., & Ceder, G. 2009 Tailoring the morphology of LiCoO$_2$: a first principles study. *Chemistry of Materials*, 21, 3799–3809.

60. Xia, L., Xia, Y., & Liu, Z. 2015 Thiophene derivatives as novel functional additives for high-voltage LiCoO$_2$ operations in lithium ion batteries. *Electrochimica Acta*, 151, 429–436.

61. Wu, L., Lee, W. H., & Zhang, J. 2014 First principles study on the electrochemical, thermal and mechanical properties of LiCoO$_2$ for thin film rechargeable battery. *Materials Today: Proceedings*, 1, 82–93.

62. Takamatsu, D., Koyama, Y., Orikasa, Y., Mori, S., Nakatsutsumi, T., Hirano, T., et al. 2012 First in situ observation of the LiCoO$_2$ electrode/electrolyte interface by total-reflection X-ray absorption spectroscopy. *Angewandte Chemie International Edition*, 51, 11597–11601.

63. Moriwake, H., Kuwabara, A., Fisher, C. A., Huang, R., Hitosugi, T., Ikuhara, Y. H., et al. 2013 First-principles calculations of lithium-ion migration at a coherent grain boundary in a cathode material, LiCoO$_2$. *Advanced Materials*, 25, 618–622.

64. Graetz, J., Hightower, A., Ahn, C. C., Yazami, R., Rez, P., & Fultz, B. 2002 Electronic structure of chemically-delithiated LiCoO$_2$ studied by electron energy-loss spectrometry. *The Journal of Physical Chemistry B*, 106, 1286–1289.

65. Huang, X., Bennett, J. W., Hang, M. N., Laudadio, E. D., Hamers, R. J., & Mason, S. E. 2017 Ab initio atomistic thermodynamics study of the (0 0 1) surface of LiCoO$_2$ in a water environment and implications for reactivity under ambient conditions. *The Journal of Physical Chemistry C*, 121, 5069–5080.

66. Liu, B., Xu, B., Wu, M. S., & Ouyang, C. Y. 2015 First-principles GGA+U study on structural and electronic properties in $LiMn_{0.5}Ni_{0.5}O_2$, $LiMn_{0.5}Co_{0.5}O_2$ and $LiCo_{0.5}Ni_{0.5}O_2$. *International Journal of Electrochemical Science*, 11, 432–445.

67. Verde, M. G., Baggetto, L., Balke, N., Veith, G. M., Seo, J. K., Wang, Z., & Meng, Y. S. 2016 Elucidating the phase transformation of $Li_4Ti_5O_{12}$ lithiation at the nanoscale. *ACS Nano*, 10, 4312–4321.

68. Vandenberg, A., & Hintennach, A. 2015 A comparative microwave-assisted synthesis of carbon-coated $LiCoO_2$ and $LiNiO_2$ for lithium-ion batteries. *Russian Journal of Electrochemistry*, 51, 310–317.

69. Yoon, C. S., Jun, D. W., Myung, S. T., & Sun, Y. K. 2017 Structural stability of $LiNiO_2$ cycled above 4.2 V. *ACS Energy Letters*, 2, 1150–1155.

70. Shiraki, S., Takagi, Y., Shimizu, R., Suzuki, T., Haruta, M., Sato, Y., et al. 2016 Orientation control of $LiCoO_2$ epitaxial thin films on metal substrates. *Thin Solid Films*, 600, 175–178.

71. Okamoto, Y., Matsumoto, R., Yagihara, T., Iwai, C., Miyoshi, K., Takeuchi, J., et al. 2017 Electronic structure and polar catastrophe at the surface of Li_xCoO_2 studied by angle-resolved photoemission spectroscopy. *Physical Review B*, 96, 125147.

72. Hara, K., Yano, T. A., Hata, J., Hikima, K., Suzuki, K., Hirayama, M., et al. 2017 Nanoscale optical imaging of lithium-ion distribution on a $LiCoO_2$ cathode surface. *Applied Physics Express*, 10(5), 052503.

73. da Costa, D. R., Zarenia, M., Chaves, A., Pereira Jr, J. M., Farias, G. A., & Peeters, F. M. 2016 Hexagonal-shaped monolayer-bilayer quantum disks in graphene: a tight-binding approach. *Physical Review B*, 94, 035415.

74. Rademaker, L., & Mellado, P. 2018 Charge-transfer insulation in twisted bilayer graphene. *Physical Review B*, 98, 235158.

75. Campos, L. C., Taychatanapat, T., Serbyn, M., Surakitbovorn, K., Watanabe, K., Taniguchi, T., et al. 2016 Landau level splittings, phase transitions, and nonuniform charge distribution in trilayer graphene. *Physical Review Letters*, 117, 066601.

76. Kresse, G., & Furthmüller, J. 1996 Software VASP, Vienna (1999). *Physical Review B*, 54, 169.

77. Parr, R. G. 1980 Density functional theory of atoms and molecules. In K. Fukui and B. Pullman (Eds.), *Horizons of Quantum Chemistry*. Springer, Dordrecht, 5–15.

78. Perdew, J. P., Burke, K., & Ernzerhof, M. 1998 Perdew, Burke, and Ernzerhof reply. *Physical Review Letters*, 80, 891.

79. Kresse, G., & Joubert, D. 1999 From ultrasoft pseudopotentials to the projector augmented-wave method. *Physical Review B*, 59, 1758.

80. Wisesa, P., McGill, K. A., & Mueller, T. 2016 Efficient generation of generalized Monkhorst-Pack grids through the use of informatics. *Physical Review B*, 93, 155109.

81. Tran, N. T. T., Lin, S. Y., Lin, C. Y., & Lin, M. F. 2017 *Geometric and Electronic Properties of Graphene-Related Systems: Chemical Bonding Schemes*. CRC Press, Boca Raton, FL.

82. Lin, S. Y., Tran, N. T. T., Chang, S. L., Su, W. P., & Lin, M. F. 2018 *Structure-and Adatom-Enriched Essential Properties of Graphene Nanoribbons*. CRC Press, Boca Raton, FL.

83. Seifert, G. 2007 Tight-binding density functional theory: an approximate Kohn–Sham DFT scheme. *The Journal of Physical Chemistry A*, 111, 5609–5613.

84. Huang, Y. C., Chang, C. P., & Lin, M. F. 2007 Magnetic and quantum confinement effects on electronic and optical properties of graphene ribbons. *Nanotechnology*, 18, 495401.

85. Ramaniah, L. M., & Nair, S. V. 1993 Optical absorption in semiconductor quantum dots: a tight-binding approach. *Physical Review B*, 47, 7132.

86. Klimeck, G., Bowen, R. C., Boykin, T. B., & Cwik, T. A. 2000 sp3s* Tight-binding parameters for transport simulations in compound semiconductors. *Superlattices and Microstructures*, 27, 519–524.

87. Eithiraj, R. D., Jaiganesh, G., & Kalpana, G. 2009 First-principles study of electronic structure and ground-state properties of alkali-metal selenides and tellurides (M2A) [M: Li, Na, K; A: Se, Te]. *International Journal of Modern Physics B*, 23, 5027–5037.

88. Lin, C. L., Arafune, R., Kawahara, K., Kanno, M., Tsukahara, N., Minamitani, E., et al. 2013 Substrate-induced symmetry breaking in silicene. *Physical Review Letters*, 110, 076801.

89. Tahir, M., & Schwingenschlögl, U. 2015 Photoinduced quantum magnetotransport properties of silicene and germanene. *The European Physical Journal B*, 88, 285.

90. Chen, S. C., Wu, C. L., Wu, J. Y., & Lin, M. F. 2016 Magnetic quantization of *sp*^{3} bonding in monolayer gray tin. *Physical Review B*, 94, 045410.

91. Chen, S. C., Wu, J. Y., & Lin, M. F. 2018 Feature-rich magneto-electronic properties of bismuthene. *New Journal of Physics*, 20, 062001.

92. Voon, L. L. Y., Lopez-Bezanilla, A., Wang, J., Zhang, Y., & Willatzen, M. 2015 Effective Hamiltonians for phosphorene and silicene. *New Journal of Physics*, 17, 025004.

93. Li, G., Luican, A., Dos Santos, J. L., Neto, A. C., Reina, A., Kong, J., & Andrei, E. Y. 2010 Observation of Van Hove singularities in twisted graphene layers. *Nature Physics*, 6, 109.

94. Kataura, H., Kumazawa, Y., Maniwa, Y., Umezu, I., Suzuki, S., Ohtsuka, Y., & Achiba, Y. 1999 Optical properties of single-wall carbon nanotubes. *Synthetic Metals*, 103, 2555–2558.

95. Mizokawa, T., Tjeng, L. H., Steeneken, P. G., Brookes, N. B., Tsukada, I., Yamamoto, T., & Uchinokura, K. 2001 Photoemission and X-ray-absorption study of misfit-layered (Bi, Pb)-Sr-Co-O compounds: electronic structure of a hole-doped Co-O triangular lattice. *Physical Review B*, 64, 115104.

96. Kim, B. J., Yu, J., Koh, H., Nagai, I., Ikeda, S. I., Oh, S. J., & Kim, C. 2006 Missing *xy*-band Fermi surface in 4*d* transition-metal oxide Sr$_2$ RhO$_4$: effect of the octahedra rotation on the electronic structure. *Physical Review Letters*, 97, 106401.

97. Evtushinsky, D. V., Inosov, D. S., Zabolotnyy, V. B., Viazovska, M. S., Khasanov, R., Amato, A., et al. 2009 Momentum-resolved superconducting gap in the bulk of Ba$_{1-x}$K$_x$Fe$_2$As$_2$ from combined ARPES and μSR measurements. *New Journal of Physics*, 11, 055069.

98. Zhou, X., Zhang, Q., Gan, L., Li, H., Xiong, J., & Zhai, T. 2016 Booming development of group IV–VI semiconductors: fresh blood of 2D family. *Advanced Science*, 3, 1600177.

99. Skibowski, M., & Kipp, L. 1994 Inverse combined with direct photoemission: momentum resolved electronic structure of 2D systems. *Journal of Electron Spectroscopy and Related Phenomena*, 68, 77–96.

100. Lomba, E., Molina, D., & Alvarez, M. 2000 Hubbard corrections in a tight-binding Hamiltonian for Se: effects on the band structure, local order, and dynamics. *Physical Review B*, 61, 9314.

101. Okuno, S. N., Kishi, T., & Tanaka, K. 2002 Spin-polarized tunneling spectroscopy of Co (0001) surface states. *Physical Review Letters*, 88, 066803.

8 Graphene as an Anode Material in Lithium-Ion Battery

Sanjaya Brahma, Alex Chinghuan Lee, and Jow-lay Huang
National Cheng Kung University

CONTENTS

8.1 INTRODUCTION

Energy conversion and storage is the major purpose of the current research around the world to overcome the regular demands posed by the present civilization that uses a huge chunk of energy to operate the portable household/consumer electronic devices, let alone the electric hybrid vehicles that need high energy/power density (Figure 8.1). The energy demand is going to increase further due to the increase in the production and usage of these devices, and therefore, a lot of research (theoretical/experimental) is ongoing to find out a new source of energy either by the modification of the existing material or by the discovery of the completely new material/technology that has the potential to overcome the issues related to the energy crisis.

The LIB [1–4] is considered as the future energy storage device because of several advantages including high energy density, high open-circuit voltage, long cycle

149

FIGURE 8.1 LIBs used in smartphones as well as electric hybrid vehicles, https://www.statisticbrain.com/hybrid-electric-vehicle-statistics/.

life, and minor self-discharge. Graphite (specific capacity $= 372$ mAh g^{-1}) is the most commonly used anode material in LIB where the Li-ions react with the carbon and form LiC_6 intercalation compounds, and during intercalation, Li transfers its $2s$ electrons to the carbon and sits between the carbon sheets. Graphite has high Coulombic efficiency and excellent cyclic stability but the low theoretical capacity obstructs its possible application in high capacity devices. Lithium intercalation compounds such as $LiCoO_2$, $LiMn_2O_4$, and $LiFePO_4$ are considered as the cathode materials in LIBs. However, the electrochemical performance of current LIBs does not meet the required charge–discharge rate, energy density, and safety, which is needed for all-electric vehicles. The energy density and LIB performance strongly depend on the physical and chemical properties of the cathode and anode materials. The development of the cathode materials is confined because of several difficult requirements including high potential, structural/thermal/chemical stability, and insertion of Li-ions in the structure. However, the anode material research in LIBs is an ongoing process and several metal oxides along with their composites with carbon (carbon nanotube, graphene) are investigated because of their high theoretical specific capacity, environmental friendly, and easy synthesis process. Some alternative anode materials such as Si (4,200 mAh g^{-1}), Sn (994 mAh g^{-1}), and SnO_2 (782 mAh g^{-1}) have also been developed. But, the practical application is still limited due to the low specific capacity/rate capability after many cycles and serious volume expansion during the alloying and de-alloying with Li-ion that damages the whole electrode.

Recently, graphene [5–10], where sp^2 bonded carbon atoms are arranged in a 2D honeycomb lattice (Figure 8.2), has attracted enormous interest around the world because of several outstanding properties such as excellent thermal/electrical conductivity, ultrahigh electron mobility (>15,000 cm^{2}V^{-1}s^{-1}) at room temperature, superior stability to harsh chemicals, high surface area (2,600 m^{2}g^{-1}), large surface to volume ratio, significant flexibility, high Young's modulus, and environmental friendly. Graphene is the basic building block of all graphitic materials that can be wrapped, rolled, stacked to 0D fullerenes, 1D nanotubes, and 3D graphite. These interesting properties of graphene lead to several industrial applications in nanoelectronics, energy storage materials (LIBs, supercapacitor), and hydrogen storage. Although graphene-based electronics have shown some serious expectations for electronics industry, the real application will be far away at least for one decade. Therefore, the major research on graphene is more focused on the bulk graphene powders for possible applications as energy storage materials in LIBs [11] and supercapacitors,

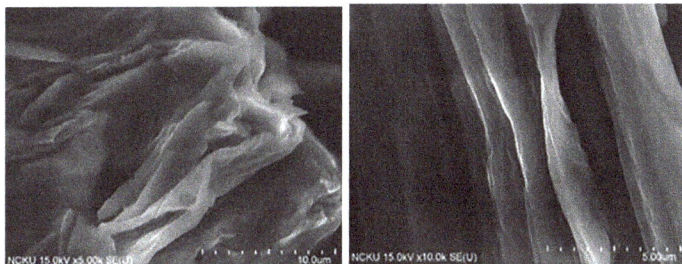

FIGURE 8.2   SEM images of graphene.

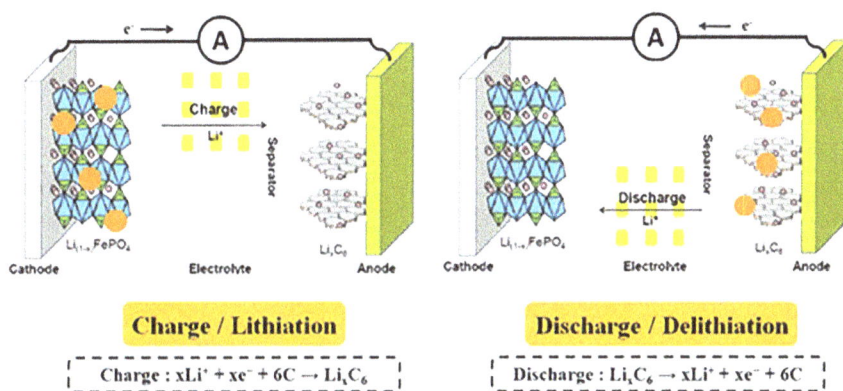

Charge / Lithiation	Discharge / Delithiation
Charge : $xLi^+ + xe^- + 6C \rightarrow Li_xC_6$	Discharge : $Li_xC_6 \rightarrow xLi^+ + xe^- + 6C$

FIGURE 8.3   Mechanism of LIBs.

catalysis [6,12,13], hydrogen storage, and so on. It is more advantageous for LIBs as the lithium ions have multiple active sites (Figure 8.3) on both sides and edges of the highly conducting graphene sheets having a large surface-to-volume ratio.

8.2   SYNTHESIS OF GRAPHENE

Graphene can be prepared by several methods such as exfoliation of graphite for hours, chemical modification, chemical vapor deposition, hydrothermal method.

Exfoliation is usually done by ultrasonication of graphite in a solution for a few hours to form individual atomic layers of graphene. However, the graphene concentration produced by this method is relatively low as the sheets again restack upon contact. Addition of a surfactant to the solution may prevent the restacking of the graphene sheets but the removal of the surfactant from the graphene is another issue to overcome.

Chemical modification involves time-consuming multiple process steps that include the oxidation of the graphite powder to graphite oxide by modified Hummer's method and then heat treatment at elevated temperature (~1,000°C) for exfoliation in air/nitrogen atmosphere. Some of these processes also use chemical reduction procedure either in hydrazine hydrate or in sodium borohydride before the heat treatment to control the concentration of oxygen-containing functional groups. Exfoliation by chemical modification has a strong potential for the large-scale production of graphene to meet the requirements for LIB applications. The above procedure can also be used to synthesize doped graphene nanosheets (GNSs) where the as-prepared graphene oxide is exfoliated at high temperature (1,050°C) in N/NH_3 atmosphere for N-doping and BCl_3 for boron doping.

Hydrothermal/solvothermal method usually involves the dispersion of graphene oxide or modified graphene oxide in a solvent or water that is then transferred to an autoclave and allowed to heat at a certain temperature (100–200°C) over a duration of time (10–20h). The brown-colored GO solution mixture then converts into black color indicating the conversion of graphene oxide to graphene.

8.3   BASIC CHARACTERIZATIONS OF GRAPHENE

8.3.1   Structure and Microstructure Analysis

Figure 8.4 illustrates the X-ray diffraction pattern of MLGN that shows a strong diffraction peak at 26.53° ($d=0.33$ nm) corresponding to the (002) peak of graphene that matches well with graphite-2H (JCPDS card no. 41–1,487). Graphene oxide (GO) shows a strong 001 peak at 10.98° ($d=0.80$ nm) that indicates a high level of oxidation of graphite. The d-spacing of GO is much larger than the MLGN because of the presence of a large concentration of oxygen-containing functional groups attached on both sides of GO sheets and the atomic-scale roughness due to the defects (sp^3 bonding) generated on the originally atomically flat graphene sheet [14]. The corresponding FESEM (field emission scanning electron microscopy) images of GO and MLGN are shown in Figure 8.5. GO shows layer-like morphology (>10 μm), and flake-like morphology is observed for MLGN and its lateral size is 2-5 μm for a single piece with laminar structure. The FETEM (field emission transmission electron microscopy) images reveal that the GO sheets are quite large in dimension (length ~5–7 nm, width ~3–5 nm) with fairly flat edges that suggest that the number of layers of GO aggregates.

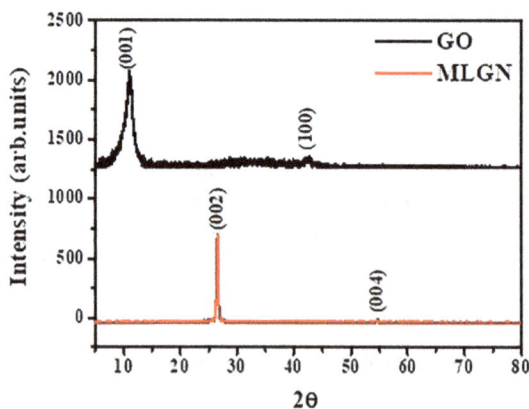

FIGURE 8.4   X-ray diffraction pattern of GO and MLGN.

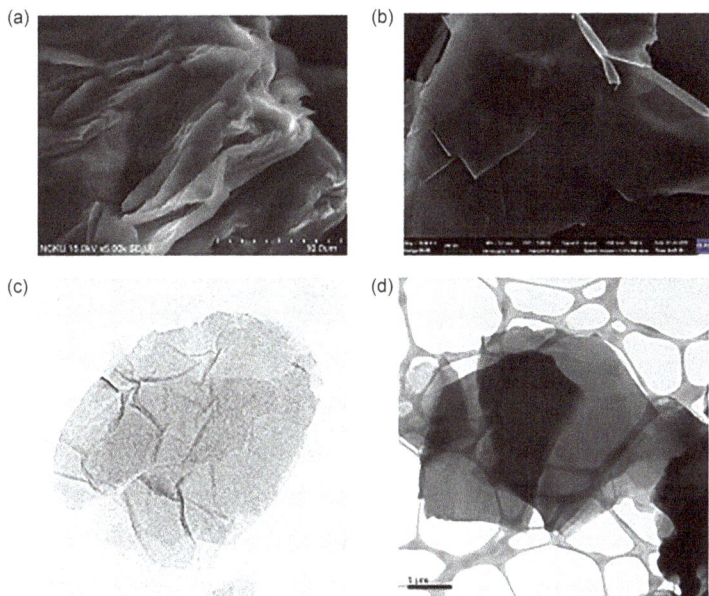

FIGURE 8.5   FESEM and FETEM images of (a, b) GO and (c, d) MLGN.

8.3.2   BONDING/BINDING ENERGY/FUNCTIONAL GROUPS AND PHONON MODES

Figure 8.6 shows the C1s XPS spectrum for MLGN (multilayer graphene) and GO (graphene oxide) where the C1s spectrum of MLGN shows peaks at 284.74, 285.84, 287.64, and 289.74 eV corresponding to C–C, C–O, C=O, and (C(O)O) functional groups, respectively. Similarly, the C1s spectrum of GO shows characteristic peaks at 284.8, 286.9, and 288.7 eV that are attributed to C–C, C=C, C–H, C–OH, and C=O bonds, respectively (Figure 5b) [15].

Figure 8.7 describes the Fourier transform infrared spectra (FTIR) of MLGN and GO. GO shows several absorption peaks at 3,400, 1,732, 1,620, 1,375, 1,252, and 1,058 cm^{-1} which are attributed to OH hydroxyl and carboxyl group [16], COOH [17], C=O [13,14], unoxidized graphitic domain [13], C–OH [18], C–O–C [13], and C–O stretching vibrations [14,19]. This indicates that the graphene oxide is associated with a large concentration of oxygen-containing functional group. Similarly, FTIR analysis of MLGN also shows different functional groups such as OH (3,400 cm^{-1}), C=C (1,620 cm^{-1}), and C=O (1,732 cm^{-1}); stretching vibrations and the bands at 1,252 and 1,071 cm^{-1} are attributed to CO stretching vibrations of carboxylic groups [20,21].

The Raman analysis of GO and MLGN is shown in Figure 8.8 that shows three peaks in GO corresponding to D band (1,342 cm^{-1}) that is assigned to the defect peak within the carbon structure, G band (1594.8 cm^{-1}) that is usually attributed to the sp^2 bonded carbon atoms in the honeycomb lattice, and the 2D band (2,700 cm^{-1}) that is a characteristic signature of the graphene, respectively [22–24], and is usually used to determine the number of layers of graphene in the sample. MLGN also shows similar peaks at 1,333, 1,580, and 2,680 cm^{-1} corresponding to D band, G band, and 2D band, respectively (Figures 8.8 and 8.9).

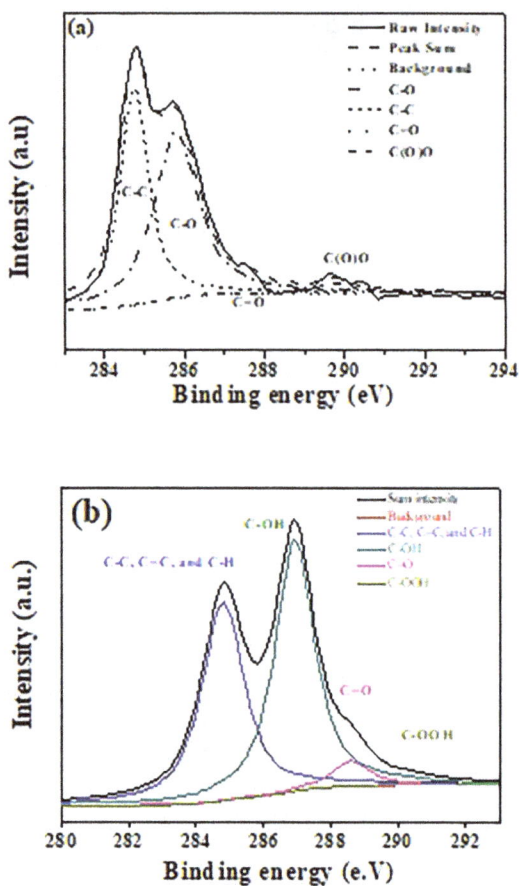

FIGURE 8.6 C1s XPS spectrum of (a) MLGN and (b) GO.

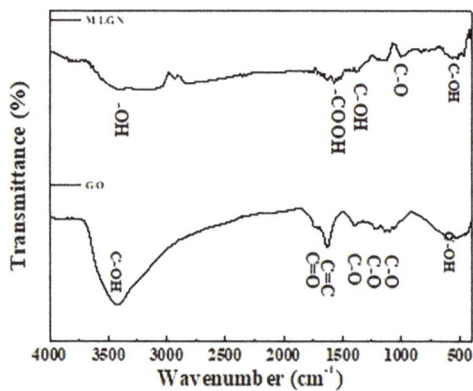

FIGURE 8.7 FTIR spectrum of GO and MLG.

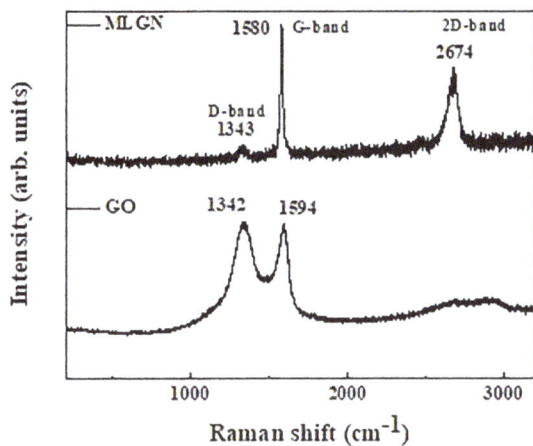

FIGURE 8.8   Raman spectrum of GO and MLGN.

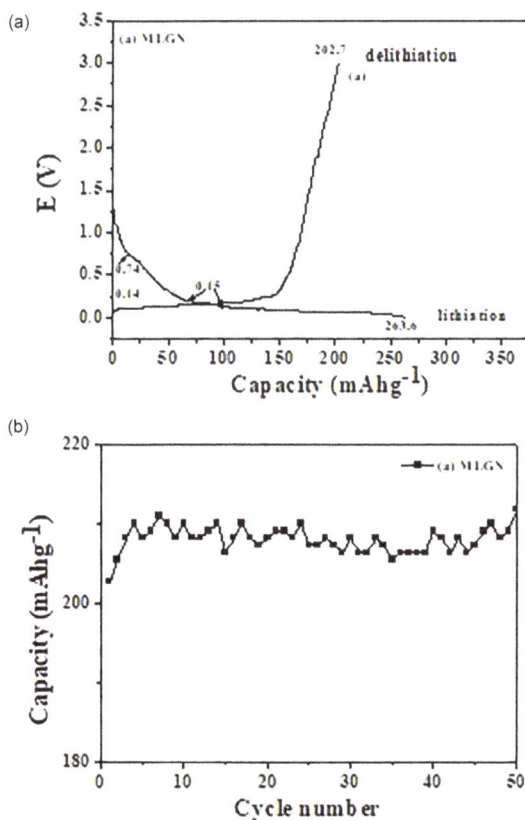

FIGURE 8.9   Lithiation and de-lithiation of: (a) MLGN, (b) discharge capacity versus cycle number of MLGN.

8.4   GRAPHENE AS ANODE IN LITHIUM-ION BATTERIES

8.4.1   GRAPHENE

Yoo et al. [25] have prepared GNSs by exfoliation of bulk graphite and used it as anode material in LIBs. GNS having wrinkled/crumpled paper-like structure is comprised of 10–20 layers of stacked graphene sheets and the thickness varies from 3 to 7 nm. Comparative investigation about the Li-storage properties has been done by incorporating carbon nanotube (CNT) and fullerenes (C_{60}) in GNS. The GNSs could achieve a reversible capacity of 540 mAh g^{-1} (current density = 0.5 A g^{-1}) that increases to 730 mAh g^{-1} and 784 mAh g^{-1} by the incorporation of CNT and C_{60}, respectively, and the capacity is much higher than the theoretical capacity of graphite. The capacity reaches 290 mAh g^{-1} (GNS), 480 mAh g^{-1} (GNS + CNT), and 600 mAh g^{-1} (GNS + C_{60}), respectively, after 20 cycles with the retention of 54%, 66%, and 77% for GNS, GNS + CNT, and GNS + C_{60}, respectively. The enhancement in the capacity is not only due to the formation of the LiC_6 compounds but also attributed to the increase in the d-spacing of the GNS ($d = 0.365$ nm) by 19% in comparison with GNS + CNT (0.40 nm) and GNS + C_{60} (0.40 nm). This increase in the d-spacing may create more active sites leading to a large change in the structural and electronic changes in GNS and the composites for the insertion of the lithium ions. However, this is the initial result of graphene as a negative electrode in LIBs and the reversible capacity/the Li-storage capability are significantly enhanced by the change in the synthesis of graphene, engineering of the oxygen-containing functional groups, surface/microstructure modification, doping, or by making composite with metal oxides.

The GNSs synthesized by the three-step procedure as described by Guo et al. [26] have shown relatively good lithium-storage capability as compared to the earlier procedure of exfoliation. The GNSs are prepared first by the oxidation of graphite by Hummer's method, rapidly heating at 1,050°C in the air to form expanded graphite and ultrasonication for 4 h. The first charge–discharge capacities of GNSs are increased to 1,233 and 672 mAh g^{-1}, respectively, at a current density of 0.2 mA cm^{-2}. The high irreversible capacity is attributed to the reaction of Li-ion with oxygen-containing functional groups and the formation of the solid electrolyte interface (SEI). The GNSs maintain the capacity of 554 mAh g^{-1} and have shown significant potential as anode materials in LIBs.

Slight variation in the synthesis procedure yield few-layer GNS having high surface area for better Li-ion adsorption/desorption and consequently the electrochemical performance. Graphite oxide exfoliation at high temperature in a different atmosphere seems to yield varying results. Lian et al. [27] have prepared graphite oxide by chemically modified procedure but exfoliated the graphite oxide at 1,050°C in nitrogen atmosphere. The first discharge and charge capacities of graphene sheets were as high as 2,035 and 1,264 mAh g^{-1} (current density = 100 mA g^{-1}), respectively, and the capacity is still maintained at 848 mAh g^{-1} after 40 cycles. This high electrochemical performance is attributed to the high-quality graphene sheets (~4 layers) with a curled/wrinkled morphology and a large specific surface area (492.5 m^2g^{-1}).

Wang et al. [28] have synthesized flower-like GNSs by chemical modification procedure that also follows a three-step procedure (oxidation of graphite/ultrasonication to disperse graphene oxide/reduction by hydrazine hydrate under reflux at 100°C). The first discharge (charge) capacity of GNSs anode is 945 mAh g^{-1} (650 mAh g^{-1}). The irreversible capacity could be associated with the formation of the SEI layer in the first cycle. The capacity is maintained at 460 mAh/g after 100 cycles and that shows good electrochemical performance. Similarly, Vargas et al. [29] have also prepared graphene by reducing graphite oxide by N_2H_4 and used it as anode with $LiNi_{0.5}Mn_{1.5}O_4$ (LNMO) as high voltage, spinel-structure cathode and achieved reasonable capacity after 50 cycles.

8.4.2 DOPED GRAPHENE

Doping of atoms such as nitrogen, boron, phosphorus seems to improve the electrochemical performance of graphene. The atomic diameter of nitrogen and carbon is almost same but differs in the electronegativity that is higher in nitrogen than carbon that enhances the interaction between a nitrogen-doped carbon material and lithium ions. Nitrogen doping enhances the electrical conductivity of graphene due to the formation of pyridine/bridgehead-like nitrogen. Wu et al. [30] have prepared nitrogen- and boron-doped graphene and used them as anode in LIBs. The doping was done at high temperature (600°C, 800°C) by using NH_3 (~99.0%) and Ar (1:2 v/v) and BCl_3 (~99.99%) and Ar (1:4 v/v) for N- and B-doping, respectively. The discharge capacity of N-doped graphene in the first cycle is 1,043 mAh g^{-1}, and after 30 cycles, it reaches 872 mAh g^{-1} (50 mA g^{-1}) that is much higher than the regular graphene. Similarly, for B-doped graphene, the discharge capacity is 1,227 mAh g^{-1} after 30 cycles (1,549 mAh g^{-1} in the first cycle). Consequently, the reversible capacity retention is 83.6% for the N-doped graphene and to 79.2% for the B-doped graphene after 30 cycles. The major advantage of these composites is the fast charging at high current rates such as 500 mA g^{-1}, 5 A g^{-1}, and 25 A g^{-1}. At a very high current rate of 25 A g^{-1}, the charge time is ~28 s/~33 s for the N-/B-doped graphene and the reversible capacity reaches 199/235 mAh g^{-1} for the N-/B-doped graphene and these results are far superior to those of the pristine graphene (~100 mAh g^{-1} at 25 A g^{-1}). The outstanding electrochemical performance of the doped graphene is due to the increase in the electrical conductivity/electrochemical activity, high thermal stability, increased disordered surface morphology, higher hydrophobicity, and better wettability toward organic electrolytes that can promote ion diffusion at the interface of electrode and electrolyte, ultrathin framework, high surface area, open porous structure, mechanical flexibility, and chemical stability.

N-doped GNS can also be obtained by exfoliating graphene oxide in N atmosphere at elevated temperature (~1,000°C) [31]. An interesting phenomenon is observed for this N-GNS, where the discharge capacity of 454 mAh g^{-1} is observed in the second cycle, continues to decrease till 17th cycle, and reverses thereafter. The capacity continues to increase to 684 mAh g^{-1} at 501 cycles showing superior electrochemical performance like that with GNS (269 mAh g^{-1} @ 100 cycles with 66% retention). The N-doping leads to the formation of pyridinic nitrogen, which in turn produces carbon vacancies in the graphene network that increases the Li^+ storage sites, resulting in the enhancement in the capacity of the N-GNS anode.

8.4.3 POROUS GRAPHENE

Recently, hierarchical porous materials have received a lot of attention for their applications as energy storage/conversion devises, photocatalysts, LIBs, and supercapacitors due to the high surface area, better interfacial transport, and reduced diffusion paths. Porous materials are comprised of porous structures with different pore sizes and depending on their dimensions, they are named as micropore (~1–2 nm) for a high concentration of ion adsorption, mesopore (2–100 nm) that facilitate the ion transfer and macropore (~400 nm) which acts as a reservoir to accommodate electrolyte [32,33]. As far as graphene is concerned, porous graphenes having pores either on the sheet or between the sheets are attributed with very high specific surface areas, excellent flexibility, electrical conductivity, specific capacity, outstanding physical and chemical stability, and significantly high concentration of defects/active sites that increase the electrochemical performance [34,35].

Fan et al. [36] have obtained porous graphene (pore size 3–8 nm) by chemical vapor deposition that presents a high reversible capacity (1,723 mAh g^{-1} at 0.1°C) and excellent rate capability (203 mAh g^{-1} at 20°C). Charge–discharge analysis of porous graphene shows very high specific capacity 4,858 mAh g^{-1} and reversible capacity of 1,827 mAh g^{-1} in the first cycle which is five times higher than those of graphite (372 mAh g^{-1}) and reduced-GO (453 mAh g^{-1}). The high capacity is due to the presence of curled edges and mesopores on the graphene sheets [37,38]. The cyclic stability investigation shows an increase of capacity from 726 to 926 mAh g^{-1} (1°C) after 100 cycles and 240-211 mAh g^{-1} (@ 20°C) after 100 cycles. The increase of capacity with the cycle is due to the activation of porous graphene during the cycling, and the significant cyclic stability is due to the effective control of the volume expansion by the curled edges of porous graphene sheets. The rate capability study also shows a great electrochemical performance of the porous graphene that exhibits reversible capacities of 955, 544, and 203 mAh g^{-1} at 0.5°C, 5°C and 20°C, respectively. This significant high rate capability and cyclic stability are partly attributed to the presence of the active sites at the edge, the mesopores in the sheets that reduce the diffusion length of lithium ions and the strongly crumpled/scrolled graphene sheet that acts as a buffer to accommodate the volume expansion during Li-insertion/Li-extraction.

Fang et al. [39] have synthesized mesoporous carbon sheets on a template by a typical monomicelle close-packing assembly approach followed by carbonization at high temperature and elimination of template by HCl. The mesoporous graphene layers have high surface area for lithium-ion adsorption and intercalation, which enables better ion transport for enhanced lithium-ion storage capacity and cyclic stability. The mesoporous graphene anode achieved excellent reversible capacity of 1,040 mAh g^{-1} (@ 100 mA g^{-1}), and the capacity is at 833 mAh g^{-1} even after 70–80 cycles. The rate capability is also good that reveals a capacity of 255 mAh g^{-1} at a large current density of 5 A g^{-1}. Therefore, the mesoporous graphene could achieve very good lithium-ion storage capacity as well as cyclic stability, and this excellent electrochemical performance is attributed to the 2D framework of the graphene sheets, large surface area, ordered mesoporous structures that are capable to withstand the high volume expansion and intercalation of the lithium ions on both sides of the graphene sheets.

Ren et al. [40] have designed a typical procedure for the production and engineer the pores in the graphene named as hierarchical porous graphene aerogel (HPGA) prepared by a hydrothermal self-assembly process and an *in situ* carbothermal reaction. HPGA having different pore sizes such as 50 nm (HPGA-50), 20 nm (HPGA-20) are synthesized that has achieved superior discharge capacity of 1100 mAh g^{-1} (HPGA-50) and 700 mAh g^{-1} (HPGA-20) at a current rate of 100 mA g^{-1} after 100 charge–discharge cycles. This high lithium storage is due to the typical structure of HPGA having pores where the lithium ions can be adsorbed and intercalated on both sides of the graphene sheets and it has also been observed that HPGA having larger mesopores has much higher reversible capacity and cyclic stability. The capacity is also measured at different charge–discharge rates that reveal ultrahigh performance (400 mAh g^{-1} @ 10 A g^{-1}, 300 mAh g^{-1} @ 20 A g^{-1}) that is again due to the presence of abundant and uniform mesopores that reduce the lithium-ion diffusion path.

The doping of porous graphene with N is an alternative way to enhance the electrochemical properties of graphene. For example, N-doped porous graphene sheets functionalized with polypyrrole by thermal activation has achieved excellent capacity (1336.3 mAh g^{-1} after 160 cycles @ 100 mA g^{-1}) and rate capability (133.2 mAh g^{-1} @ current density of 40 A g^{-1}) [41]. Jiang et al. [42] has reported the synthesis of nitrogen-doped graphene hollow microspheres by a template method that shows very good lithium storage capacity of 753.7 mAh g^{-1} (after 50 cycles, current density = 100 mA g^{-1}). Hu et al. [43] have utilized thermal reduction of graphene oxide with ammonium hydroxide and synthesized nitrogen-doped porous graphene that achieved high discharge capacity (453 mAh g^{-1} @current density of 2 A g^{-1} after 550 cycles.

Sui et al. [44] have prepared N-doped porous graphene material (NPGM) by a novel method by mixing melamine–formaldehyde resin with graphene oxide that forms a brown-colored graphene oxide/melamine–formaldehyde (GOMF) hydrogel and black-colored graphene/melamine–formaldehyde (GMF) after heating at 140°C for 12 h in an autoclave. The porous materials could be obtained by thermal treatment of GMF aerogel at 800°C for 1 h at a ramp rate of 10°C min^{-1}. Comparative investigation has been done on LIB performance between NPGM and PGM electrodes with initial discharge capacities of 1,279 and 793 mAh g^{-1}, which decrease to 742 and 621 mAh g^{-1} in the second cycle. The rate capability study shows a high discharge capacity of NPGM electrode (672 mAh g^{-1} @ 100 mA g^{-1}) at the 10th cycle, which is much higher than that of PGM electrode (450 mAh g^{-1}). At higher current density (1.5 A g^{-1}), NPGM also shows a high discharge capacity of 317 mAh g^{-1} after 70th cycle. Cyclic stability study of NPGM reveals reasonably high capacity after 200 cycles (496 mAh g^{-1} @ 400 mA g^{-1}) which is much higher than graphite as well as other potential materials. The enhanced LIB performance is due to the doping of nitrogen, large surface area, and porous structure of NPGM.

N-doped porous graphene sheets having high surface area, crumpled structure has been synthesized by Wang et al. [45] by chemical activation process by using polypyrrole to functionalize the graphene sheets with KOH. These doped graphene sheets have achieved specific capacity of 1516.2 mAh g^{-1} @100 mA g^{-1} (~4 times than commercial graphite), cyclic stability (1336.3 mAh g^{-1} after 160 cycles @100 mA g^{-1}, 1200.4 mAh g^{-1} @200 mA g^{-1} even after 300 cycles), and high rate

capability (133.2 mAh g^{-1} @ 40 A g^{-1}). This indicates that the conventional intercalation mechanism may not be the only reason for this high reversible capacity, and the
large irreversible capacity is usually due to the formation of the SEI layer, decomposition of the electrolyte by the surface defects generated due to the N-doping or due
to the high surface area of the composite or by the presence of N of the pyridinic
nitrogen atoms with a lone pair of electrons that may act as the active site.

8.4.4 CHEMICALLY MODIFIED GRAPHENE FOR FAST-CHARGING LITHIUM-ION BATTERY (LIB)

Single-layered graphene (SLG), few-layered graphene (FLG), or reduced graphene
oxide (rGO), possessing large surface-to-mass ratio, high electrical conductivity,
and high mechanical strength, are promising anode materials for high-capacity,
fast-charging LIB. It is theoretically considered that SLG can store Li-ions at both
sides and then provide a specific capacity (~740 mAh g^{-1}) two times that of graphite.
Since 2008, Itaru et al. have introduced macromolecules, like carbon nanotube or
fullerenes, into GNS matrix for improving Li-ion intercalation structure, thus significantly enhancing a specific capacity of GNSs [25]. It was not only found that
the charging capacity of the assembled battery monotonically increased with the
expanding d-spacing perpendicular to the basal plane of graphene, suggesting a possible electrochemically intercalating mechanism, yielding a high specific capacity of
~780 mAh g^{-1}, but also opened a new avenue for investigation of defect structure on
Li-ion adsorbing/diffusion phenomena in defective grapheme [25]. The basal plane
of pristine SLG is more difficult to adsorb Li-ion than graphite with van der Waals
interaction. Li-ion usually adsorbs on the edge of SLG and forms intercalated compounds of LiC$_6$ or Li$_{0.3}$C$_6$. Up to now, the best specific capacity of ~1,500 mAh g^{-1}
was reached from a graphene-based LIB that was mainly composed of graphene
nanoflake with a lateral dimension of <100 nm [46]. Itaru et al. tried to synthesize
zigzag-edge-rich graphene for possible LIB material [47]. It could be noticed that a
high-energy-density LIB using folded graphene films has shown high packing density up to 0.64 g cm^{-3}, high C-rate up to 50 A g^{-1}, and long cyclic stability up to
50,000 cycles [48]. There remain other graphene-based batteries made from FLG or
rGO and show moderate specific capacity. Therefore, these progresses suggest that it
is important to take an advantage of the edge of SLG for lithium uptake, thus significantly increasing gravimetric capacity two to four times than that of commercially
available graphite.

Another processing strategy to enhance Li-ion coverage is to activate the surface
of SLG with defects. It is reported that Stone–Wale defect, double vacancy, or higher
order defects on graphene have a tendency to capture Li-ion and allow Li-diffusion
through the basal plane [49]. Yi Fang et al. [39] and Long Ren et al. [40] separately
synthesized mesoporous SLG nanosheets and assembled half-cell exhibiting excellent reversible capacities of 1,040 and 1,100 mAh g^{-1} at a charging rate of 0.1 A g^{-1},
respectively. The former has a larger pore size of ~100 nm and the latter has a denser
distribution of pore size around 50 nm. Dohyeon Yoon et al. controlled the exfoliation
and oxidation degree of graphene oxide then produced hydrogen-enriched rGO via
supercritical fluid treatment [50]. Their assembled battery also exhibited an excellent

TABLE 8.1

Microstructures of SLG or rGO and Their Corresponding Specific Capacity at Different Characteristic Charging Conditions

Microstructure	Initial Discharge Capacity/Coulombic Efficiency (mAh g⁻¹, %)	Rate Capacity at 2 A g⁻¹	Capacity at High C-Rate (mAh g⁻¹)	References
Mesoporous graphene (pore size ~100 nm)	3,535/29%	300	255 (5 A g⁻¹)	[6]
Mesoporous graphene (pore size ~50 nm)	2,900/48%	600	300 (20 A g⁻¹)	[7]
Hydrogen-enriched	1,521/87%	450	146 (50 A g⁻¹)	[8]
HBC-OMe/3D assembly	1,260/77%	570	—	[9]

reversible capacity of 1,331 mAh g^{-1} at a charging rate of 0.05 A g^{-1}. Yen et al. developed self-assembled GNSs with controlled functionalized groups that prevented the restacking of graphene meanwhile wisely optimized geometric and electronic structure [51]. Moreover, it is necessary to balance the rate performance and cycling stability when considering the durability of graphene-based materials. Table 8.1 collects the above-mentioned remarkable researches relating to LIB performance using SLG or rGO as the anode. These chemically modified graphenes seemed to alter electronic transport and redox reaction thus increased their fast-charging performance. Hence, it can be inferred that these pores or defective structures led to charge redistribution and enhanced chemical bonding between host carbon atom and Li-ion, leading to higher Li-ion diffusion kinetics in graphene-based batteries over that in traditional graphite material.

Since Reddy et al. successfully deposited N-doped graphene on Cu-substrate and demonstrated its superior electrochemical properties [52], heteroatom-doped graphene is becoming important and is possible to strike a balance between LIB performance and industrial production. Table 8.2 summarizes the recent progresses of electrochemical performance of LIB made from heteroatom-doped graphene. The major doping element is nitrogen that could contribute to enhance electric conductivity and increase Li-ion uptake of doped-graphene. The state-of-the-art fast-charging battery is a high-quality nitrogen-doped mesoporous graphene with an excellent rate capability of 440 mAh g^{-1} at a charging rate of ~22 A g^{-1} [53]. The other dopants, including boron, sulfur, and phosphorus, appear to modify surface functionality and related physicochemical properties that can improve the electrochemical performance of graphene.

On the contrary, it seems that heteroatom-doped graphene can be a candidate material for developing fast-charging battery, as shown in Table 8.2. Their rate capacity is generally twice ~ five times that of graphite. With regard to the intercalation mechanism of these heteroatom-doped graphenes, they are inevitably suffered from the formation of secondary electrolyte interphase (SEI) and show a low Coulombic efficiency in the first cycle. Once a stable SEI film is formed, almost all the graphene-based batteries exhibit high cyclic stability. Although the electrochemical properties

TABLE 8.2

Composition and Structure of Heteroatom-Doped Graphenes and Their Corresponding Specific Capacity at Different Characteristic Charging Conditions

Dopants/ Structure of Grapheme	Initial Discharge Capacity/Coulombic Efficiency (mAh g⁻¹, %)	Rate Capacity (mAh g⁻¹) at 2 A g⁻¹	1) Capacity (mAh g⁻¹) at High C-Rate or 2) Cyclic Stability	References
N/NSt	2,128/49%	350	1) 199 (25 A g⁻¹)	[12]
B/NSt	2,750/55%	470	1) 235 (25 A g⁻¹)	
N/NSt	1,310/37%	–	2) 684 (500 cycle)	[13]
N/NSt	900	250	–	[14]
P/NSt	910/45%	200	–	[15]
N, P/NSt	7,507/30%	–	1) 450 (5 A g⁻¹)	[16]
N, S/NSt	1,636/44%	300~850	2) 900 (750 cycles)	[17]
S/NSt	1,700/51%	400	1) 285 (11 A g⁻¹) 2) 300 (500 cycles)	[18]
N/mesoporous NSt	1,444/49%	210	2) 1,078 (350 cycles)	[19]
N/NSt	895/60%	210	2) 550 (100 cycles)	[20]
N, B/3D framework	1,409/73%	400	2) 307 (4 A g⁻¹, 200 cycles)	[21]
N/mesoporous scaffold	945/76%	730	1) 440 (22 A g⁻¹) 2) 774 (500 cycles)	[11]

NSt, Nanosheet; N, nitrogen; B, boron; P, phosphorus; S, sulfur

of graphene-based batteries have been widely investigated, it is still a lack of chemical information within the cell. In order to further improve anode performance, the newly developed *in operando* analytical techniques are possible to be utilized to reveal possible lithium intercalation behavior. For instance, *in situ* Raman spectroscopy is expected to detect the profile change of disorder-induced line (at about $1,360\,cm^{-1}$) or E_{2g2} band (at $1,587\,cm^{-1}$), enabling the one to examine electrochemical phenomena within an optical skin depth (i.e., about 100 nm) [54]. *In situ* gas chromatogram and mass spectroscopy can determine gas evolution kinetics at the effluence that comes from electrolyte decomposition upon SEI formation. The chemical information can rapidly feedback to adjust the material recipe for optimizing battery performance.

8.4.5 Discussions

Although graphene can be considered as one of the very suitable materials as anode in LIBs, the problem still lies with the primary synthesis procedure and the post-annealing processes at elevated temperature (~1,000°C). The low-temperature synthesis including the chemical reduction procedure with hazardous reductants (sodium borohydride or hydrazine hydrate) is not environmental friendly and a green synthesis of graphene is a specific requirement. Exfoliation of graphene seems to be one of the choices with mass production and restacking of the individual sheets that lead to detrimental effect on the electrochemical properties of graphene. The doping with elements like B/N seems to improve the specific capacity, but the doping process needs high-temperature processing (~1,000°C) in NH_3 or BCl_3 vapor and the electrochemical performance such as cyclic stability and rate capability should be improved for industrial application. Modification of the microstructure of graphene with the introduction of porosity has been reported as one of the alternatives with similar underlying problems as described above.

8.5 CONCLUSIONS

The details about the synthesis of graphene, rGO, porous graphene, and doped GNSs are investigated. Although graphene shows significant performance as anode material in LIBs as compared to graphite, overall improvements that include the modification of synthesis procedure that avoids using hazardous materials, discovery of a perfect green synthesis of graphene at room temperature, and avoids high-temperature processing should be the future outlook.

ACKNOWLEDGEMENT

This work was financially supported by the Hierarchical Green-Energy Materials (Hi-GEM) Research Center, from The Featured Areas Research Center Program within the framework of the Higher Education Sprout Project by the Ministry of Education (MOE) and the Ministry of Science and Technology (MOST 109-2634-F-006-020) in Taiwan.

REFERENCES

1. Wu, S.; Xu, R.; Lu, M.; Ge, R.; Iocozzia, J.; Han, C.; Jiang, B.; Lin, Z. Graphene-containing nanomaterials for lithium-ion batteries, *Advanced Energy Materials*, **2015**, 1500400, 1–40.
2. Goriparti, S.; Miele, E.; Angelis, F. D.; Fabrizio, E. D.; Zaccaria, R. P.; Capiglia, C. Review on recent progress of nanostructured anode materials for Li-ion batteries, *Journal of Power Sources*, **2014**, 257, 421–443.
3. Etacheri, V.; Marom, R.; Elazari, R.; Salitra, G.; Aurbach, D. Challenges in the development of advanced Li-ion batteries: A review, *Energy and Environmental Science*, **2011**, 4, 3243.
4. Deng, D. Li-ion batteries: Basics, progress, and challenges, *Energy Science and Engineering*, **2015**, 3(5), 385–418.
5. Zhu, Y.; Ji, H.; Cheng, H.-M.; Ruoff, R. S. Mass production and industrial applications of graphene materials, *National Science Review*, **2018**, 5, 90–101.
6. Li, X.; Zhi, L. Graphene hybridization for energy storage applications, *Chemical Society Reviews*, **2018**, 47, 3189–3216.
7. Xin, H.; Li, W. A review on high throughput roll-to-roll manufacturing of chemical vapor deposition graphene, *Applied Physics Reviews*, **2018**, 5, 031105.

8. Randviir, E. P.; Brownson, D. A. C.; Banks, C. E. A decades of graphene research, production, applications and outlook, *Materials Today*, **2014**, 17(9), 426–432.

9. Allen, M. J.; Tung, V. C.; Kaner, R. B. Honeycomb carbon: A review of graphene, *Chemical Reviews*, **2010**, 110, 132–145.

10. Novoselov, K. S.; Geim, A. K.; Morozov, S. V.; Jiang, D.; Zhang, Y.; Dubonos, S. V.; Grigorieva, I. V.; Firsov, A. A. Electric field effect in atomically thin carbon films, *Science* **2004**, 306, 666–669.

11. Luo, R.-P.; Lyu, W.-Q.; Wen, K.-C.; He, W.-D. Overview of graphene as anode in lithium-ion batteries, *Journal of Electrochemical Science and Technology*, **2018**, 16(1), 57–68.

12. Qiu, B.; Xing, M.; Zhang, J. Recent advances in three-dimensional graphene based materials for catalysis applications, *Chemical Society Reviews*, **2018**, 47, 2165.

13. Zhu, Y.; Ji, H.; Cheng, H.-M.; Ruoff, R. S. Mass production and industrial applications of graphene materials, *National Science Review*, **2018**, 5, 90–101.

14. Shen, J.; Hu, Y.; Shi, M.; Lu, X.; Qin, C.; Li, C.; Ye, M. Fast and facile preparation of graphene oxide and reduced graphene oxide nanoplatelets, *Chemistry of Materials*, **2009**, 21, 3514–3520.

15. Zhu, Y. G.; Wang, Y.; Xie, J.; Cao, G.-S.; Zhu, T.-J.; Zhao, X.; Yang, H. Y. Effects of graphene oxide function groups on SnO_2-graphene nanocomposite for lithium storage application, *Electrochimica Acta*, **2015**, 154, 338–344.

16. Liu, X.; Zhong, X.; Yang, Z.; Pan, F.; Gu, L.; Yu, Y. Gram-scale synthesis of graphene-mesoporous SnO_2 composite as anode for lithium-ion batteries, *Electrochimica Acta*, **2015**, 152, 178–186.

17. Wu, G.; Wu, M.; Wang, D.; Yin, L.; Ye, J.; Deng, S.; Zhu, Z.; Ye, W.; Li, Z. A facile method for in-situ synthesis of SnO_2-graphene as a high performance anode material for lithium-ion batteries, *Applied Surface Science*, **2014**, 315, 400–406.

18. Wu, M.; Liu, J.; Tan, M.; Li, Z.; Wu, W.; Li, Y.; Wang, H.; Zheng, J.; Qiu, J. Facile hydrothermal synthesis of SnO_2/C microspheres and double layered core–shell SnO_2 microspheres as anode materials for Li-ion secondary batteries, *RSC Advances*, **2014**, 4, 25189–25194.

19. Park, S.-K.; Yu, S.-H.; Pinna, N.; Woo, S.; Jang, B.; Chung, Y.-H.; Cho, Y.-H.; Sung, Y.-E.; Piao, Y. A facile hydrazine-assisted hydrothermal method for the deposition of monodisperse SnO_2 nanoparticles onto graphene for lithium ion batteries, *Journal of Materials Chemistry*, **2012**, 22, 2520–2525.

20. Scipioni, R.; Gazzoli, D.; Teocoli, F.; Palumbo, O.; Paolone, A.; Ibris, N. Preparation and characterization of nanocomposite polymer membranes containing functionalized SnO_2 additives, *Membranes (Basel)*, **2014**, 4, 123–142.

21. Xiong, D.; Li, X.; Shan, H.; Zhao, Y.; Dong, L.; Xu, H. Oxygen-containing functional groups enhancing electrochemical performance of porous reduced graphene oxide cathode in lithium ion batteries, *Electrochimica Acta*, **2015**, 174, 762–769.

22. Wang, D.; Li, X.; Wang, J.; Yang, J.; Geng, D.; Li, R.; Cai, M.; Sham, T.-K.; Sun, X. Defect-rich crystalline SnO_2 immobilized on graphene nanosheets with enhanced cycle performance for Li ion batteries, *The Journal of Physical Chemistry C*, **2012**, 116, 22149–22156.

23. Liang, J.; Wei, W.; Zhong, D.; Yang, Q.; Li, L.; Guo, L. One-step in situ synthesis of SnO_2/graphene nanocomposite and its application as an anode material for Li-ion batteries, *ACS Applied Materials and Interfaces*, **2012**, 4, 454–459.

24. Li, Y.; Lv, X.; Lu, J.; Li, J. Preparation of SnO_2 nanocrystal/graphene nanosheets composites and their lithium storage ability, *The Journal of Physical Chemistry C*, **2010**, 114, 21770–21774.

25. Yoo, E. J.; Kim, J.; Hosono, E.; Zhou, H.-S.; Kudo, T.; Honma, I. Large reversible Li storage of graphene nanosheet families for use in rechargeable lithium ion batteries, *Nano Letter*, **2008**, 8(8), 2277–2282.

26. Guo, P.; Song, H.; Chen, X. Electrochemical performance of graphene nanosheets as anode material for lithium-ion batteries, *Electrochemistry Communications*, **2009**, 11, 1320–1324.

27. Lian, P.; Zhu, X.; Liang, S.; Li, Z.; Yang, W.; Wang, H. Large reversible capacity of high quality graphene sheets as an anode material for lithium-ion batteries, *Electrochimica Acta,* **2010**, 5, 3909–3914.

28. Wang, G.; Shen, X.; Yao, J.; Park, J. Graphene nanosheets for enhanced lithium storage in lithium ion batteries, *Carbon,* **2009**, 47, 2049–2053.

29. Vargas, O.; Caballero, A.; Morales, J.; Elia, G. A.; Scrosatiw, B.; Hassoun, J. Electrochemical performance of a graphene nanosheets anode in a high voltage lithium-ion cell, *Physical Chemistry Chemical Physics*, **2013**, 15, 20444.

30. Wu, Z.-S.; Ren, W.; Xu, L.; Li, F.; Cheng, H.-M. Doped graphene sheets as anode materials with super high rate and large capacity for lithium ion batteries, *ACS Nano*, **2011**, 5(7), 5463–5471.

31. Li, X.; Geng, D.; Zhang, Y.; Meng, X.; Li, R.; Sun, X. Superior cycle stability of nitrogen-doped graphene nanosheets as anodes for lithium ion batteries, *Electrochemical Communications*, **2011**, 13, 822–825.

32. Tao, Y.; Kong, D.; Zhang, C.; Lv, W.; Wang, M.; Li, B.; Huang, Z. H.; Kang F.; Yang, Q. H. Monolithic carbons with spheroidal and hierarchical pores produced by the linkage of functionalized graphene sheets, *Carbon*, **2014**, 69, 169–177.

33. Shu, K.; Wang, C.; Li, S.; Zhao, C.; Yang, Y.; Liu H.; Wallace, G. Flexible free-standing graphene paper with interconnected porous structure for energy storage, *Journal of Materials Chemistry A*, **2015**, 3(8), 4428–4434.

34. Han, S.; Wu, D.; Li, S.; Zhang, F.; Feng, X. Porous graphene materials for advanced electrochemical energy storage and conversion devices, *Advanced Materials*, **2014**, 26(6), 849–864.

35. Russo, P.; Hu, A.; Compagnini, G. Synthesis, properties and potential applications of porous grapheme: A review. *Nano Micro Letters*, **2013**, 5(4), 260–273.

36. Fan, Z.; Yan, J.; Ning, G.; Wei, T.; Zhi, L.; Wei, F. Porous graphene networks as high performance anode materials for lithium ion batteries, *Carbon*, **2013**, 60, 538–561.

37. Pan, D. Y.; Wang, S.; Zhao, B.; Wu, M. H.; Zhang H. J., Wang Y, et al. Li storage properties of disordered graphene nanosheets. *Chemistry Materials,* **2009**, 21(14), 3136–3142.

38. Uthaisar, C.; Barone, V. Edge effects on the characteristics of Li diffusion in graphene, *Nano Letter*, **2010**, 10(8), 2838–2842.

39. Fang, Y.; Lv, Y.; Che, R.; Wu, H.; Zhang, X.; Gu, D.; Zheng, G.; Zhao, D. Two-dimensional mesoporous carbon nanosheets and their derived graphene nanosheets: Synthesis and efficient lithium ion storage, *Journal of American Chemical Society*, **2013**, 135, 1524–1530.

40. Ren, L.; Hui, K. N.; Hui, K. S.; Liu, Y.; Qi, X.; Zhong, J.; Du, Y.; Yang, J. 3D hierarchical porous graphene aerogel with tunable meso-pores on graphene nanosheets for high performance energy storage, *Scientific Reports*, **2015**, 5, 14229.

41. Wang, H.; Wang, Y.; Li, Y.; Wan Y.; Duan, Q. Exceptional electrochemical performance of nitrogen-doped porous carbon for lithium storage, *Carbon*, **2015**, 82, 116–123.

42. Jiang, Z.; Jiang, Z. J.; Tian X.; Luo, L. Nitrogen-doped graphene hollow microspheres as an efficient electrode material for lithium ion batteries, *Electrochimica Acta*, **2014**, 146, 455–463.

43. Hu, T.; Sun, X.; Sun, H.; Xin, G.; Shao, D.; Liu C.; Lian, J. Rapid synthesis of nitrogen-doped graphene for a lithium ion battery anode with excellent rate performance and super-long cyclic stability, *Physical Chemistry Chemical Physics*, **2014**, 16(3), 1060–1066.

44. Sui, Z.-Y.; Wang, C.; Yang, Q.-S.; Shu, K.; Liu, Y.-W.; Han, B.-H.; Wallace, G. G. A highly nitrogen-doped porous graphene: An anode material for lithium ion batteries, *Journal of Materials Chemistry A*, **2015**, 3, 18229–18237.

45. Wang, H.-G.; Wang, Y.; Li, Y.; Wan, Y.; Duan, Q. Exceptional electrochemical performance of nitrogen-doped porous carbon for lithium storage, *Carbon*, **2015**, 82, 116–123.

46. Hassoun, J.; Bonaccorso, F.; Agostini, M.; Angelucci, M.; Betti, M. G.; Cingolani, R.; Gemmi, M.; Mariani, C.; Panero, S.; Pellegrini, V.; Scrosati, B. An advanced lithium-ion battery based on a graphene anode and a lithium iron phosphate cathode, *Nano Letter,* **2014**, 14(8), 4901–4906.

47. Tamura, N.; Tomai, T.; Oka, N.; Honma, I. Capacity improvement of the carbon-based electrochemical capacitor by zigzag-edge introduced graphene, *Applied Surface Science,* **2018**, 428, 986–989.

48. Liu, T. Y.; Kim, K. C.; Kavian, R.; Jang, S. S.; Lee, S. W. High-density lithium-ion energy storage utilizing the surface redox reactions in folded graphene films, *Chemistry of Materials,* **2015**, 27(9), 3291–3298.

49. Yildirim, H.; Kinaci, A.; Zhao, Z. J.; Chan, M. K. Y.; Greeley, J. P. First-principles analysis of defect-mediated Li adsorption on graphene, *ACS Applied Materials and Interfaces,* **2014**, 6 23, 21141–21150.

50. Yoon, D.; Chung, K. Y.; Chang, W.; Kim, S. M.; Lee, M. J.; Lee, Z.; Kim, J. Hydrogen-enriched reduced graphene oxide with enhanced electrochemical performance in lithium ion batteries, *Chemistry of Materials,* **2015**, 27(1), 266–275.

51. Yen, H.-J.; Tsai, H.; Zhou, M.; Holby, E. F.; Choudhury, S.; Chen, A.; Adamska, L.; Tretiak, S.; Sanchez, T.; Iyer, S.; Zhang, H.; Zhu, L.; Lin, H.; Dai, L.; Wu, G.; Wang, H.-L. Structurally defined 3D nanographene assemblies via bottom-up chemical synthesis for highly efficient lithium storage, *Advanced Materials,* **2016**, 28(46), 10250–10256.

52. Reddy, A. L. M.; Srivastava, A.; Gowda, S. R.; Gullapalli, H.; Dubey, M.; Ajayan, P. M. Synthesis of nitrogen-doped graphene films for lithium battery application, *ACS Nano,* **2010**, 4(11), 6337–6342.

53. Mo, R. W.; Li, F.; Tan, X. Y.; Xu, P. C.; Tao, R.; Shen, G. R.; Lu, X.; Liu, F.; Shen, L.; Xu, B.; Xiao, Q. F.; Wang, X.; Wang, C. M.; Li, J. L.; Wang, G.; Lu, Y. F. High-quality mesoporous graphene particles as high-energy and fast-charging anodes for lithium-ion batteries, *Nature Communications*, **2019,** 10, 10.

54. Inaba, M.; Yoshida, H.; Ogumi, Z.; Abe, T.; Mizutani, Y.; Asano, M. In-situ Raman-study on electrochemical Li-intercalation into graphite, *Journal of the Electrochemical Society*, **1995**, 142(1), 20–26.

9 Liquid Plasma

A Synthesis of Carbon/ Functionalized Nanocarbon for Battery, Solar Cell, and Capacitor Applications

Masahiro Yoshimura
National Cheng Kung University

Jaganathan Senthilnathan
Indian Institute of Technology Madras

Anupama Surenjan
National Institute of Technology Karnataka

CONTENTS

9.1   INTRODUCTION

As discussed in Chapter 1, nanocarbon is one of the most exciting materials investigated for their potential applications in various fields, including composite materials, hydrogen storage, electronic nanodevices, micro-electromechanical systems, and templates for nanorods and nanowires [1–3]. The high-temperature and toxic chemicals used in microwave plasma deposition, solvothermal synthesis, segregation growth, thermal annealing, arc discharge, chemical vapor deposition, thermal exfoliation, gas plasma or glow discharge plasma deposition, etc., are few factors that make the synthesis of nanocarbons complicated [4]. Recently, the generation of plasma in liquid has emerged as the most attractive and economically viable technique for the formation of nanocarbon materials such as graphene or reduced graphene oxide (rGO), fullerene (C_{60}), and single- and multiwall carbon nanotubes (CNTs and MWCNTs) [5–6]. Different forms of liquid plasma, such as spark plasma, arc plasma, streamer plasma, and glow discharge plasma, have been used to generate plasma in aqueous and organic solvents [7–8]. In an aqueous solution, free electrons in the plasma split the water molecules and produce H, •OH, and •O radicals, whereas the low conductive nature of organic solvents allows the partial discharge of plasma when compared to an aqueous solution [9]. Nonthermal plasma or cold plasma, generated inside the nonpolar solvents, displays a high electron temperature when compared to ions and neutrals [9], whereas the electrons and ions in thermal plasma are in thermal equilibrium with each other at the same temperature [2]. In liquid plasma, a plasma discharge produces electrons, ions (positive ions), neutral, and free radicals, radiation (UV range), and shock waves, which alter the chemical composition of the liquid [10–12]. One of the main advantages of liquid plasma is that the plasma glow can be controlled by altering the repetition frequency and pulse width [13]. Various forms of nanocarbon synthesized by the liquid plasma process in aqueous and organic solvents are given in Table 9.1.

9.2   FORMATION OF VARIOUS FORMS OF NANOCARBON IN THE LIQUID PLASMA PROCESS

There are different forms of carbon nanomaterials such as graphene, rGO, MWCNTs, CNTs, nano-onions, nanohorns, nitrogen-functionalized graphene (NFG), fullerene, and diamond-like carbon, which have been successfully synthesized by the liquid plasma process [14–21]. The carbon nanosphere (CNS) was synthesized by generating plasma (power supply 25–65 kHz) in benzene (500 mg of CNS was produced from 100 mL of carbon precursor benzene) [22]. Figure 9.1 shows the HR-TEM images of the nanosphere synthesized with 25–65 kHz power supply [22].

Quality graphite was deposited on the electrode surface, and amorphous carbons were suspended in the solution from the plasma generated with the carbon precursor benzene/pyridine [23]. Several organic solvents and liquids such as oil (cooking, lubricating, and waste oil), hexafluoroethane, benzene, pyridine, acetonitrile, ethanol, propanol, trichloroethylene, and n-heptane have been used for the formation of different forms of nanocarbon [10,24–31]. Similarly, microwave plasma (2.5 GHz) and ultrasonic s(19 kHz) were used for the formation of graphitic and amorphous

TABLE 9.1

Synthesis of Different Forms of Nanocarbon by the Liquid Plasma Process

S. No.	Plasma	Source	Precursor Solution	Nanocarbon	References
1	Pulsed plasma	Graphite	Ethanol	Diamond-like carbon	[7]
2	Pulsed plasma	Graphite	Acetonitrile	Nitrogen polymer	[10]
3	Pulsed discharge	Graphite	Toluene	Fullerene (C_{60})	[20]
4	Pulsed plasma	Graphite	Ethanol	Nanocarbon	[21]
5	Pulsed plasma	Graphite sheets	Benzene	Carbon nanospheres	[22]
6	Pulsed plasma	Graphite	Acetonitrile	Nanoclay/graphene hybrids	[24]
7	Electric discharge	Carbon rod	Liquid N_2/H_2O	MWCNTs	[38]
8	Arc discharge	Carbon electrode	Water	CNTs	[40]
9	Arc discharge	Graphite	Toluene	CNTs	[41]
10	Arc discharge	Graphite	Water	Onion/CNTs	[42]
11	Arc discharge	Graphite	Water	MWCNTs	[43]
12	Arc/laser ablation	Graphite	Liquid N_2/water	Graphene/CNTs	[44]
13	Arc discharge	Graphite	Water	CNTs	[45]
14	Arc discharge	Graphite	Water	Carbon onion/ fullerene	[47]
15	Pulsed plasma	Graphene oxide	Acetonitrile	Au/graphene	[57]
16	Arc discharge	Graphite	Water	Carbon onion	[59]
17	Pulsed plasma	Graphene oxide	Acetonitrile	NFG	[60]
18	Arc discharge	Graphite	Water/liquid nitrogen	CNTs/ carbon onion	[61]
19	Pulsed discharge	Graphite	Ethanol/butanol	Graphene	[62]
20	Pulsed discharge	Graphite	Ethanol	Graphene	[63]
21	Arc discharge	Graphite	Ethanol/alkane/ aromatic	Graphene	[64]
22	Arc discharge	Graphite	Water	Graphene	[65]
23	Pulsed discharge	Carbon soots	Water	Co-CNTs	[66]
24	Arc discharge	Graphite	Water	MWCNTs	[67]

Note: NFG: nitrogen-functionalized graphene; CNTs: carbon nanotubes; MWCNTs: multiwall carbon nanotubes.

carbon from n-dodecane [25]. Morishita et al. demonstrated the formation of nanocarbons from cyclohexane, hexadecane, hexane, and benzene in the liquid plasma process [32], whereas the nanocarbon produced from aliphatic linear molecules is more crystalline (graphitic structure) than the one produced from aromatic benzene compound [32]. Further, in saturated organic compounds such as hexane and hexadecane, a C–H dissociation is the initial step, followed by the formation

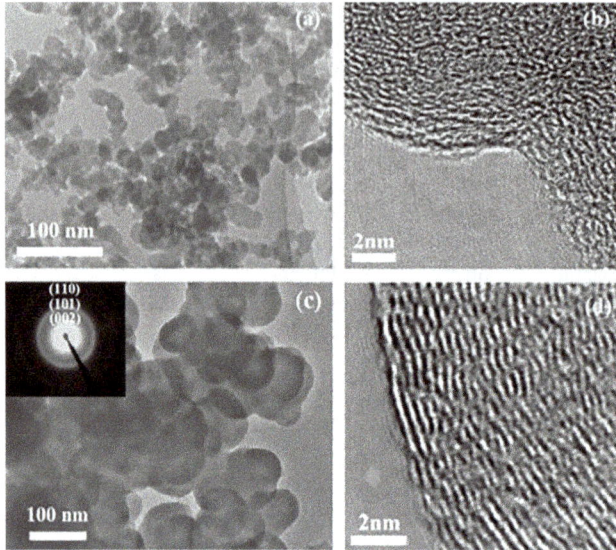

FIGURE 9.1   (a–c) TEM/HRTEM image of CNSs synthesized with 25 kHz (d) synthesized with 65 kHz. Reproduced from Kang et al. [22] with permission from Elsevier Ltd.

of unsaturated ring molecules in the subsequent steps [10,32]. However, unsaturated compounds such as naphthalene, acetonitrile, ethylene, perfluorocarbons, acrylonitrile, pyridine, ferrocene, 1,2-dicyanoethylene, and tetracyanoethylene undergo polymerization through radical reaction depending on the condition [33–36]. A diamond-like carbon was directly deposited on a silica wafer counter electrode (working electrode tungsten) using ethanol as a carbon precursor and by applying a high potential across the electrodes [21,37]. An arc discharge plasma was used to produce CNTs, MWCNTs, and nano-onions by using graphite electrode in water and toluene [38–45]. Similarly, fullerene (C_{60}) was produced by the discharge of plasma in liquid toluene using graphite electrodes [46]. A very high specific surface area of $984.3\,m^2g^{-1}$ displayed by carbon onion produced by arc discharge in aqueous solution using a graphite electrode [47]. Similarly, Au-supported CNT was synthesized by a one-pot gas–liquid interfacial plasma method to convert hydrosilane to silanols in aqueous solution [48]. The Au-supported nanotubes were synthesized by the liquid plasma process for the selective oxidation of silanes in water, and the Au catalyst was reused without any loss in the catalytic activity [48].

9.2.1   FORMATION OF UNCONVENTIONAL POLYMERS IN THE LIQUID PLASMA PROCESS

Polymerization of naphthalene, styrene, benzonitrile, fluorocarbon, acrylonitrile, ethylene, 1,2-dicyanoethylene, and tetracyanoethylene with the gas plasma process has been extensively studied and well documented [33–36], whereas only a few studies have been done on polymerization of organic compounds in liquid plasma conditions. Yoshimura and his research groups have been successful in developing carbon and polymerized products from different organic solvents [10]. In liquid plasma, solvents such as methanol, ethanol, hexane, and propanol form an ionic species rather than radical species and polymerization is not favorable in such a condition [10,21,37,49,50]. Sentilnathan et al. reported that the liquid plasma reaction in the acetonitrile compound initiates H^+ detachment in the first step and forms a highly reactive free radical monomer, $\cdot CH_2C \equiv N$ [10,51]. Initial expulsion of H^+ ion rather than N from $-C \equiv N$ might be due to the low bond energy of C–H (413 kJ mol^{-1}) compared to that of $C \equiv N$ (891 kJ mol^{-1}) which favors the H^+ discharge and forms $\cdot CH_2 \equiv CN$ radical monomers (Figure 9.2) [10]. A polymer that is produced by the liquid plasma process is completely different from the commercially available polymers [52]. Polymers produced in the liquid plasma process contain an unconventional repetitive unit that has potential application in the areas of light-harvesting, solar cell, and light-emitting diode applications.

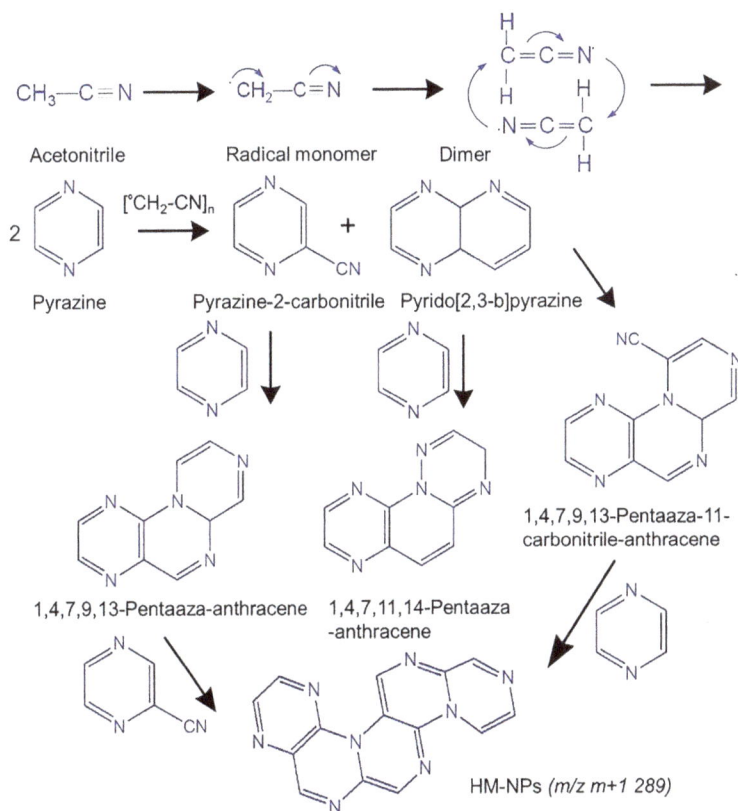

FIGURE 9.2 Proposed reaction mechanism for the formation of nitrogen polymer in liquid plasma process in acetonitrile solvent. Reproduced from Senthilnathan et al., Sci. Reps. 3, 2414 (2013), with permission from the Nature group.

9.2.2 DIRECT FUNCTIONALIZATION OF GRAPHENE IN THE LIQUID PLASMA PROCESS

Functionalized graphene has been considered to be one of the most emergent materials; however, as described in Chapter 8 of this book, it is not well established to date. Senthilnathan et al. demonstrated the nitrogen functionalization of graphene in the liquid plasma process at ambient conditions [53]. In this study, a high potential was applied across the electrodes (graphite and platinum electrodes) in acetonitrile and radicalized graphene and punctured graphene sheets were exfoliated from the graphite electrode. The radicalized acetylene monomer $\bullet CH_2C\equiv N$ present in the acetonitrile readily reacts with punctured graphene and forms nitrogen-functionalized graphene [10,53]. A schematic representation of micro-plasma discharge, graphene, and acetonitrile radical reaction, and the formation of nitrogen-functionalized graphene are given in Figure 9.3. Similarly, the liquid plasma process was used to prepare the nitrogen and boron co-doped graphene with acetonitrile containing sodium tetraphenylborate solution [54]. The HR-TEM images show that the amorphous-like nanocarbons were produced with 5 mM concentration of boron (Figure 9.4a–c), whereas a sponge-like graphene network was observed when the boron concentration was reduced from 5 to 0.1 mM (Figure 9.4d–f). Further, graphene nanosheets

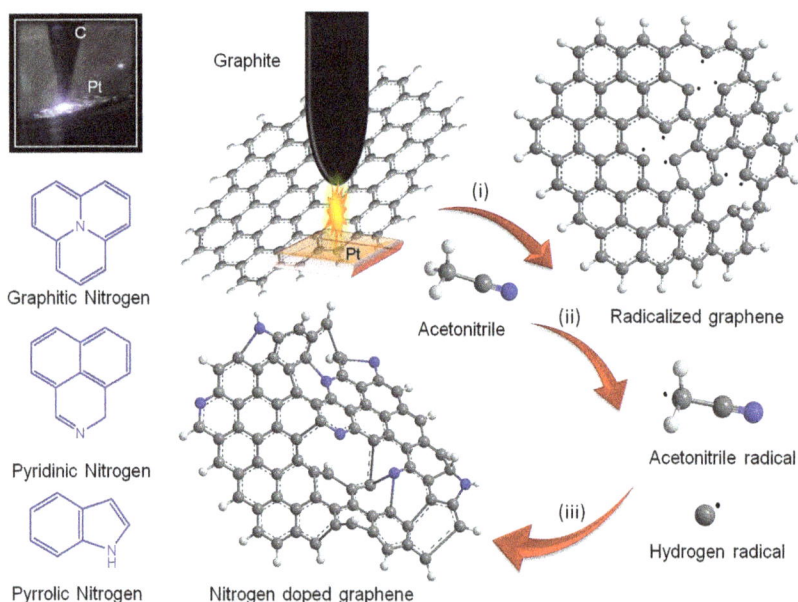

FIGURE 9.3 Proposed mechanism for the formation of nitrogen-functionalized graphene in submerged liquid plasma process (i) micro plasma discharge facilitates the exfoliation of radicalized graphene layer in the acetonitrile solution (ii) electron generated in micro plasma discharge collision with acetonitrile and forms nascent hydrogen and acetonitrile radicals (iii) formation of nitrogen functionalized graphene. Reproduced from Senthilnathan et al. [10], with permission from The Royal Society of Chemistry.

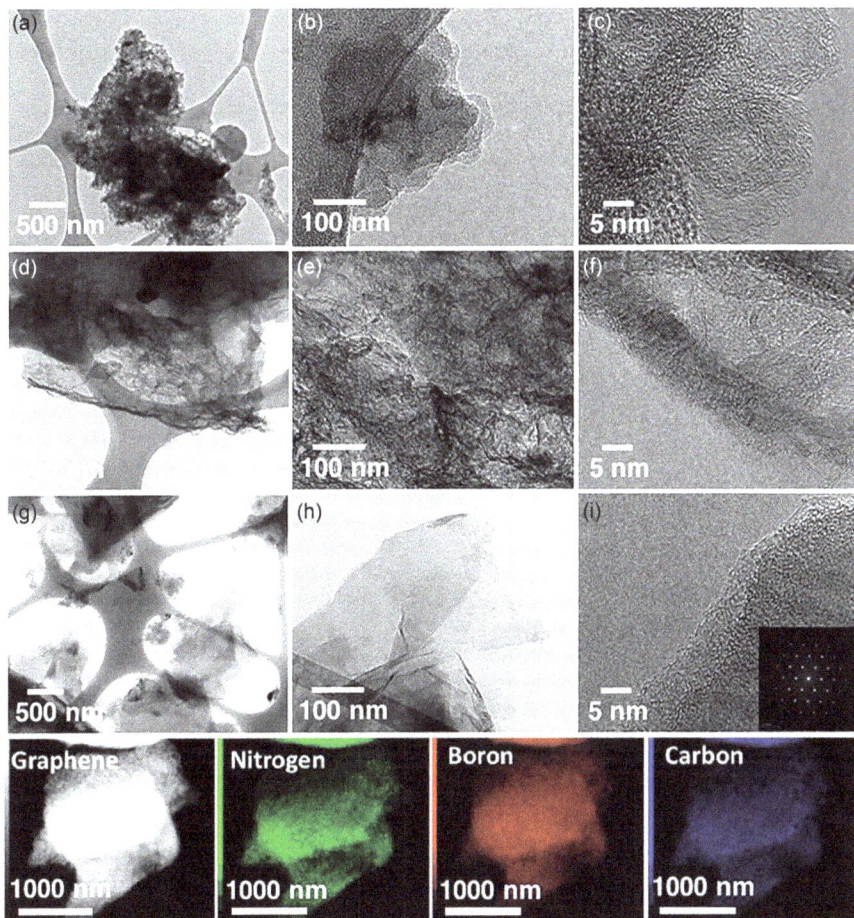

FIGURE 9.4 The low and corresponding high-resolution TEM images of few-layered NB-GNs synthesized by SLPE in the presence of 5 mM (A–C), 2.5 mM (D–F), and 0.1 mM sodium tetraphenylborate-NaBPh4 (G–I) in 0.77 mol ACN. The EDX mapping analysis was showing the presence of N, B, and C in NB-GNs by SLPE, with a precursor containing 0.1 mM NaBPh4 in 0.77 mol ACN. Reproduced from Elumalai et al. [54], with permission from Frontiers.

(few-layered) were obtained when the 1 and 0.1 mM boron were used, and the presence of N, B, and C has been monitored by EDX mapping analysis (Figure 9.4g–i). Hyun and Saito successfully demonstrated the formation of nitrogen–carbon nanosheets from 2-pyrrolidone, 1-methylpyrrolidine, pyrrolidine, pyrrole, cyclopentanone, and cyclohexanone in a liquid plasma condition [55].

Lee et al. reported that the silicon- and nanoclay-supported nanocarbon shows the highest theoretical energy density and non-toxic, eco-friendly material [24,56]. Senthilnathan et al. studied the direct formation of nanoclay–nitrogen-doped graphene using HB and 4B rod pencil as the working electrode (Pt counter electrode;

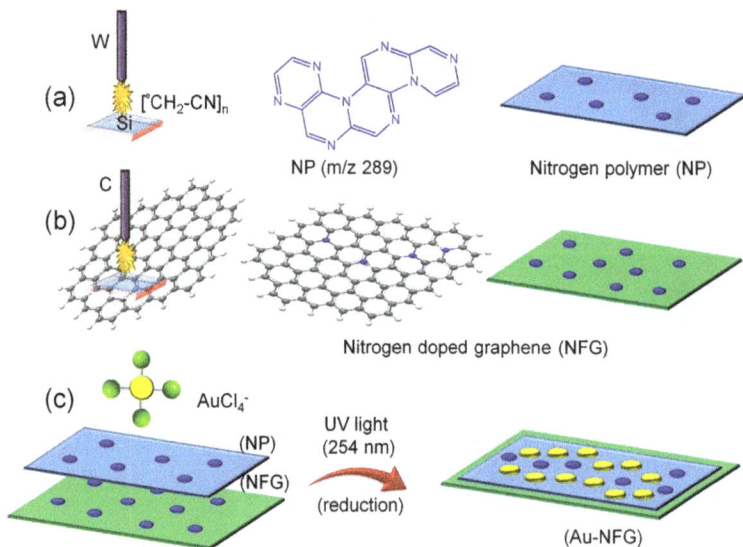

FIGURE 9.5 Proposed mechanism for the formation of Au-nitrogen functionalized graphene under UV light (a) Formation of nitrogen polymer in acetonitrile solution by a liquid plasma process, (b) formation of nitrogen functionalized grapheme in liquid plasma process, and (c) formation of Au-nitrogen polymer and nitrogen functionalized graphene UV light. Reproduced from Senthilnathan et al. [57], with permission from Royal Society of Chemistry.

discharge voltage 3.1 eV; repetition rate 10 kHz; pulse delay 500 μs; pulse width 5 ms) in acetonitrile [56].

Similarly, Au-decorated nitrogen-doped graphene was synthesized by the liquid plasma process in the three-step process (Figure 9.5). Acetonitrile polymers or oligomers produced in liquid plasma were used for the reduction of Au^{3+} under UV light [57]. The nitrogen polymers produced in the liquid plasma process provide excellent control over the size and shape of the Au nanoparticles [57]. Yang and Li prepared metal nanocarbon-functionalized materials (Pd-CNTs, Au-CNTs) by the gas–liquid ionic plasma process. The Pd-CNTs and Au-CNTs showed high catalytic performance for the organic molecule transformation reactions (Suzuki reaction) [58]. The Au-CNTs displayed high stability, high recyclability, and good reactivity for the selective oxidation of silanes in H_2O [58].

9.3 APPLICATIONS OF NANOCARBONS SYNTHESIZED FROM THE LIQUID PLASMA PROCESS

9.3.1 APPLICATION NANOCARBON HYBRIDS/COMPOSITES FOR FUEL CELL APPLICATIONS

Nanocarbon materials such as graphene, rGO, CNTs, MWCNTs, fullerene, and carbon onion play a substantial role in the development of alternative, clean and sustainable energy technologies [68,69]. Graphene nanocomposite with different inorganic moieties is one of the most attractive materials that can be used for different fields of applications, including photovoltaic cells, supercapacitors, and fuel cells [70,71]. Panomsuwan et al. [72] demonstrated the formation of nitrogen-doped carbon nanoparticles (specific surface area $210–250\,m^2g^{-1}$) with the particle size in the rage of $20–40\,nm$ in a liquid plasma process using cyano-aromatic molecules [72,73]. An electrochemical study of nitrogen-doped graphene showed enhanced oxygen reduction potential and current density in alkaline and acidic conditions [72]. Alsaeedi and Show demonstrated the formation of nanocarbon (graphitic carbon ~100 nm size) from ethanol in the liquid plasma process [74]. The nanocarbon was used as the support material for Pt nanoparticles. The Pt-supported nanocarbon catalyst was used as the catalyst for fuel cell application, and the maximum output power of $65\,mW\ cm^{-2}$ was obtained from fuel cell [74,75]. Similarly, Pt-supported graphene was synthesized and used in a membrane electrode assembly as the catalytic layer. The energy generation capacity of the membrane electrode assembly was evaluated to be $240\,mA\ cm^{-2,}$ and the observed value was much higher when compared to the carbon block [76]. Similarly, Pt-decorated CNTs were prepared by dispersing Pt nanoparticle from Pt sheet in a liquid plasma condition. The platinum-incorporated CNTs used in proton exchange membrane fuel cell as a catalyst displayed a maximum output power of $108\,mW\ cm^{-2}$ [77].

9.3.2 APPLICATION NANOCARBON HYBRIDS/COMPOSITES FOR SPECIFIC CAPACITANCE APPLICATIONS

The nitrogen-functionalized graphene–nanoclay composite prepared in the liquid plasma process displayed distinct redox peaks in cyclic voltameter, with the specific capacitances of 40 and 111 F g^{-1}, respectively, obtained at the scan rate of $5\,mV\ s^{-1}$ [24]. The specific capacitance obtained in this method was very low when compared to nanocarbon synthesized by other methods of synthesis. In the same way, a simultaneous reduction and functionalization of GO were performed in the liquid plasma process using acetonitrile and graphene oxide solution. The nitrile and amine groups present in reduced nitrogen-functionalized GO (r-FGO) show high cyclic stability with a specific capacitance value of 349 F g^{-1} at the scan rate of $10\,mV\ s^{-1}$. Only <10% reduction of specific capacitance was observed at the end of 1000 cycles [57].

Senthilnathan et al. demonstrated the high specific capacitance use of rGO–aminopyridine (rGO-$(AmPy)_n$) hybrid in the liquid plasma process [78]. Distinct redox peaks were observed at the scan rates of 5, 10, and $25\,mV\ s^{-1}$. Also, rGO-$(AmPy)_n$ showed excellent cyclic stability with a high specific capacitance of 418, 400, and 381 F g^{-1} for 0.1, 0.2, and 0.3 A g^{-1} current density, respectively (Figure 9.6a–c) [78]. The high specific capacitance value is due to the presence of pyridinic and pyrrolic groups in rGO-$(AmPy)_n$ (Table 9.2). Further, the electrochemical generation of H_2O_2 at the rGO-$(AmPy)_n$ surface was evaluated and found to be 0.42 mM for the fixed reaction time of 30 min. An H_2O_2/•OH radical generation capacity of rGO-$(AmPy)_n$ electrode was evaluated in the electrochemical cell by using methyl parathion as the target pollutant (Figure 9.6d). Lee et al. [23] demonstrated the formation of nitrogen-doped CNT–supported iron oxide for supercapacitor applications [23]. The electrochemical

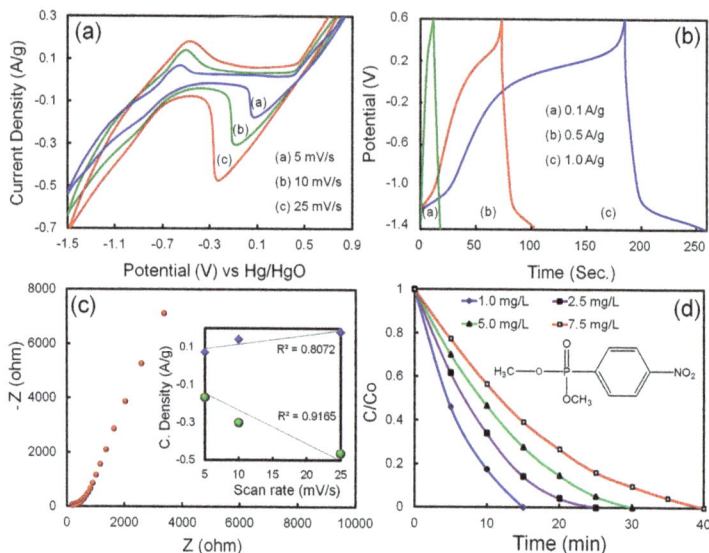

FIGURE 9.6 (a) Electrochemical CV curves of rGO-(AmPy)n hybrid electrode in 1 M NaOH obtained at the scan rate of 5, 10 and 25 mV/s. (b) Galvanostatic charge/discharge curves of a rGO-(AmPy)n based capacitor under different constant currents (0.1, 0.2 and 0.3 A/g). (c) Nyquist plot of rGO-(AmPy)n hybrid electrode. Inset shows peak current dependence on the square root of scan rate. Note: C. density = Current density. (d) Pseudofirst-order degradation patterns of various concentrations of methyl parathion (1.0, 2.5, 5.0 and 7.5 mg/L) in the modified electro Fenton process with rGO-(AmPy)n hybrid electrode. Reproduced from Senthilnathan et al. [78], with permission from Elsevier.

performance of the iron oxide–deposited nitrogen-doped CNTs was higher than that of pristine MWCNTs and nitrogen-doped CNTs [23]. The pristine MWCNTs displayed exhibited a specific capacitance of 19.10 F g^{-1}, and the cyclic stability study showed 16.5 F g^{-1} at the end of the 100th cycle [23]. The comparison of the specific capacitance value of nanocarbon prepared from the liquid plasma process with other methods is given in Table 9.2.

9.4 FUTURE PROSPECTIVE

In summary, this review intends to provide an overview of studies focusing on non-thermal plasma interactions with liquid plasma in a nonaqueous solution. However, there are many unsolved queries in the liquid plasma process in nonaqueous solution, and more profound research is needed in this area. The following are the important conclusions and possible perspectives obtained based on the nanocarbon synthesized from a nonaqueous liquid in the liquid plasma process: (i) The fast-moving electrons generated at the interface are effectively quenched by the nonaqueous compound and new materials formed, (ii) minimal surface damage was observed, (iii) large-scale production of nanocarbons is very much possible, and (iv) no toxic chemicals or high temperature required. The formation of nanocarbons in nonaqueous solutions

TABLE 9.2

Comparison of Specific Capacitance of Nitrogen-Functionalized rGO with Other Nanocarbons Synthesized by Other Methods

S. No.	Materials	Process	Faradic/ Non-faradic	Groups	Capacitance (F g⁻¹)	References
1	[a]f-G	SLP	Faradic	Nitrogen	291	[53]
2	r-FGO	Liquid plasma	Faradic	Nitrogen	363	[60]
3	rGO-(AmPy)	Liquid plasma	Faradic	Nitrogen	418	[78]
4	[b]f-GO	Chemical	Faradic	Nitrogen	525	[79]
5	[b]f-GO	Chemical	Faradic	Sulfur	201	[80]
6	[a]f-G	Thermal	Non-faradic	Oxygen	417	[81]
7	[c]rGO	Thermal	Non-faradic	Oxygen	315	[82]
8	[c]rGO	Solvothermal	Faradic	Oxygen	276	[83]
9	[a]f-G	Hydrothermal	Non-faradic	Nitrogen	326	[84]
10	[a]f-G	Solvothermal	Non-faradic	Nitrogen	301	[85]
11	[c]rGO	Thermal	Faradic	Oxygen	261	[86]
12	[c]rGO	Chemical	Non-faradic	Oxygen	238	[87]
13	[c]rGO	Chemical	Non-faradic	Oxygen	255	[88]
14	[a]f-G	Plasma	Non-faradic	Nitrogen	280	[89]

[a] f-G: functionalized graphene.
[b] f-GO: functionalized graphene oxide.
[c] rGO: reduced graphene oxide.
Reproduced from Senthilnathan et al. [53], with permission from Nature group.

using the liquid plasma process is an important and powerful tool for a low-cost and eco-friendly synthesis. Furthermore, our research group has been successful in the preparation of multilayer graphene sheets by an electrochemical exfoliation under ambient pressure and temperature by applying a volt in H_2O_2+NaOH solution [90,91]. Moreover, the preparations were all based on low energy production and scalable processing. Therefore, continuous or, at the least, successive production of functionalized graphene and their hybrids with Au and other nanoparticles and their inks is possible. Furthermore, this method of processing demonstrates great potential and perspective.

ACKNOWLEDGMENT

The authors thank Professor Yury Gogotsi (Drexel University) for his encouragement and support. The authors are grateful to Professor Jaw-Lay Huang and Professor Wen-Ta Tsai, Department of Materials Science and Engineering, National Cheng Kung University, Tainan, Taiwan, for the discussions and support. The authors gratefully acknowledge the support of Professor Jiunn-Der Liao, Department of Material Science and Engineering, and Professor Jih-Jen Wu, Department of Chemical Engineering, National Cheng Kung University, for the help and discussions regarding experiments.

REFERENCES

1. Fridman, G. Brooks, A. D. Balasubramanian, M. Fridman, A. Gutsol, A. Vasilets, V. N. Ayan, H. Friedman. G. *Plasma Process Polym.* 4, 370–375 (2007).
2. Fridman, G. Peddinghaus, M. Ayan, H. Fridman, A. Balasubramanian, M. Gutsol, A. Brooks, A. Friedman, G. *Plasma Chem. Plasma Process.* 26, 425 (2006).
3. Locke, B. R. Shih, K.-Y. *Plasma Sources Sci. Technol.* 20, 034006 (2011).
4. Choi, W. Lahiri, I. Seelaboyina, R. Kang, Y.-S. *Crit. Rev. Solid State.* 35, 52–71 (2010).
5. Dobrynin, D. Fridman, G. Friedman, G. Fridman, A. *New J. Phys.* 11, 115020 (2009).
6. Tan, H. Wang, D. Guo, Y. *Materials.* 12, 2279–2289 (2019).
7. Wang, H. Li, J. Quan, X. *J. Electrostat.* 64, 416–421 (2006).
8. Sugiarto, A. T. Ohshima, T. Sato, M. *Thin Solid Films.* 407, 174–178 (2002).
9. Grande, S. Tampieri, F. Nikiforov, A. Giardina, A. Barbon, A. Cools, P. Morent, R. Paradisi, C. Marotta, E. De-Geyter, N. *Front. Chem.* 7, 344 (2019).
10. Senthilnathan, J. Weng, J. C.C. Liao, J.-D. Yoshimura, M. *Sci. Reps.* 3, 2414–2420 (2013).
11. Georgakilas, V. Otyepka, M. Bourlinos, A. B. Chandra, V. Kim, N. Kemp, K. C. Hobza, P. Zboril, R. Kim, K. S. *Chem. Rev.* 112, 6156–6214 (2012).
12. Bruggeman, P. Leys, C. *J. Phys. D. Appl. Phys.* 42, 053001 (2009).
13. Lin, L. Wang, Q. *Plasma Chem. Plasma Process.* 35, 925–962 (2015).
14. Hyun, K. Ueno, T. Li, O. L. Saito, N. *RSC Adv.* 6, 6990 (2016).
15. Cobley, C. M. Chen, J. Cho, E. C. Wang, L. V. Xia, Y. *Chem. Soc. Rev.* 40, 44–56 (2010).
16. Li, O. L. Chiba, S. Wada, Y. Panomsuwan, G. Ishizaki, T. *J. Mater. Chem. A.* 5, 2073 (2017).
17. Hyun, K. Ueno, T. Saito, N. *Jpn. J. Appl. Phys.* 55, 01AE18 (2016).
18. Li, O. L. Hayashi, H. Ishizaki, T. Saito, N. *RSC Adv.* 6, 51864 (2016).
19. Lee, S. H. Heo, Y. K. Bratescu, M. A. Ueno, T. Saito, N. *Phys. Chem. Chem. Phys.* 19, 15264 (2017).
20. Bratescu, M. A., Cho, S. P., Takai, O. Saito, N. *J. Phys. Chem. C.* 115, 24569–24576 (2011).
21. Watanabe, T. Wang, H. Yamakawa, Y. Yoshimura M. *Carbon.* 44, 799–823 (2006).
22. Kang, J. Li, O.-L. Saito, N. *Carbon.* 60, 292–298 (2013).
23. Lee, H. Wada, Y. Kaneko, A. Li, O.-L. Ishizaki, T. *Jpn. J. Appl. Phys.* 57, 0102BD (2018).
24. Senthilnathan, J. Rao, K. S. Lin, W.-H., Liao, J.-D. Yoshimura, M. *Carbon.* 78, 446–454 (2014).
25. Nomura, S. Toyota, H. *Appl. Phys. Lett.* 83, 4503 (2003).
26. Nomura, S. Toyota, H. Mukasa, S. Yamashita, H. Maehara, T. *Appl. Phys. Lett.* 8, 211503 (2006).
27. Nomura, S. Toyota, H. Mukasa, S. Yamashita, H. Maehara, T. Kawashima, A. *J. Appl. Phys.* 106, 073306 (2009).
28. Ishijima, T. Hotta, H. Sugai, H. *Appl. Phys. Lett.* 91, 121501 (2007).
29. Ishijima, T. Sugiura, H. Satio, R. Toyada, H. Sugai, H. *Plasma Sources Sci. Technol.* 19, 015010 (2010).
30. Egiza, M. Murasawa, K. Ali, A.-M. Fukui, Y. Gonda, H. Sakurai, M. Yoshitake, T. *Jpn. J. Appl. Phys.* 58, 075507 (2019).
31. Shimoeda, H. Kondo, H. Ishikawa, K. Hiramatsu, M. Sekine, M. Hori, M. *Jpn. J. Appl. Phys.* 53, 040305 (2014).
32. Morishita, T. Ueno, T. Panomsuwan, G. Hieda, J. Yoshida, A. Bratescu, M. A. Saito, N. *Sci. Rep.* 6, 36880 (2016).

33. Neira-Velazquez, M. G. Ramos-deValle, L. F. Hernández-Hernández, E. Ponce-Pedraza, A. Solís-Rosales, S. G. Sánchez-Valdez, S. Bartolo-Pérez, P. González-González, V. A. *Plasma Process Polym.* 8, 842–849 (2011).
34. Qingsong, Y. U. Mu, Y. E. Lizhen, L. U. Jie, C. Fosong, W. Osada, Y. *Chinese J. Polym. Sci.* 6(2), 172–177 (1998).
35. Herbert, P. A. F. O'Neill, L. Jaroszynska-Wolinska, J. *Chem. Mater.* 21, 4401–4407 (2009).
36. Inagaki, N. *Plasma Surface Modification and Plasma Polymerization*, Technomic Publishing Company, Lancaster, PA, 1996.
37. Wang, H. Yoshimura, M. *Chem. Phys. Lett.* 348, 7–10 (2001).
38. Antisari, M. V. Marazzi, R. Krsmanovic, R. *Carbon.* 41, 2393–2401 (2003).
39. Tan, H. Wang, D. Guo, Y. *Materials.* 12(14), 2279 (2019).
40. Biró, L. P. Horváth, Z. E. Szalmás, L. Kertész, K. Wéber, F. Juhász, G. Radnóczi, G. Gyulai, J. *Chem. Phys. Lett.* 372, 399–402 (2003).
41. Okada, T. Kaneko, T. Hatakeyama, R. *Thin Solid Films*, 515, 4262–4265 (2007).
42. Lange, H. Sioda, M. Huczko, A. Zhu, Y. Q. Kroto, H. W. Walton, D. R. M. *Carbon.* 41(8), 1617–1623 (2003).
43. Sano, N. Naito, M. Chhowalla, M. Kikuchi, T. Matsuda, S. Iimura, K. *Chem. Phys. Lett.* 378(1–2), 29–34 (2003).
44. Scuderi, V. Bongiorno, C. Faraci, G. Scalese, S. *Carbon.* 50, 2365–2369 (2012).
45. Zhu, H. W. Li, X. S. Jiang, B. Xu, C. L. Zhu, Y. F. Wu, D. H. Chen, X. H. *Chem. Phys. Lett.* 366(5–6), 664–669 (2002).
46. Beck, M. T. Dinya, Z. Kéki, S. Papp, L. *Tetrahedron.* 49, 285–290 (1993).
47. Sano, N. Wang, H. Alexandrou, I. Chhowalla, M. Teo, K. B. K. Amaratunga, G. A. J. *J. Appl. Phys.* 92(5), 2783–2788 (2002).
48. Liu, T. Yang, F. Li, Y. Ren, L. Zhang, L. Xu, K. Wang, X. Xu, C. Gao, J. *J. Mater. Chem. A.* 2, 245–250 (2014).
49. Wang, H. Shen, M.-R. Ning, Z.-Y. Ye, C. Cao, C.-B. Dang, H.-Y. Zhu, H.-S. *Appl. Phys. Lett.* 69, 1074–1076 (1996).
50. Chen, Q. Kitamura, T. Saito, K. Haruta, K. Yamano, Y. Ishikawa, T. Shirai, H. *Thin Films.* 516, 4435–4440 (2008).
51. Inagaki, N. Tasaka, S. Yamada, Y. *J. Polym. Sci. Part A Polym. Chem.* 30(9), 2003–2010 (1992).
52. Yasuda, H. *J. Polym. Sci. Macromol. Rev.* 16, 199–293 (1981).
53. Senthilnathan, J. Rao, K. S. Yoshimura, M. *J. Mater. Chem. A.* 2, 3332–3337 (2014).
54. Elumalai, S. Su, C.-Y. Yoshimura, M. *Front. Mater.* 6, 216 (2019).
55. Hyun, K. Saito, N. *Sci. Rep.* 7, 3825 (2017).
56. Lee, J. K. Smith, K. B. Hayner, C. M. Kung, H. H. *Chem. Commun.* 46, 2025–2027 (2010).
57. Senthilnathan, J. Rao, K. S. Lin, W.-H. Ting, J.-M. Yoshimura, M. *J. Mater. Chem. A.*, 3, 3035–3043 (2015).
58. Yang, F. Li, Y. *Int. J. Chem. Eng. Appl.* 6(1), 49–52 (2015).
59. Sano, N. Wang, H. Chhowalla, M. Alexandrou, I. Amaratunga, G. A. J. *Nature.* 414, 506–507 (2001).
60. Senthilnathan, J. Liu, Y.-F. Rao, K. S. Yoshimura, M. *Sci. Rep.* 4, 4395 (2014).
61. Alexandrou, I. Wang, H. Sano, N. Amaratunga, G. A. J. *J. Chem. Phys.* 120(2), 1055–1058 (2004).
62. Hagino, T. Kondo, H. Ishikawa, K. Kano, H. Sekine, M. Hori, M. *Appl. Phys. Express.* 5(3), 035101 (2012).
63. Matsushima, M. Noda, M. Yoshida, T. Kato, H. Kalita, G. Kizuki, T. Uchida, H. Umeno, M. Wakita, K. *J. Appl. Phys.* 113(11), 114304 (2013).

64. Muthakarn, P. Sano, N. Charinpanitkul, T. Tanthapanichakoon, W. Kanki, T. *J. Phys. Chem. B*. 110(37), 18299–18306 (2006).
65. Sano, N. Charinpanitkul, T. Kanki, T. Tanthapanichakoon, W. *J. Appl. Phys*. 96(1), 645–649 (2004).
66. Hsin, Y. L. Hwang, K. C. Chen, F.-R. Kai, J.-J. *Adv. Mater*. 13, 830–833 (2001).
67. Roslan, M. S. Chaudhary, K. T. Doylend, N. Agam, A. Kamarulzaman, R. Haider, Z. Mazalan, E. Ali, J. J. *Saudi Chem. Soc*. 23(2), 171–181 (2019).
68. Jenkins, G. Manz, A. *J. Micromech. Microeng*. 12, N19 (2002).
69. Horikoshi, S. Serponec, N. *RSC Adv*. 7, 47196 (2017).
70. Stankovich, S. Dikin, D. A. Dommett, G. H. B. Kohlhaas, K. M. Zimney, E. J. Stach, E. A. Piner, R. D. Nguyen, S. T. Ruoff, R. S. *Nature*. 442, 282–286 (2006).
71. Ramanathan, T. Abdala, A. A. Stankovich, S. Dikin, D. A. Herrera-Alonso, M. Piner, R. D. Adamson, D. H. Schniepp, H. C. Chen, X. Ruoff, R. S. Nguyen, S. T. Aksay, I. A. Prud'Homme, R. K. Brinson, L. C. *Nat. Nanotechnol*. 3, 327–331 (2008).
72. Panomsuwan, G. Saito, N. Ishizaki, T. *Carbon*. 98, 411–420 (2016).
73. Panomsuwan, G. Saito, N. Ishizaki, T. *Electrochem. Commun*. 59, 81–85 (2015).
74. Alsaeedi, Y. Show, A. *Nanomater. Nanotech*. 9, 1–6 (2019).
75. Bruggeman, P. J. Kushner, M. J. Locke, B. R. Gardeniers, J. G. E. Graham, W. G. Graves, D. B., Hofman-Caris, R. C. H. M. Maric, D. Reid, J. P. Ceriani, E. Rivas, D. F. *Plasma Sources Sci. Technol*. 25, 053002 (2016).
76. Gamaleev, V. Kajikawa, K. Takeda, K. Hiramatsu, M. *J. Carbon Res*. 4, 65, 2–9 (2018).
77. Show, Y. Hirai, A. Almowarai, A. Ueno, Y. *Thin Solid Films*. 596, 198–200 (2015).
78. Senthilnathan, J. Yoshimura, M. *J. Hazmat. Mater*. 340, 26–35 (2017).
79. Liu, Y. Deng, R. Wang, Z. Liu, H. *J. Mater. Chem*. 22, 13619–13624 (2012).
80. Kumar, N. A. Choi, H. J. Bund, A. Baek, J. B. Jeong, Y. T. *J. Mater. Chem*. 22, 12268–12274 (2012).
81. Fang, Y. Luo, B. Jia, Y. Li, X. Wang, B. Song, Q. Kang, F. Zhi, L. *Adv. Mater*. 24, 6348–6355 (2012).
82. Ye, J. Zhang, H. Chen, Y. Cheng, Z. Hu, L. Ran, Q. *J. Power Sources*. 212, 105–110 (2012).
83. Lin, Z. Liu, Y. Yao, Y. Hildreth, O.-J. Li, Z. Moon, K. Wong, C.-P. *J. Phys. Chem. C*. 115, 7120–7125 (2011).
84. Sun, L. Wang, L. Tian, C. Tan, T. Xie, Y. Shi, K. Li, M. Fu, M. *RSC Adv*. 2, 4498–4506 (2012).
85. Lu, Y. Lu, Y. Zhang, F. Zhang, T. Leng, K. Zhang, L. Yang, X. Ma, Y. Huang, Y. Zhang, M. Chen, Y. *Carbon*. 63, 508–516 (2013).
86. Zhao, B. Liu, P. Jiang, Y. Pan, D. Y. Tao, H. H. Song, J. S. Fang, T. Xu, W. W. *J. Power Sources*. 198, 423–427 (2012).
87. Zhang, D. Zhang, X. Chen, Y. Wang C. Ma, Y. *Electrochim. Acta*. 69, 364–370 (2012).
88. Lei, Z. Lu, L. Zhao, X. S. *Energy Environ. Sci*. 5, 6391–6399 (2012).
89. Jeong, H. M. Lee, J. W. Shin, W. H. Choi, Y. J. Shin, H. J. Kang, J. K. Choi, J. W. *Nano Lett*. 11(6), 2472–2477 (2011).
90. Rao, K. S. Senthilnathan, J. Cho, H.-W. Wu, J. J. Yoshimura, M. *Adv. Funct. Mater*. 25(2), 298–305 (2014).
91. Rao, K. S. Senthilnathan, J. Ting, J.-M. Yoshimura, M. *Nanoscale*. 6, 12758–12768 (2014).

10 Ionic Liquid-Based Electrolytes

Synthesis and Characteristics and Potential Applications in Rechargeable Batteries

Linh T. M. Le, Thanh D. Vo, Hoang V. Nguyen, Man V. Tran, and Phung M. L. Le
University of Science
Viet Nam National University

CONTENTS

10.1 OVERVIEW

Electrolyte is a medium responsible for ionic transport, typically for Na–ion/Li–ion transport in rechargeable batteries, and hence controls the power density. Ionic conductivity is the topmost important property for an electrolyte, and at the same time, it should have negligible electronic conductivity in order to avoid short circuits. Electrolytes should also have good mechanical, thermal, and electrochemical stability along with good interfacial properties. Thermal and voltage stability ranges of batteries depend on the lowest unoccupied molecular orbital (LUMO) and highest occupied molecular orbital (HOMO) energy levels of the electrolyte as shown in Figure 10.1. In order to have better thermal stability, redox energies E_a (anode) and E_c (cathode) should lie within the band gap (E_g) of the electrolyte [1].

10.1.1 DEFINITION

Ionic liquids (ILs) are composed of ions, exhibit negligible vapor pressure, like solid salts, and are considered as nonflammable. Compared to other organic solvents, they do not vaporize unless heated to the point of thermal decomposition, typically 200–300°C or more. One of the most important applications of IL utilization is for energy systems, especially for energy storage and conversion materials and devices, because of the continuously increasing demand for clean and sustainable energy.

In this chapter, the applications of ILs are reviewed by focusing on their use as electrolyte medium for Li–ion/Na–ion batteries due to their intrinsic properties such as nonvolatility, high thermal stability, and high ionic conductivity regarding various applications. These properties are, however, much affected by interactions between ions (electrostatic or van der Waals interactions), and even the directionality of ions.

On the other hand, for further specific development, we need to have a proper orientation for the application of ILs and find ways to optimize their properties.

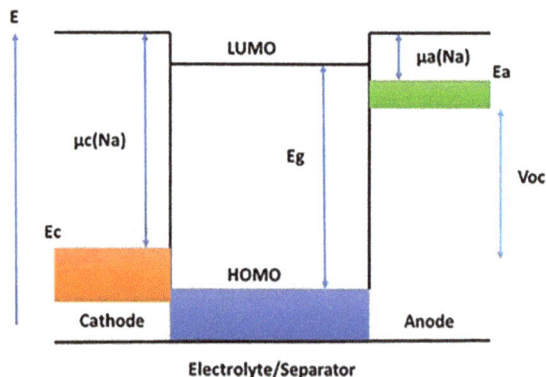

FIGURE 10.1 Schematic energy diagram of Na cell at open circuit showing the thermodynamic stability conditions, requiring the redox energies of the cathode (E_c) and anode (E_a) to lie within the band gap (E_g) of the electrolyte.

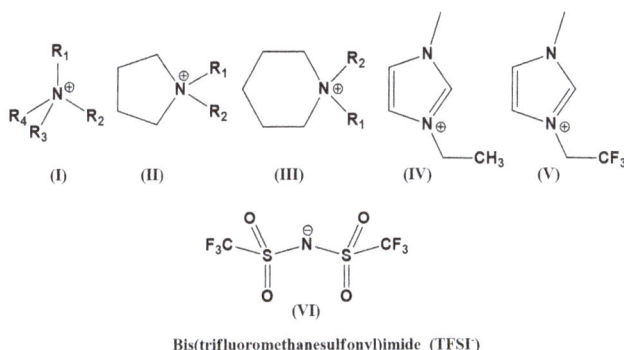

(I) (II) (III) (IV) (V)

(VI)

Bis(trifluoromethanesulfonyl)imide (TFSI⁻)

FIGURE 10.2 Typical anion and cations comprising ILs [6,7].

Lithium/sodium batteries and fuel cells are examples, for which commercial application mainly in hybrid, electric, and fuel cell vehicles as well as for stationary use in homes, buildings, and up to large power grid scale has already commenced. Carbonate-based electrolytes have been used in the highest energy density secondary batteries ever developed Li-ion batteries [2–4]. However, organic carbonates are still considered unsafe related to flammability, side reactions, dissolution of electroactive materials, and evaporation of the solvents in the case of using devices at high temperature, especially when Li–ion batteries are used on large scale applications such as in electric vehicles and power grids. These issues are prompting research on new electrolyte materials based on ILs [5]. Typical cations and anions comprising ILs were synthesized and characterized in the work of M.L.P. Le and co-workers [6,7] (Figure 10.2).

10.1.2 CLASSIFICATION

Among rechargeable batteries, lithium–ion batteries not only have higher working voltage and energy density, but also have long cycle life. Such superior characteristics enable lithium–ion batteries to fulfill complex requirements for diversified growth in devices. Lithium–ion batteries still have some existed problem need to be improved for eco-friendly applications such as power storage, health care.

The potential of $LiMO_2$ (M: transition metals)-layered cathode materials is typically working in the range of 3.5–4.2 V versus Li/Li⁺. On the other hand, the potential of carbonaceous negative electrodes is as low as 0–0.3 V versus Li/Li⁺; therefore, aprotic electrolytes are used to avoid significant side reactions at the negative electrodes. Nowadays, lithium salts dissolved in aprotic molecular solvents such as ethylene carbonate (EC) and dimethyl carbonate (DMC) or propylene carbonate (PC) are used as electrolytes in practical Li–ion batteries [2,4]. To resolve flammability and volatility problems of the organic solvents, improvement of electrolyte safety for Li–ion batteries is strongly desired, especially for large-scale energy storage systems. Room-temperature Na–ion batteries using aprotic electrolytes have also been investigated as alternative power sources [5]. Thermally stable electrolytes have been extensively investigated to improve the thermal stability of these batteries [2,5,8–12].

TABLE 10.1

Characteristics of Different Electrolytes Used for Lithium–Ion Batteries (LIBs)

	Liquid Electrolytes	Ionic Liquid Electrolytes	Solid Polymer Electrolytes	Gel Polymer Electrolytes
Composition	Organic solvents + lithium salts	Ionic liquids + lithium salts	Polymer + lithium salts	Organic solvents + polymer + lithium salts
Ion conductivity	High	High	Low	Relatively high
Low-temperature performance	Relatively good	Poor	Poor	Relatively good
Thermal stability	Poor	Good	Excellent	Relatively good

During the last decades, some promising designed ILs with different anion and cation structures have been developed as prospective electrolytes for the batteries (Table 10.1). All of them exhibit low volatility and fire-retardant ability, and high thermal stability. Good Li^+ and Na^+ ion-conducting ability and electrochemical stability, respectively, are requirements for a good electrolyte used in Li/Na batteries. The discharge/charge becomes less efficient if the negative and positive electrodes are highly reducing and oxidizing agents. Therefore, a wide electrochemical window is desirable for Li–ion/Na–ion battery electrolytes. In this section, IL electrolytes for Li–ion/Na–ion batteries are being discussed.

10.2 SOME CONCEPTS OF IL-BASED ELECTROLYTES FOR LI–ION/NA–ION BATTERIES

10.2.1 LOW-MELTING ALKALINE SALTS

10.2.1.1 Low-Melting Lithium Salts

Many conventional lithium salts were combined with other ingredients to use as electrolytes for Li–ion batteries with the melting point typically above 200°C, for example, 236°C for $LiClO_4$, 296.5°C for $LiBF_4$, and 200°C for $LiPF_6$. This is mainly because of the strong electrostatic attraction between the Li^+ ion and the anion, ligand structure, or asymmetric structure of the salts. The charge density of Li^+ ion is high, because of its small ionic radius, resulting in a strong electrostatic attraction between the Li^+ ion and the anion. Sometimes, the factors described above make us difficult to predict the melting temperatures of lithium salts because interactions between Li^+ and anions are relatively weak due to Lewis acidity of Li^+; thus, it is difficult to lower the melting point of a lithium salt down to room temperature. However, even the lithium salt of a large aluminate anion which is weak Lewis base has a melting point much higher than room temperature [13,14]. Fujinami et al. reported that the introduction of ether groups into the aluminate structure gave liquid-like Li salts [15,16].

The ether groups act as Li$^+$ coordinating ligands, dissociating the Li$^+$ cations from the central atom of the anion's centers. Watanabe and co-workers reported ILs consisting of Li$^+$ cation and borates anion having electron-withdrawing groups, to reduce the basicity, as well as Li$^+$ coordinating ether ligands, to dissociate the lithium cations from the central atom of the anions [17–19].

10.2.1.2 Mixtures of Alkaline Imide Salts

The melting point of A[TFSI] (A: alkali metal; TFSI: bis(trifluoromethanesulfonyl) imide = [(CF$_3$SO$_2$)$_2$N$^-$) is relatively high, about 236°C. The melting points of the imide salt as well as the ionic conductivity of the molten salts are low at room temperature because of their very high viscosity ion pairs. The strong Lewis-acidic nature of the Li$^+$ ion favors ion pairing even in the molten state. However, to increase the ionic conductivity, the temperature should be elevated.

10.2.2 ALKALINE SALTS DISSOLVED IN ORGANIC IONIC LIQUIDS

There are numerous aprotic ILs composed of organic cations and counter anions (Table 10.2), having melting points lower than room temperature without containing Li$^+$/Na$^+$ ions; therefore, lithium or sodium salts should be mixed with the organic ILs to form mixtures that are used as electrolytes for Li–ion/Na–ion batteries, and the ILs are used as electrolyte solvents for the lithium or sodium salts. The physicochemical properties of lithium salt/IL binary mixtures are significantly affected by the structures of cations and anions of the organic IL. Herein, we highlight the characteristics of lithium salt/IL binary mixtures [2,5].

TABLE 10.2

Abbreviations of Typical Cations and Anions Comprising Aprotic ILs

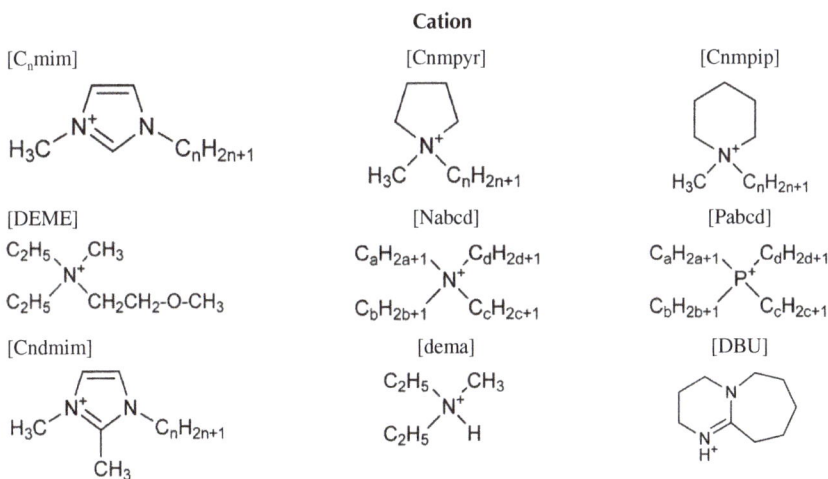

Cation

[C$_n$mim]	[Cnmpyr]	[Cnmpip]

[DEME]	[Nabcd]	[Pabcd]

[Cndmim]	[dema]	[DBU]

(*Continued*)

TABLE 10.2 *(Continued)*

Abbreviations of Typical Cations and Anions Comprising Aprotic ILs

Anion

[TFSA]   [FSA]   [FTA]

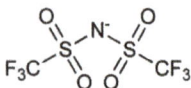   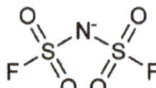   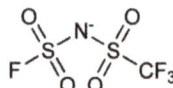

[BETA]   [TSAC]   [FAP]

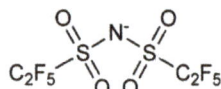   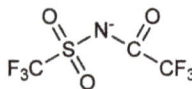   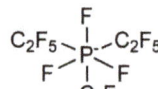

[TfO]   [MS]   [DFOB]

$CF_3SO_3^-$ $CH_3SO_3^-$ 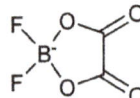

[DCA]

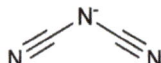

10.2.2.1   Effects of Cation Structure

The oxidation stability of the organic ILs is strongly affected by the chemical nature of the cations. It is widely known that aliphatic quaternary ammonium (AQA) cations and aliphatic quaternary phosphonium (AQP) cations have excellent oxidation stability [20,21]. Hence, many researchers have investigated AQA- and AQP-based ILs as electrolyte solvents for Li–ion/Na–ion batteries [7].

Besides, within the same cation structure, the changes in alkyl chain length affect the oxidation stability of ionic liquids [6] (Figure 10.3).

The electrochemical reactions of negative electrode materials are affected by not only the cation species but also the anion species. Therefore, the combination of cation and anion of an IL and the lithium salt or sodium salt should be selected carefully for battery applications. The organic cations, especially ionic radius, the anion size, and the flexibility of cation and anion as well as functional groups contained in the cation/anion structures also affected many properties of lithium salt/IL binary mixtures such as transport property of Li$^+$, viscosity, and conductivity. For example, organic cations containing ether groups tend to result in lower viscosities for the Li salt/IL binary mixtures owing to the interaction between the Li$^+$ and the ether group (Figure 10.4) [22–28] (Table 10.3).

For some mixtures of ILs in different cation types, P. Le et al. [2] reported the comparison between the theoretical and practical viscosity of IL mixtures to demonstrate the effect of cation type on viscosity. The work shows that if the presence of solvent EC in electrolyte mixtures created an ideal solution (i.e., there is no interaction between molecules), theoretical viscosity will be calculated by the formula:

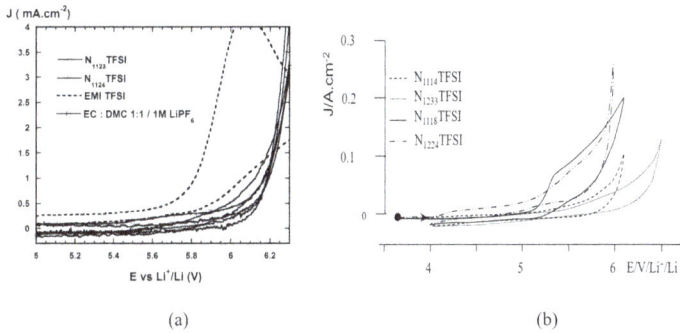

FIGURE 10.3 CV curves (a) different pure ionic liquids compared with conventional electrolyte 1 M $LiPF_6$/EC-DMC (1:1) [2]; (b) Variation of alkyl chains length of quaternary ammonium cation [6]. (Platinum working and counter electrodes. Scan rate 0.1 mV/s, room temperature).

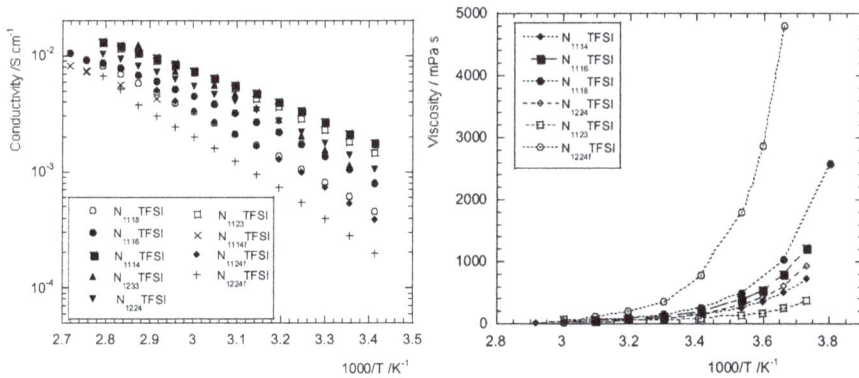

FIGURE 10.4 Ionic conductivity and viscosity as a function of temperature for several ionic liquids [6], Reproduced with permission. Copyright 2010, ACS.

$\text{Ln}\,\eta_{\text{mix}} = X_1 \ln \eta_1 + X_2 \ln \eta_2$ [2], where X_1 and X_2 are mole fractions of solution 1 (ionic liquid) and solution 2 (polar solvent), respectively. The difference between the theoretical value and the real value is relatively high, up to 51% (Table 10.4). It can be demonstrated that these solutions are not ideal and some interactions between solvent with ionic liquid exist. The larger the cations are, the stronger the interaction between the solvent and the IL is shown.

10.2.2.2   Effects of Anion Structure

The oxidative stability of anions mainly determines the anodic limit of the ILs. Fluorinated anions such as BF_4^- and PF_6^-, and amide-type anions, such as [TFSA]$^-$ and [FSA]$^-$, or imide-type anions, such as [TFSI]$^-$, are frequently used in the electrolytes owing to their good electrochemical stability. The fluorinated anions interact weakly with organic cations in the liquids because they are Lewis bases. When a lithium salt is dissolved in solvents (e.g., propylene carbonate), the Li^+ ion is stabilized

TABLE 10.3

Properties of Quaternary Ammonium Imide Ionic Liquids [6]

ILs	Cation Structure	T_g °C	T_m °C	T_d °C	Density g⁻¹mL¹/20°C	Conductivity (σ) 10⁴ S⁻¹cm¹ 20°C	Viscosity (η) mPa⁻¹s⁻¹ 20°C (50°C)
N_{1114}TFSI	$(CH_3)_3C_4H_9N^+$	−78	8/17	400	1.41	17	148 (30)
N_{1114f}TFSI	$(CH_3)_3CF_3(CH_2)_3N^+$		90	408		4 (70°C)	
N_{1233}TFSI	$CH_3C_2H_5(C_3H_7)_2N^+$		15/29	405	1.32	8	155
N_{1123}TFSI	$(CH_3)_2C_2H_5(C_3H_7)N^+$	−95	−9	402	1.39	14	82 (40)
N_{1224}TFSI	$CH_3(C_2H_5)_2(C_4H_9)N^+$	−89	11	398	1.34	10	161 (41)
N_{1224f}TFSI	$CH_3(C_2H_5)_2(CF_3C_3H_6)N^+$	−70	36	420	1.46	2	774 (119)
N_{1124f}TFSI	$(CH_3)_2C_2H_5(CF_3C_3H_6)N^+$		51	410		4	
N_{1116}TFSI	$(CH_3)_3C_6H_{13}N^+$	−76	32	395	1.32	8	205 (52)
N_{1118}TFSI	$(CH_3)_3C_8H_{17}N^+$	−77	7	380	1.26	5	257 (65)

T_g = glass transition temperature; T_m = melting point; T_d = degradation temperature.

TABLE 10.4

Theoretical and real viscosity of ionic liquids mixed with different percentages of ethylene carbonate [2]

	N_{1123}TFSI		N_{1124}TFSI		EMITFSI	
% EC	Theory	Reality	Theory	Reality	Theory	Reality
10	19.8	15.8	23.3	22.1	11.3	–
15	13.8	11.1	15.9	15.6	8.6	–
20	10.4	9.3	11.7	12.3	6.9	14.1
25	8.1	7.0	9.0	10.1	5.8	–

Reproduced with permission. Copyright 2010, ACS

by solvation and the salt dissociates into a solvated cation $[Li(solvent)_x]^+$ and counter anion. In the case of a binary mixture of lithium salt and IL, the Li^+ is solvated by anions and forms complexes $[Li − (X)_x](x − 1)^-$ (X: anion).

For example, P. Le et al. reported that Li^+ and $[TFSI]^-$ may form complexes of $[Li(TFSI)_2]^-$ in certain binary mixtures [2]. The hydrodynamic radius of Li^+ is increased due to the complex formation, and the viscosity of the mixture is increased with increasing lithium salt concentration, resulting in an ionic conductivity decrease. In comparison with the conventional electrolyte (lithium salt and organic solvent), the quaternary ammonium cation–based ILs (QAILs) showed a good oxidation stability with a potential of more than 6.0 V versus Li/Li^+. The oxidation potential of QAILs is slightly higher than that of $LiPF_6$/EC:DMC (2:1), an electrolyte for high-performance lithium–ion batteries. The imidazolium ionic liquid exhibits a lower oxidation stability than others. The potential limitation of 1.5 V is found to be comparable for all the ILs with TFSI anion. The reduction wall is associated with the imide anion reduction and the lack of passive layer formation [29–32]. It is the fact that the addition of lithium salt increases the reduction stability with the formation of a passive layer derived from an insoluble salt.

In the binary mixture of Li salt/IL, complexes between anion and Li+ make Li+ relatively low, resulting in low limiting current density in the Li–ion cells. Matsumoto et al. reported that the viscosity of ILs is significantly affected by the anion structure; thus, the design of anion structure is critical.

The Li+ mobility in the electrolyte was enhanced by low viscosity. Li salts were dissolved in various ILs. The advantages of [TFSI]-based ILs as electrolytes for lithium batteries have been reported by many researchers [2,3,7,33]. The electrode/electrolyte interfacial charge transfer process is also greatly affected by the anion structure of the ILs. At the interface, the decomplexation (desolvation) of $[Li(X)_x](x - 1)^-$ occurs and anions are liberated. Therefore, the interaction between Li+ and anions has a significant effect on the interfacial charge transfer process [34]. Graphite is a representative, commonly employed as negative electrode material for Li–ion batteries. During the charging of a Li–ion cell, the reduction of graphite occurs and Li+ intercalates into the layered graphite structure to maintain electrical neutrality. In the case of lithium salt/IL electrolytes (except for [FSA]-based ILs), it is known that the organic cation is inserted into the graphite instead of Li+, and destruction of the layered structure of graphite takes place. However, in the [FSA]-based ILs, the intercalation of organic cation into graphite is suppressed and reversible Li+ intercalation occurs [5].

Figure 10.5 shows first charge and cycling performance of hard carbon electrode in binary mixtures of Li[TFSI]/ILs. In the case of Li[TFSI]/[C$_2$mim][TFSI], the [C$_2$mim]+ cation is intercalated into the graphite at an electrode potential of ca. 1 V versus Li/Li+ during the first cathodic scan, and this intercalation is irreversible. On the other hand, in the case of [FSA]-based ILs, the Li+ intercalation reaction takes place reversibly in the potential range of 0.2–0 V. In addition, in the case [FSA]-based ILs, the reductive decomposition of [C$_2$mim]+ cation at the negative electrode is also prohibited. At present, it is not clear why the intercalation

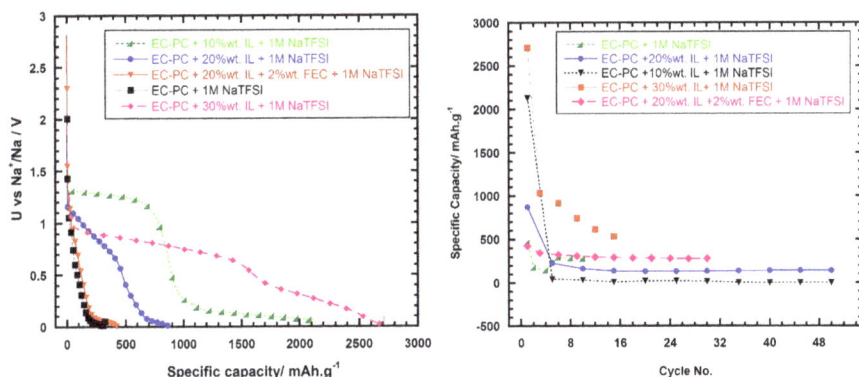

FIGURE 10.5   (a) 1st discharge capacity of half-cell (-) Na | EC-PC (1:1) + x %wt. EMI-TFSI +1M NaTFSI | Hard carbon, (b) Cycling discharge capacity as a function cycle number.

of organic cation and decomposition of $[C_2mim]^+$ cation are prevented in the [FSA]-based ILs. A small amount of $[FSA]^-$ anion decomposition during the initial charging stabilizes passivation layer on the electrode surface. To support this hypothesis, another mechanism was also proposed recently [35]. As well as lithium salt/organic IL binary systems, the sodium salt/organic ILs can also be used as the electrolytes for Na and Na–ion batteries [5].

10.2.2.3   Effect of Organic Solvent Added to ILs

The effects of the incorporation of ethylene carbonate (EC) or dimethyl carbonate (DMC) on the physicochemical and electrochemical properties of ionic liquids (ILs) based on aliphatic quaternary ammonium and imide anion were studied [4]. The evolution of the melting point, glass transition, ionic conductivity, diffusion coefficient, and electrochemical stability were evaluated. The addition of a low amount of solvent, that is, 20 wt% resulted in significantly improve the conductivity values, reaching 12 mS cm^{-1} at 40°C. The incorporation of a polar solvent, EC, has no positive effect on the IL dissociation. Moreover, the incorporation of EC in ILs improves the electrochemical stability toward reduction, whereas the high anodic stability is maintained. The addition of LiTFSI in IL + solvent mixed electrolytes reduces the ionic conductivity, but still higher than pure ILs, showing the beneficial effect of the additive solvent [4].

Phung Le and co-workers [5] conducted a calculation based on VTF fit of ionic conductivity in the range of ~25°C–~60°C for supercooled liquids and glasses rather than a straight-line Arrhenius behavior to explain the conduction mechanism of the complex electrolyte:

$$\sigma = A \cdot \exp\left(-\frac{E_a}{R(T - T_o)}\right)$$

where T_o is referred to as the Vogel temperature, equal to the glass transition in ideal glasses, the effects of charge carrier concentration, often related to the refactor, A, and segmental motion, related to the activation energy, E_a, on overall conductivity, σ, at a given temperature T. Fits were performed by linearizing the data according to the form of the equation. The solver tool within Microsoft Excel and verified by the fminbnd function of MATLAB® R2016b was used with manually varying the value of T_o confirming the local maximum in R-squared to maximize the linearity of the resulting data.

Table 10.5 showed that the activation energy of EMI-TFSI has a decreasing tendency with an increase in the percentage of EC. This is due to the fact that dilution of solutions partially holds up ionic bond slightly, which is easy to them extract and become non-electrical charged particles. On the contrary, an opposite tendency for electrolytes with an IL used as a co-solvent was observed with an increase in the IL addition amount. This result is due to the increase in viscosity as well as the presence of significant ionic interactions between anion–cation and ion–dipole molecules. However, the degree of activation energy values deduced are gently lower than those of an IL used as the main solvent (Table 10.5).

TABLE 10.5

Activation Energy of the Mixed Electrolytes: EMI-TFSI – $x$wt% EC and EC-PC (1:1) + x wt% EMI-TFSI [5]

Electrolytes (EMI-TFSI as Main Solvent)	E_a (J mol^{-1})	Electrolytes (EMI-TFSI as Co-solvent)	E_a (J mol^{-1})
Pure IL	2714	EC-PC + 1 M NaTFSI	1812
IL + 0.5 M NaTFSI	3267	EC-PC + 10% IL + 1 M NaTFSI	1927
IL + 5% EC + 0.5 M NaTFSI	3009	EC-PC + 20% IL + 1 M NaTFSI	2097
IL + 10% EC + 0.5 M NaTFSI	2868	EC-PC + 25% IL + 1 M NaTFSI	2147
IL + 15% EC + 0.5 M NaTFSI	2726	EC-PC + 30% IL + 1 M NaTFSI	2219
IL + 20% EC + 0.5 M NaTFSI	–	EC-PC + 20% IL + 2% FEC + 1 M NaTFSI	1832
IL + 25% EC + 0.5 M NaTFSI	2477	EC-PC + 25% IL + 2% FEC + 1 M NaTFSI	1928
IL + 30% EC + 0.5 M NaTFSI	2564	EC-PC + 30% IL + 2% FEC + 1 M NaTFSI	2011
		EC-PC + 50% IL + 1 M NaTFSI	2320

Reproduced with permission. Copyright 2010, ACS.

10.2.3 SOLVENT-IN-SALT ELECTROLYTES

We can also prepare electrolytes using various solvents to enhance the solvation of lithium salt rather than ILs [2–4,7]. In these electrolytes mixed with alkaline salt, the Li$^+$ and solvent form a solvate [Li(solvent)$_x$]$^+$ cation. The coordination number of Li$^+$ in liquids is typically 4–5. Therefore, in extremely dilute electrolytes with the amount of solvent higher than that of salt, all the solvent molecules can be assumed to be in the first solvation shell of Li$^+$, and free solvents thus scarcely exist in the solution. However, the extremely concentrated electrolytes possess unique properties, some of them either very similar to those of solvate ILs or different from those of common electrolyte solutions with salt concentrations less than 2 M [36]. Regarding the organic solvent-in-salt electrolytes for Li batteries in a review article by Yamada et al., the advantages of the electrolytes, such as the wide electrochemical windows, suppression of volatility, suppression of Al corrosion, and highly reversible reactions of Li metal and graphite electrodes, were well highlighted [36]. Recently, the solvent-in-salt concept was extended to the aqueous electrolytes [37–41] for Li–ion cells. The anodic limit expansion can be attributed to the decrease in solvent in the solution. The cathodic limit expansion is attributed to the formation of a LiF-based passivation layer, which is a decomposition product of the [TFSI]$^-$ anion. The layer passivates the electrode surface and suppresses the further reductive decomposition of electrolyte.

10.2.4 LI$^+$-CONDUCTING POLYMER ELECTROLYTES CONTAINING IONIC LIQUIDS

ILs can be mixed with polymers [42], resulting in gels, hereafter termed "ion-gels," which can be used as electrolytes for batteries. In ion-gels (polymer electrolyte), ILs are confined in the polymer matrix, and this is advantageous to avoid the leakage of

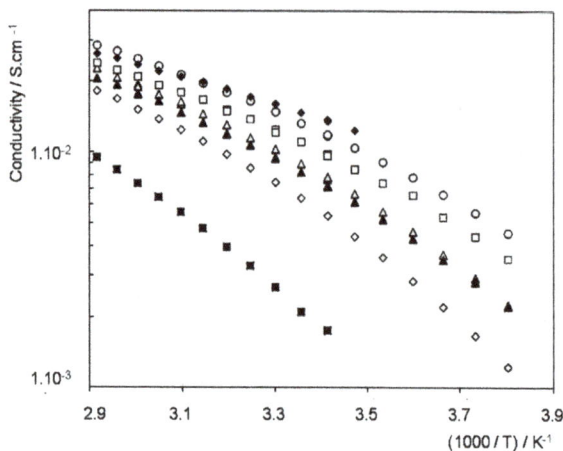

FIGURE 10.6 Ionic conductivity as a function of temperature for N_{1114}TFSI + solvent: (■) N_{1114}TFSI, (◊) N_{1114}TFSI + 20 wt % EC, (Δ) N_{1114}TFSI + 30 wt % EC, (○) N_{1114}TFSI + 50 wt % EC, (▲) N_{1114}TFSI + 20 wt % DMC, (□) N_{1114}TFSI + 30 wt % DMC, (◆) N_{1114}TFSI + 50 wt % DMC [4], Reproduced with permission. Copyright 2010, ACS.

electrolytes and resolve the safety issue of batteries. Ion-gels, however, are clearly different from conventional polymer electrolytes. In the case of L^{i+-}-conducting conventional polymer electrolytes, Li salts are dissolved in polymers such as poly(ethylene oxide) (PEO), and the ionic conduction is coupled with the segmental motion of the polymer chains [43]. Therefore, the ionic conductivities of conventional polymer electrolytes are as low as 10^{-6}–10^{-4} S cm^{-1} at room temperature. On the other hand, in the case of ion-gels, the liquid-state salts such as Li salt/organic ILs are mixed with polymers and the ILs behave as both charge carriers and plasticizers in the gels. By using ILs, we can prepare "polymer-in-salt" [44]-type electrolytes having relatively high ionic conductivity of ca. 10^{-4}–10^{-3} S cm^{-1} at room temperature. Polyethylene oxide (PEO) [45], poly(methyl methacrylate) (PMMA) [46], and copolymer of poly(vinylidene fluoride–co-hexafluoropropylene) (PVDF-HFP) [47] have been reported to be compatible with ILs. It is known that the PVDF-HFP has relatively good mechanical strength due to the partially crystalline nature even if some plasticizer is included in the matrix [42]. The ionic conductivity is enhanced when increasing the content of IL in an ion-gel; however, the gel also becomes mechanically weaker. To achieve both high ionic conductivity and sufficient mechanical strength, the polymer cross-linking is effective [46].

10.3   SYNTHESIS OF IONIC LIQUIDS

10.3.1   TYPICAL IONIC LIQUID SYNTHETIC ROUTE

The general synthesis of ILs involves a consecutive quaternization–metathesis reaction procedure as shown in Figure 10.7. Many ILs are prepared by a metathesis reaction from a halide or similar salt of the desired cation. We also scanned the different routes using differing energy sources to promote IL syntheses, via conductive heating (conventional heating), microwave irradiation, and ultrasonic irradiation [48].

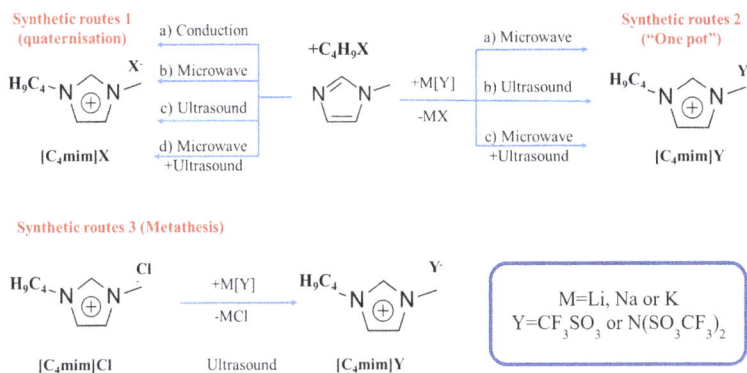

FIGURE 10.7 Typical IL synthesis routes.

10.3.1.1 Synthetic Route 1 (Quaternization)

Scheme 10.1 shows that most of the 1-alkyl-3-imidazolium halide salt preparations are executed using traditional heating under reflux, different from nowadays with nitrogen or argon used, since it prevents moisture in ILs and makes it more colorless [49].

A slightly excess amount of 1-haloalkane and long reaction time are required to achieve good yields and promote the completion of the reactions at a reasonable rate (Scheme 10.2) in the preparation of 1-alkyl-3-methylimidazolium halide salts.

The purification of 1-alkyl-3-methylimidazolium halide salts (Figure 10.8) shows a way to remove the excess 1-haloalkane, unconverted 1-methylimidazole, and any solvent remaining during the preparation.

The synthesis of ionic liquids was promoted by microwave irradiation because of the ability to control reaction conditions very specifically, monitoring temperature,

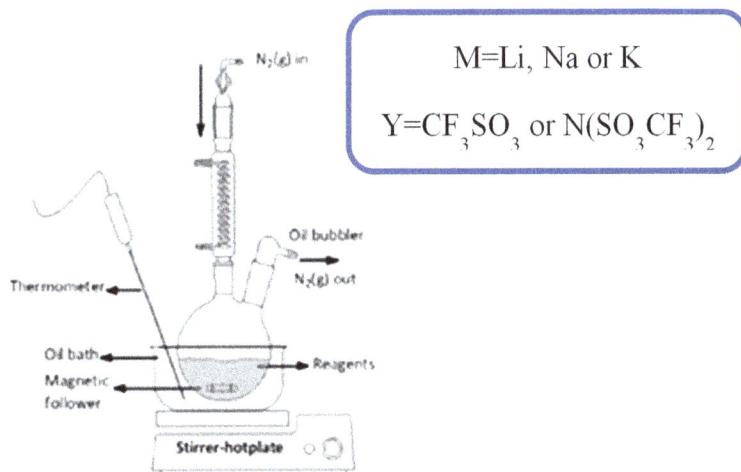

$$M=Li, Na \text{ or } K$$

$$Y=CF_3SO_3 \text{ or } N(SO_3CF_3)_2$$

SCHEME 10.1 1-Alkyl-3-imidazolium halide salt preparations with conventional heating under reflux.

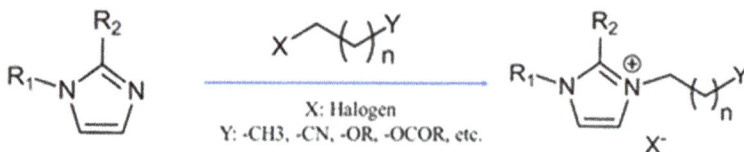

SCHEME 10.2 Preparations of l-alkyl-3-methylimidazolium halide salts using an excess of 1-haloalkane and long reaction time.

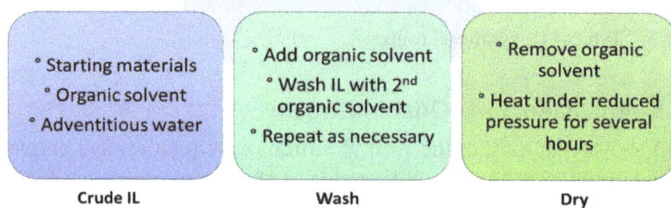

Crude IL	Wash	Dry
° Starting materials ° Organic solvent ° Adventitious water	° Add organic solvent ° Wash IL with 2nd organic solvent ° Repeat as necessary	° Remove organic solvent ° Heat under reduced pressure for several hours

FIGURE 10.8 Purification of 1-alkyl-3-methylimidazolium halide salts.

pressure, and reaction time [50]. The promotion of 1-alkyl-3-methylimidazolium halide syntheses using microwave irradiation is favored by an ionic conduction heating mechanism as ILs well absorb microwave irradiation (Scheme 10.3).

Synthesis of ionic liquids could be promoted by ultrasonic irradiation at about the same time, appear ultrasound-promoted preparations of ILs (Scheme 10.4) [51,52].

Synthesis of ionic liquids was promoted via halide-free route. Most ionic liquids are usually prepared by quaternization of imidazoles, alkyl amines, or phosphines, often employing alkyl halides as the alkylating agents, followed by anion metathesis. The anion metathesis methods produce many good-quality ILs, but the contamination by residual halide could limit the production of high-purity materials. The halide traces in the resulting ILs drastically change the physical and electrochemical properties of the neat ILs [53].

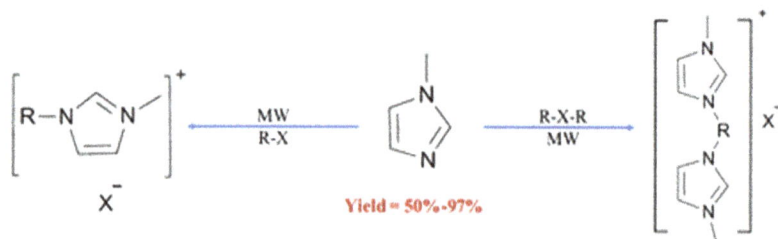

SCHEME 10.3 Microwave-promoted preparation of 1-alkyl-3-methylimidazolium halide.

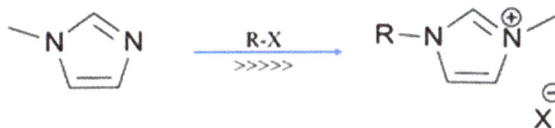

SCHEME 10.4   Ultrasound-promoted preparation of 1-alkyl-3-methylimidazolium halide.

Therefore, various synthetic strategies have been devised to synthesize via a halide-free route [54]. For example, the direct synthesis of 1,3-disubstituted imidazolium tetrafluoroborate ILs achieved in a one-step procedure (Scheme 10.5) affords a mixture of ILs.

10.3.1.2   Metathesis Reaction

Most of the common ILs are prepared via metathesis with a metal salt or an acid base neutralization reaction producing a stoichiometric amount of waste MX or HX (Scheme 10.6) [6]. Hydrophobic ILs are easily separated from reaction media if the

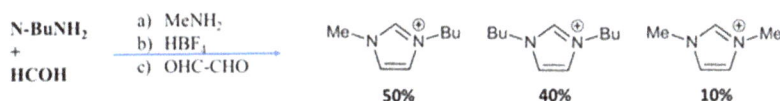

SCHEME 10.5   Halogen-free synthesis of ILs.

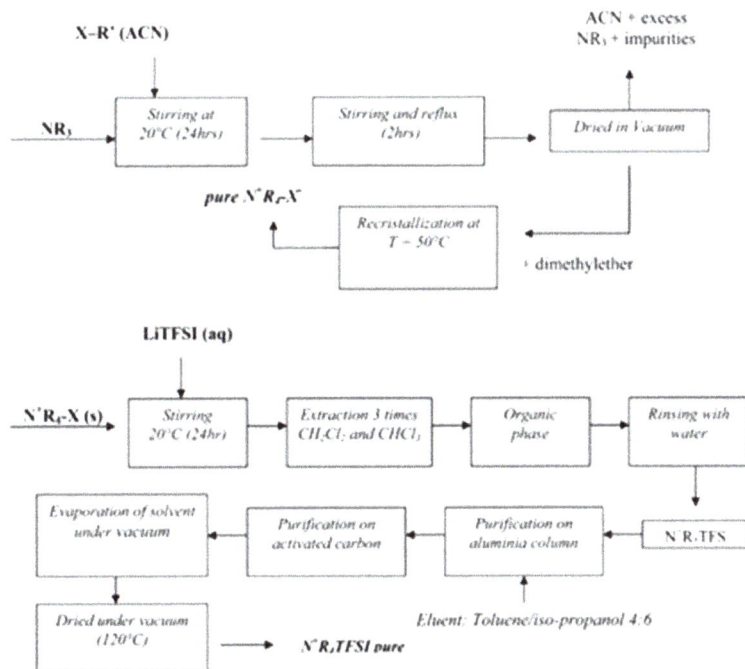

SCHEME 10.6   Synthesis and preparation of IL TFSI salts [6].

metathesis reaction occurs in water. However, the slow decantation and removal of water require a large energy input. For the water-miscible ionic liquids, the separation of the by-products from the desired IL becomes challenging. To make feasible this separation, the exchange reaction of the halide anion is generally performed in the presence of silver salt associated with numerous anions, e.g., dicyanamide, thiocyanate, and tetrafluoroborate. Consequently, the metathesis reactions are generally occur in an aqueous solution with either the free acid of the appropriate anion or its ammonium or alkali metal salt. The resulting ILs are extracted from the aqueous solution by an organic solvent. However, ILs are contaminated by appreciable amounts of chloride or bromide ions. This lowers the yield of the final product since their elimination requires several aqueous washings. The use of small water volume and lowering the temperature to ~0°C can reduce the loss of ILs affording yields in the range of 70%–80%. The use of molecular solvent, of silver salt, and the generated amounts of salts require removal and subsequent disposal, which do not comply with the first principle of green chemistry.

10.4 APPLYING ILs FOR LI–ION/NA–ION BATTERIES

Low-melting temperature ionic liquids (ILs), based on aliphatic quaternary ammonium N_{1xxx} and 1-ethyl-3-methylimidazolium cations with the imide anion, were prepared and characterized [3]. The physicochemical and electrochemical properties of these ILs, including melting point, degradation temperature, viscosity, density, ionic conductivity, and electrochemical stability, were determined. The literature reported that fluorinated imidazolium cation has good electrochemical stability compared to the non-fluorinated form. The addition of an organic solvents to the ionic liquids affected the high viscosity of aliphatic ammonium-based ILs without detriment to the interesting properties of ILs (i.e., nonvolatility, non-flammability, high thermal stability, electrochemical stability), resulting in good cycling performance of the mixtures IL–ethylene carbonate was characterized using $LiMn_2O_4$-based electrode materials in the voltage range of 3.5–4.3 V (Figure 10.9) [3].

The prepared gel polymer electrolyte (GPE) films using the polyvinylidene fluoride–hexafluoropropylene (PVDF-HFP) matrix achieved good sodium–ion conductivity at room temperature. GPEs of PVDF-HFP were formed by microporous PVDF-HFP membranes filled and swollen in different liquid electrolytes: $NaClO_4$, $NaPF_6$, and NaTFSI salts dissolved in mixtures of carbonate solvents (propylene carbonate (PC), fluoroethylene carbonate (FEC)) or ionic liquids (1-ethyl-3-methylimidazolium bis(trifluoromethanesulfonyl)imide and 1-butyl-3-methyl bis(trifluoromethanesulfonyl) imide). GPEs showed a "sponge-like" structure with a high porosity of 80–85% and a large pore size. The ionic conductivity of GPEs reached 1–2 mS cm^{-1} at room temperature. Sodium intercalation into a $Na_{0.44}MnO_2$ electrode was performed in coin-cell type by using GPEs as the conduction media. Cycling data of sodium cell using PVDF-HFP/$NaClO_4$ 1 M-PC:2%FEC exhibited an excellent stable specific capacity of 100 mAh g^{-1} at room temperature (Figure 10.10) [42].

The use of ionic liquid–based electrolyte can solve the safety problems associated with lithium–ion/sodium–ion batteries, especially for large-scale applications due to their thermal stability and inflammability. The main challenge of ionic liquids is the

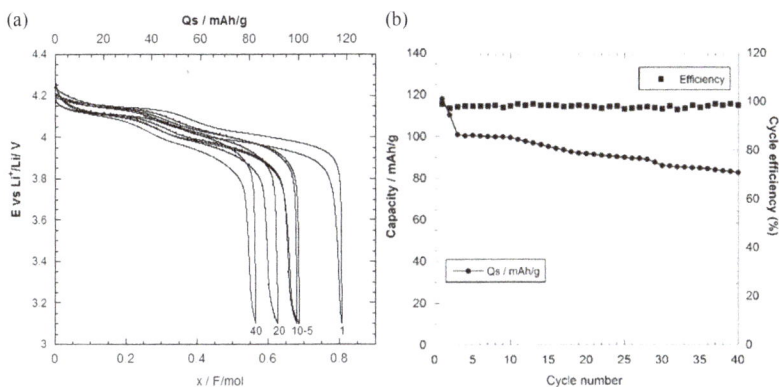

FIGURE 10.9   (a) Discharge curve of the cells using electrode $LiMn_2O_4$ in the electrolyte IL+ 20% EC with different lithium percentage intercalated and its (b) cycling performance.

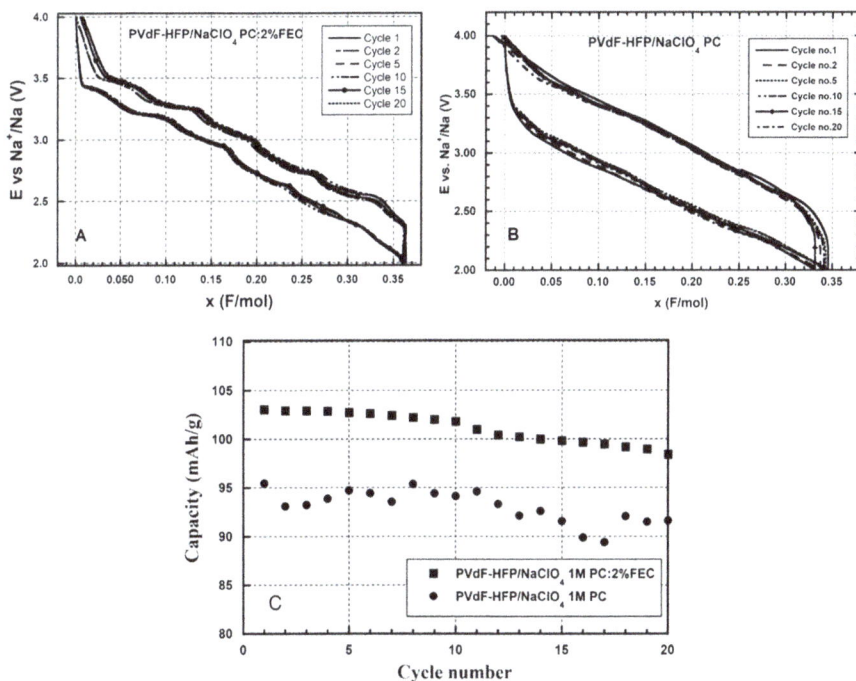

FIGURE 10.10   Charge discharge curves at various cycles in the electrolyte (a) with FEC additive (b) without using FEC of the cells using the electrode $Na_{0.44}MnO_2$ and PVDF-HFP separator and (c) cycling performance of the cells.

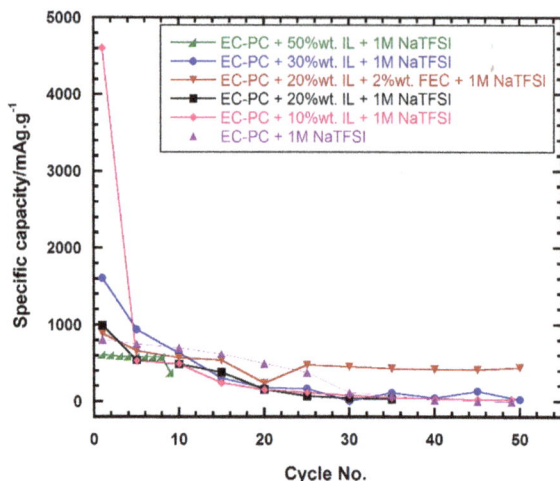

FIGURE 10.11 Specific capacity versus cycle number of SnS/C anode material based sodium half-cell in different electrolytes.

relatively high viscosity, and the use of organic solvent at suitable content could be an alternative solution for reducing viscosity; meanwhile, the electrochemical properties as well as the good thermal stability of the mixed solutions remain similar to those of the pure ionic liquid. Electrolyte mixtures based on EMI-TFSI with organic solvents (EC and PC) have good electrochemical compatibility in half-cell configuration with respect to sodium metal anode and the electrode materials used: SnS/C, hard carbon (HC), and $Na_{0.44}MnO_2$. Moreover, the thermal stability, the flammability, and the conduction mechanism of electrolyte mixtures were also explored and discussed [5].

With increasing amounts of IL in the binary electrolytes, the irreversible capacity is also climbed in the first cycle of discharge due to the formation of SEI layer on the anode material (Figure 10.11). In the following cycles, the discharge capacity will be stabilized. However, with increasing of ILs amount in EC–PC mixture, first irreversible increase significantly and the battery lost completely performance in following cycles, especially up to 30 wt% IL. The stable performance is only maintained within a good value of capacity when using FEC additive due to stabilizing SEI layer to prevent the electrolyte reduce or oxidize of electrolyte so far after first discharge cycle.

REFERENCES

1. Vignarooban, K.; Kushagra, R.; Elango, A.; Badami, P.; Mellander, B. E.; Xu, X.; Tucker, T. G.; Nam, C.; Kannan, A. M. Current trend and future challenges of electrolytes for sodium-ion batteries. *International Journal of Hydrogen Energy*, 41, 2829–2846, 2016.
2. Le, L. T.; Vo, T. D.; Ngo, K. H.; Okada, S.; Alloin, F.; Garg, A.; Le, P. M. Mixing ionic liquids and ethylene carbonate as safe electrolytes for lithium-ion batteries. *Journal of Molecular Liquids*, 271, 769–777, 2018.

3. Le, M. L. P.; Tran, N. A.; Ngo, H. P. K.; Nguyen, T. G. Liquid electrolytes based on ionic liquids for lithium-ion batteries. *Journal of Solution Chemistry*, 44(12), 2332–2343, 2015.

4. Le, M. P.; Cointeaux, L.; Strobel, P.; Leprêtre, J. C.; Judeinstein, P.; Alloin, F. Influence of solvent addition on the properties of ionic liquids. *The Journal of Physical Chemistry C*, 116(14), 7712–7718, 2012.

5. Le, L. T. M.; Vo, T. D.; Nguyen, Q. D.; Okada, S.; Alloin, F.; Le, P. M. L. High performance electrolyte using mixtures of ionic liquid–solvent for sodium-ion batteries. *ECS Transactions*, 85(13), 215–226, 2018.

6. Le, M. L. P.; Alloin, F.; Strobel, P.; Leprêtre, J. C.; Pérez del Valle, C.; Judeinstein, P. Structure–properties relationships of lithium electrolytes based on ionic liquid. *The Journal of Physical Chemistry B*, 114(2), 894–903, 2009.

7. Alloin, F.; Strobel, P.; Leprêtre, J. C.; Cointeaux, L.; del Valle, C. P. Electrolyte based on fluorinated cyclic quaternary ammonium ionic liquids. *Ionics*, 18(9), 817–827, 2012.

8. Tarascon, J. M.; Armand, M. Issues and challenges facing rechargeable lithium batteries. Nature, 414, 171–179, 2011.

9. Armand, M.; Tarascon, J. M. Building better batteries. *Nature*, 451(7179), 652, 2008.

10. Goodenough, J. B.; Kim, Y. Challenges for rechargeable Li batteries. *Chemistry of Materials*, 22(3), 587–603, 2009.

11. Xu, K. Nonaqueous liquid electrolytes for lithium-based rechargeable batteries. *Chemical Reviews*, 104(10), 4303–4418, 2004.

12. Xu, K. Electrolytes and interphases in Li-Ion batteries and beyond. *Chemical Reviews*, 114, 11503–11618, 2014.

13. Tokuda, H.; Watanabe, M. Characterization and ionic transport properties of nano-composite electrolytes containing a lithium salt of a superweak aluminate anion. *Electrochimica Acta*, 48, 2085–2091, 2003.

14. Tokuda, H.; Tabata, S. I.; Susan, M. A. B. H.; Hayamizu, K.; Watanabe, M. Design of polymer electrolytes based on a lithium salt of a weakly coordinating anion to realize high ionic conductivity with fast charge-transfer reaction. *The Journal of Physical Chemistry B*, 108, 11995–12002, 2004.

15. Fujinami, T.; Buzoujima, Y. Novel lithium salts exhibiting high lithium ion transference numbers in polymer electrolytes. *Journal of Power Sources*, 119–121, 438–441, 2003.

16. Tao, R.; Miyamoto, D.; Aoki, T.; Fujinami, T. Novel liquid lithium borates characterized with high lithium ion transference numbers. *Journal of Power Sources*, 135, 267–272, 2004.

17. Shobukawa, H.; Tokuda, H.; Tabata, S.; Watanabe, M. Preparation and transport properties of novel lithium ionic liquids. *Electrochimica Acta*, 50, 305–309, 2004.

18. Shobukawa, H.; Tokuda, H.; Susan, M. A. B. H.; Watanabe, M. Ion transport properties of lithium ionic liquids and their ion gels. *Electrochimica Acta*, 50, 3872–3877, 2005.

19. Tokuda, H.; Watanabe, M. Physicochemical properties and structures of ionic liquids and their utilization for lithium rechargeable batteries. *Battery Technology (Committee of Battery Technology, Electrochemical Society of Japan)*, 20, 65–71, 2008.

20. Tsunashima, K.; Sugiya, M. Physical and electrochemical properties of low-viscosity phosphonium ionic liquids as potential electrolytes. *Electrochemistry Communications*, 9, 2353–2358, 2007.

21. Girard, G. M. A.; Hilder, M.; Zhu, H.; Nucciarone, D.; Whitbread, K.; Zavorine, S.; Moser, M.; Forsyth, M.; MacFarlane, D. R.; Howlett, P. C. Electrochemical and physicochemical properties of small phosphonium cation ionic liquid electrolytes with high lithium salt content. *Physical Chemistry Chemical Physics*, 17, 8706–8713, 2015.

22. Tokuda, H.; Hayamizu, K.; Ishii, K.; Susan, M. A. B. H.; Watanabe, M. Physicochemical properties and structures of room temperature ionic liquids. 2. Variation of alkyl chain length in imidazolium cation. *The Journal of Physical Chemistry B*, 109, 6103–6110, 2005.

23. Matsumoto, H.; Sakaebe, H.; Tatsumi, K. Li/LiCoO$_2$ cell performance using ionic liquids composed of N, N-diethyl-N-methyl-N-(2-methoxyethyl)ammonium – effect of anionic structure. *ECS Transactions*, 16, 59–66, 2008.

24. Sato, T.; Maruo, T.; Marukane, S.; Takagi, K. Ionic liquids containing carbonate solvent as electrolytes for lithium ion cells. *Journal of Power Sources*, 138, 253–261, 2004.

25. Tsuzuki, S.; Hayamizu, K.; Seki, S.; Ohno, Y.; Kobayashi, Y.; Miyashiro, H. Quaternary ammonium room-temperature ionic liquid including an oxygen atom in side chain/lithium salt binary electrolytes: ab initio molecular orbital calculations of interactions between ions. *The Journal of Physical Chemistry B*, 112, 9914–9920, 2008.

26. Seki, S.; Kobayashi, Y.; Miyashiro, H.; Ohno, Y.; Mita, Y.; Usami, A.; Terada, N.; Watanabe, M. Reversibility of lithium secondary batteries using a room-temperature ionic liquid mixture and lithium metal. *Electrochemical and Solid-State Letters*, 8, A577–A578, 2005.

27. Seki, S.; Kobayashi, Y.; Miyashiro, H.; Ohno, Y.; Usami, A.; Mita, Y.; Watanabe, M.; Terada, N. Highly reversible lithium metal secondary battery using a room temperature ionic liquid/lithium salt mixture and a surface-coated cathode active material. *Chemical Communications*, 42, 544–545, 2006.

28. Seki, S.; Ohno, Y.; Miyashiro, H.; Kobayashi, Y.; Usami, A.; Mita, Y.; Terada, N.; Hayamizu, K.; Tsuzuki, S.; Watanabe, M. Quaternary ammonium room-temperature ionic liquid/lithium salt binary electrolytes: electrochemical study. *Journal of the Electrochemical Society*, 155, A421–A427, 2008.

29. Matsumoto, H.; Yanagida, M.; Tanimoto, K.; Kojima, T.; Tamiya, Y.; Mizayaki, Y. Improvement of ionic conductivity of room temperature molten salt based on quaternary ammonium cation and imide anion. In: P.C. Trulove, et al. (Eds.), *Molten Salt XII*, Electrochemical Society, Pennington, NJ, pp. 186–192, 2000.

30. Sakaebe, H.; Matsumoto, H. N-methyl-N-propylpiperidinium bis (trifluoromethane-sulfonyl) imide (PP13–TFSI) – novel electrolyte base for Li battery, *Electrochemistry Communications*, 5, 594–598, 2003.

31. Kiani, M. A.; Mousavi, M. F.; Rahmanifar, M. S. Synthesis of nano- and micro-particles of LiMn2O4: electrochemical investigation and assessment as a cathode in Li battery, *International Journal of Electrochemical Science*, 6, 2581–2595, 2011.

32. Bolloli, M.; Kalhoff, J.; Alloin, F.; Bresser, D.; Phung Le, M. L.; Langlois, B., Passerini, S.; Sanchez, J.-Y. Fluorinated carbamate as suitable solvent for LiTFSI-based lithium-ion electrolytes: physicochemical properties and electrochemical characterization. *The Journal of Physical Chemistry C*, 119(39), 22404–22414, 2015.

33. Tran, A. N.; Van Do, T. N.; Le, L. P. M.; Le, T. N. Synthesis of new fluorinated imid-azolium ionic liquids and their prospective function as the electrolytes for lithium-ion batteries. *Journal of Fluorine Chemistry*, 164, 38–43, 2014.

34. Zhou, Q.; Henderson, W. A.; Appetecchi, G. B.; Passerini, S. (). Phase behavior and thermal properties of ternary ionic liquid–lithium salt (IL– IL– LiX) electrolytes. *The Journal of Physical Chemistry C*, 114(13), 6201–6204, 2010.

35. Yamagata, M.; Nishigaki, N.; Nishishita, S.; Matsui, Y.; Sugimoto, T.; Kikuta, M.; Higashizaki, T.; Kono, M.; Ishikawa, M. Charge–discharge behavior of graphite nega-tive electrodes in bis(fluorosulfonyl)imide-based ionic liquid and structural aspects of their electrode/electrolyte interfaces. *Electrochimica Acta*, 110, 181–190, 2013.

36. Yamada, Y.; Yamada, A. Review of superconcentrated electrolytes for lithium batteries. *Journal of the Electrochemical Society*, 162, A2406–A2423, 2015.

37. Suo, L.; Borodin, O.; Gao, T.; Olguin, M.; Ho, J.; Fan, X.; Luo, C.; Wang, C.; Xu, K. "Water-in-salt" electrolyte enables high-voltage aqueous lithium-ion chemistries. *Science*, 350, 938–943, 2015.

38. Suo, L.; Han, F.; Fan, X.; Liu, H.; Xu, K.; Wang, C. "Water-in-salt" electrolytes enable green and safe Li-ion batteries for large scale electric energy storage applications. *Journal of Materials Chemistry A*, 4, 6639–6644, 2016.

39. Suo, L.; Borodin, O.; Sun, W.; Fan, X.; Yang, C.; Wang, F.; Gao, T.; Ma, Z.; Schroeder, M.; von Cresce, A.; Angell, C.; Xu, K.; Wang, C. Advanced high-voltage aqueous lithium-ion battery enabled by "water-in-bisalt" electrolyte. *Angewandte Chemie – International Edition*, 55, 7136–7141, 2016.

40. Miyazaki, K.; Shimada, T.; Ito, S.; Yokoyama, Y.; Fukutsuka, T.; Abe, T. Enhanced resistance to oxidative decomposition of aqueous electrolytes for aqueous lithium-ion batteries. *Chemical Communications*, 52, 4979–4982, 2016.

41. Yamada, Y.; Usui, K.; Sodeyama, K.; Ko, S.; Tateyama, Y.; Yamada, A. Hydrate-melt electrolytes for high-energy-density aqueous batteries. *Nature Energy*, 1, 16129, 2016.

42. Vo, D. T.; Do, H. N.; Nguyen, T. T.; Nguyen, T. T. H.; Okada, S.; Le, M. L. P. Sodium ion conducting gel polymer electrolyte using poly (vinylidene fluoride hexafluoropropylene). *Materials Science and Engineering: B*, 241, 27–35, 2019.

43. Armand, M. Polymer solid electrolytes – an overview. *Solid State Ionics*, 9–10, 745–754, 1983.

44. Angell, C. A.; Liu, C.; Sanchez, E. Rubbery solid electrolytes with dominant cationic transport and high ambient conductivity. *Nature*, 362, 137–139, 1993.

45. Shin, J.-H.; Henderson, W. A.; Passerini, S. An elegant fix for polymer electrolytes. *Electrochemical and Solid-State Letters*, 8, A125–A127, 2005.

46. Susan, M. A. B. H.; Kaneko, T.; Noda, A.; Watanabe, M. Ion gels prepared by in situ radical polymerization of vinyl monomers in an ionic liquid and their characterization as polymer electrolytes. *Journal of the American Chemical Society*, 127, 4976–4983, 2005.

47. Fuller, J.; Breda, A. C.; Carlin, R. T. Ionic liquid–polymer gel electrolytes from hydrophilic and hydrophobic ionic liquids. *Journal of Electroanalytical Chemistry*, 459, 29–34, 1998.

48. Abdul-Sada, A. A. K.; Elaiwi, A. E.; Greenway, A. M.; Seddon, K. R. Evidence for the clustering of substituted imidazolium salts via hydrogen bonding under the conditions of fast atom bombardment mass spectrometry. *European Mass Spectrometry*, 3, 245–247, 1997.

49. Tsuzuki, S.; Hayamizu, K.; Seki, S.; Ohno, Y.; Kobayashi, Y.; Miyashiro, H. Quaternary ammonium room-temperature ionic liquid including an oxygen atom in side chain/ lithium salt binary electrolytes: ab initio molecular orbital calculations of interactions between ions. *The Journal of Physical Chemistry B*, 112(32), 9914–9920, 2008.

50. Rogers, R. D.; Seddon, K. R. *Ionic Liquids: Progress and Prospects* (ACS Symposium Series, Vol. 856), American Chemical Society, Washington, DC, pp. 100–107, 2003.

51. Namboodiri, V. V.; Varma, R. S. Solvent-free sonochemical preparation of ionic liquids. *Organic Letters*, 4(18), 3161–3163, 2002.

52. Lévêque, J. M.; Luche, J. L.; Pétrier, C.; Roux, R.; Bonrath, W. An improved preparation of ionic liquids by ultrasound. *Green Chemistry*, 4(4), 357–360, 2002.

53. Seddon, K. R.; Stark, A.; Torres, M. J. Influence of chloride, water, and organic solvents on the physical properties of ionic liquids. *Pure and Applied Chemistry*, 72(12), 2275–2287, 2000.

54. Dupont, J.; de Souza, R. F.; Suarez, P. A. Ionic liquid (molten salt) phase organometallic catalysis. *Chemical Reviews*, 102(10), 3667–3692, 2002.

11 Imidazolium-Based Ionogels via Facile Photopolymerization as Polymer Electrolytes for Lithium–Ion Batteries

Yu-Chao Tseng and Jeng-Shiung Jan
National Cheng Kung University

CONTENTS

11.1 INTRODUCTION

Batteries have been developed for more than 100 years and have a great impact on our daily life. Recently, due to the use of mobile devices, a battery with high capacity and high safety is especially in demand. Apart from mobile devices, electric vehicles are now ready to market and also in need of batteries with high capacity as well as high energy density. Li–ion batteries possess the above properties, and they are now the most promising candidate for commercialization.[1–4] Electrolytes are the most important in batteries, as they carry ions between a pair of electrodes to complete a circuit of charges. At least the following requirements should be met with the electrolyte materials: good ionic conductivity, high thermal stability, high mechanical property, a wide electrochemical window, inertness to other components, and great durability. The commercial electrolytes are mixtures of carbonates and lithium salts; however, being volatile and being flammable are big problems to these kinds of liquid electrolytes.[5–7] To solve these problems, it is worth researching other materials to replace liquid electrolyte, such as polymer electrolyte. Polymer electrolytes are suitable alternatives to solve security problems since polymer matrix can afford high thermal, mechanical, chemical, and electrochemical stability, but some major problems still limit their utilization, including low ionic conductivity and incompatibility of the interfaces between electrodes and electrolytes. Combining the aforementioned discussion, a moderate addition of ionic liquid (IL) may have the possibility to promote the performance of the electrolyte system.

Ionic liquids (ILs) have attracted lots of attention to be the suitable solution of these disadvantages due to their negligible flammability, broad electrochemical window, and vapor pressure, which are especially in demand of the electrolytes for next-generation batteries.[8–15] Poly(ionic liquids) (PILs), which are macromolecular analogs of ILs, have emerged as a new class of polymer electrolytes due to their combined properties emanating from the IL units and their intrinsic polymeric nature.[16–21] PILs can be of cationic type, anionic type, or zwitterionic type. In the latter case, the polymer backbone carries both a cationic and anionic functionality.[22] Currently, the cationic-type PILs exhibit more potential for development and dominate the PIL research,[23–26] while the zwitterionic-type PILs still remain problems to be studied.[27–32] In addition, 1-ethyl-3-methylimidazolium bis(trifluoromethanesulfon ylimide) (1E3m-TFSI) is one of the typical ionic liquids used as the additive of electrolyte membrane. 1E3m-TFSI possesses high thermal stability and can be employed as a solvent to dissolve together with lithium salts, prepolymer, and poly(ethylene glycol) diacrylate (PEGDA, cross-linker). For IL-based electrolytes, there is a balance between their ionic conductivities and mechanical properties depending on the composition ratio of the polymer phase and additive phase. Herein, we report a series of quasi-solid polymer electrolytes (Q-SPEs) based on 1-ethyl-3-vinylimidazo-lium bis(trifluoromethanesulfonylimide) (1E3V-TFSI), PEGDA (Mn = 700 g mol⁻¹), PEGME (Mn = 500 g mol⁻¹), 1E3m-TFSI, and LiTFSI by photopolymerization and their application as electrolyte system for Li/LiFePO₄ cells. Though few literatures have reported the preparation of electrolytes using poly(ionic liquid) as the mechanical matrix, the electrolyte system of poly(ionic liquid)/PEGDA/PEGME is currently not evaluated. Additionally, the method of photopolymerization is still an innovative process of preparation as it combines the steps of polymerization and solvent casting without the use of any organic solvent, which helps shorten the preparation time and increase the cost-effectiveness.[29,30,32–38]

11.2   EXPERIMENT

11.2.1   MATERIALS

1-Vinylimidazole (Vim) (99%), PEGDA (Mn = 700 g mol^{-1}), poly(ethylene glycol) methyl ether methacrylate (PEGME) (Mn = 500 g mol^{-1}), lithium bis(trifluoromethanesulfonyl)imide (LiTFSI), and 1-ethyl-3-methylimidazolium bromide (1E3m-Br, 98%) were supplied from Sigma-Aldrich. Bromoethane was supplied by Alfa Aesar. 2-Hydroxy-2-methylpropiophenone was supplied from Merck. Lithium metal and aluminum foil were provided by UBIQ company. Super P and LiFePO$_4$ were supplied by Timcal and Aleees, respectively. All of the organic solvents used in the research were ACS reagent grade.

11.2.2   SYNTHESIS OF PREPOLYMER, 1-ETHYL-3-VINYLIMIDAZOLIUM BIS (TRIFLUOROMETHANESULFONYLIMIDE) (1E3V-TFSI)

1E3V-TFSI was synthesized by refluxing the solution of Vim (1 equiv.) and bromoethane (1.2 equiv.) in acetone overnight. The bromide intermediate would be collected from the mixture at the end of the reaction. After stirring with acetone (three times) to remove any unreacted impurity, the intermediate was then dissolved in water followed by adding excessive LiTFSI into the solution and the mixtures were stirred for 5 h. After the reaction, the clarified liquid precipitated from water was washed several times until the residual (e.g., LiBr) could not be detected by 0.1 M AgNO$_3$ aqueous solution. Finally, the liquid was dried under vacuum at 80°C overnight to yield the desired prepolymer, 1E3V-TFSI.

11.2.3   ANION SUBSTITUTION OF IL ADDITIVE

The ionic compound 1-ethyl-3-methylimidazolium bromide from Sigma-Aldrich was dissolved in water, and the excessive molar amount of LiTFSI was added. The mixture was then mechanically stirred and sonicated to facilitate the replacement of Br$^-$ ion with TFSI$^-$ ion. After centrifugation at 4000 rpm for 10 min, the solution was phase-separated into aqueous layer and oil layer. The upper aqueous layer containing halide was removed, and the oil layer was collected and dried at 100°C for 1 day under vacuum to give a colorless liquid.

11.2.4   PREPARATION OF ELECTROLYTES

The mixtures with desired ratios of 1E3V-TFSI/PEGDA/PEGME/1E3m-TFSI/LiTFSI were prepared in an argon-filled glove box. The weight amount of LiTFSI was kept constant, while the weight percentage of added EMIM-TFSI was between 38 and 80 wt% based on the total mass. And the molar ratios of 1E3V-TFSI/PEGDA/PEGME were set as 1:0.7:0.3, 1:0.5:0.5, and 1:0.3:0.7 for YC1, YC2, and

YC3 series, respectively. It is worth noting that there are two different methods that can be adopted for polymerization. The difference is that we use petri dishes as the template at the beginning. To increase the degree of polymerization as well as to improve the defects that glass may absorb some wavelength range of light, we further optimize the step. The membrane containing the highest degree of cross-linking and the lowest addition of ionic liquid (YC1-38%), for instance, was prepared as follows: 1E3V-TFSI (1182.5 mg), PEGDA (1025.0 mg), PEGME (292.5 mg), LiTFSI (598.0 mg), 1E3m-TFSI (1902.0 mg), and initiator 2-hydroxy-2-methylpropiophenone (25 mg) were mixed together for 3 h to become a homogeneous solution. Then, the solution was dropped on the cathode directly. The photopolymerization took place upon UV light at a wavelength of 350 nm for 20 min in the glove box. Finally, the transparent electrolyte sticking to the cathode was thus obtained. The experimental process and the electrolyte photographs are shown in Figure 11.1. Other membranes were obtained through the procedure similar to that of YC1-38% except the individual contents were adjusted to the corresponding proportion and the compositions of all prepared materials are recorded in Table 11.1.

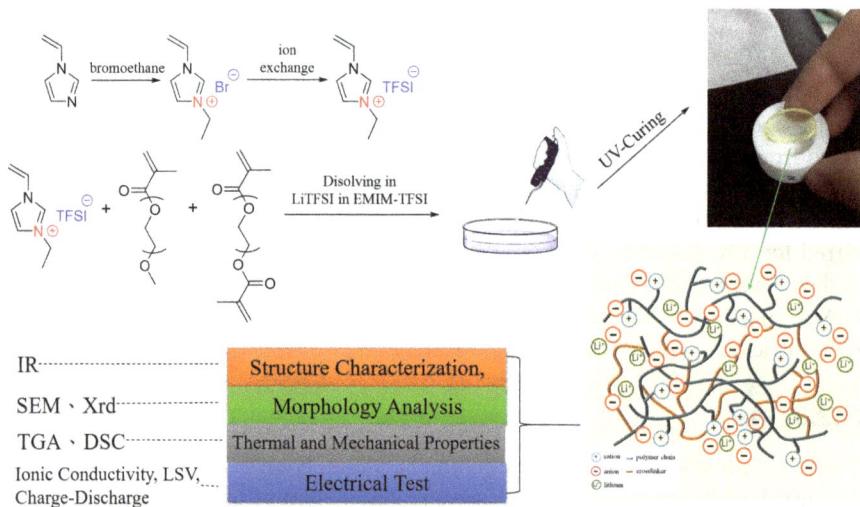

* Improvement of the photopolymerization

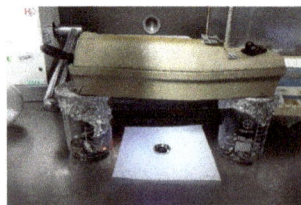

UV-Curing for 20 min.
(350 nm)

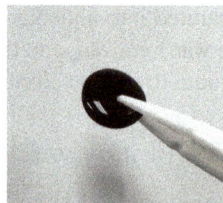

LiFePO$_4$ electrode attached with the electrolyte

FIGURE 11.1 Experimental process of photopolymerization as well as schematic diagram of the electrolyte membranes.

TABLE 11.1
Composition, Decomposition Temperature, and Ionic Conductivity of the Prepared Electrolyte Membranes

| | Weight (mg) | | | | | | Ionic |
Sample	Vim-Et-TFSI	PEG Methyl Ether Methacrylate	PEG Diacrylate	LiTFSI	EMIM-TFSI	$T_{d5\%}$ (°C)	Conductivity (log(S/cm))
YC1-38%	1182.5	292.5	1025.0	598.0	1902.0	354	−3.73
YC1-60%	1182.5	292.5	1025.0	598.0	4647.0	358	−2.85
YC1-80%	1182.5	292.5	1025.0	598.0	12392.0	376	−2.77
YC2-38%	1225.0	531.0	744.0	598.0	1902.0	356	−3.65
YC2-60%	1225.0	531.0	744.0	598.0	4647.0	361	−2.78
YC3-38%	1271.0	788.0	441.0	598.0	1902.0	355	−3.48
YC3-60%	1271.0	788.0	441.0	598.0	4647.0	359	−2.81

11.2.5 SAMPLE CHARACTERIZATION

Fourier transform infrared (FT-IR) spectroscopic measurements were recorded on Thermo Nicolet Nexus 6700 FTIR within the 600–3600 cm⁻¹ range of wave number. X-ray diffraction (XRD) analysis was performed on a Rigaku Ultima IV-9407F701 X-ray spectrometer using Cu Kα (0.154 nm) radiation (50 kV, 250 mA).The samples were mounted on a sample holder and scanned from $2\theta = 5°$ to 40° at a speed of 10° min⁻¹. Thermal gravimetric analysis (TGA) of the samples was performed on a TGA7 instrument (Perkin Elmer). The range of the measurement was set from 150°C to 600°C, and the temperature ramping rate was 15°C min⁻¹. The sample membranes were heated to 100°C and kept at the temperature for 15 min before starting. $T_{d5\%}$ is the temperature when the weight loss of sample reaches its 5%. Field-emission scanning electron microscopy (FE-SEM) analysis of the polymer membranes was performed on a Hitachi SU8010 microscope (voltage: 1–10 kV), and the samples were mounted on the platform by carbon tape for characterization.

11.2.6 IONIC CONDUCTIVITY AND LINEAR SWEEP VOLTAMMETRY (LSV) OF MEASUREMENT

To obtain the ionic conductivity, the sample thickness was measured at first. Then, it was sandwiched between two stainless steels and assembled into a coin cell. The measurement was performed by AC impedance analysis with the frequency set as 0.1–10⁶ s⁻¹. Extrapolation was performed based on the five points at the high-frequency end. The x-axis value of the intersection was defined as the resistance of the electrolyte (R). Then, the ionic conductivity can be calculated by the following equation:

$$\sigma\left(\frac{S}{cm}\right) = \frac{d}{RA}$$

where R is the bulk resistance, d is the membrane thickness denoted in cm, and A is the contact area of the disk and the membrane (1.840 cm^2).

LSV was used to determine the durability of each electrolyte on cathodic potential. The electrolytes were sandwiched between a Li metal and a stainless steel disk in a coin cell. The potential range was set from 0 to 6 V, and the scan rate was 0.005 V s^{-1}.

11.2.7 BATTERY CELL ASSEMBLY

A cathode slurry was made in N-methyl-2-pyrrolidone solvent at a weight ratio of 80:10:10 corresponding to $LiFePO_4$ powder/Super P/polyvinylidene difluoride (PVDF) and further spread on aluminum foil. The coated foil was dried at 80°C under vacuum overnight. Then, the cathode electrode was roll-pressed to tight the structure. Further, coin cell assembly was performed inside the glove box. Li metal and the prepared electrolytes were used to be anode and ion transfer medium, respectively.

11.2.8 CHARGE–DISCHARGE PERFORMANCE AND CYCLE LIFE

The charge–discharge tests of the as-prepared half cells were conducting using the BAT-700 battery testing system. The operation voltages for the cells were in the range from 2.5 to 4.0 V. Galvanostatic charge–discharge cycling performance was also taken on the half coin cell at a constant current rate of 0.2 C. The rest time during each cycle was 20 s.

11.3 RESULTS AND DISCUSSION

11.3.1 PREPARATION AND CHARACTERIZATION

The electrolytes with varying ILs and degrees of cross-linking were prepared by UV photopolymerization. The lithium salt concentration is kept constant, while the weight percentage of added 1E3m-TFSI is between 38 and 80 wt% based on the total mass. And the molar ratios of 1E3V-TFSI/PEGDA/PEGME are set as 1:0.7:0.3, 1:0.5:0.5, and 1:0.3:0.7 for YC1, YC2, and YC3 series, respectively. The final membranes are named as AX-Y%, in which X represents the degree of cross-linking (YC1 > YC2 > YC3) while Y is the weight percent of IL additive. All the products prepared as above are found to be self-standing and translucent appearance (Figure 11.1). After membrane formation, four peaks assigned to the vibration sorption of TFSI anions ($v_a(SO_2)$, $v_a(CF_3)$, $v_s(SO_2)$, and $v_a(S-N-S)$) appear at around 1344, 1177, 1128, and 1050 cm^{-1} in FT-IR spectra (Figure 11.2a). The result of polymerization for YC1-38% is also characterized by FT-IR, during which the intensity of the characteristic peak of the vinyl group (1638 cm^{-1}) significantly decreases as shown in Figure 11.2b, suggesting that most of the vinyl moieties had been reacted and incorporated into

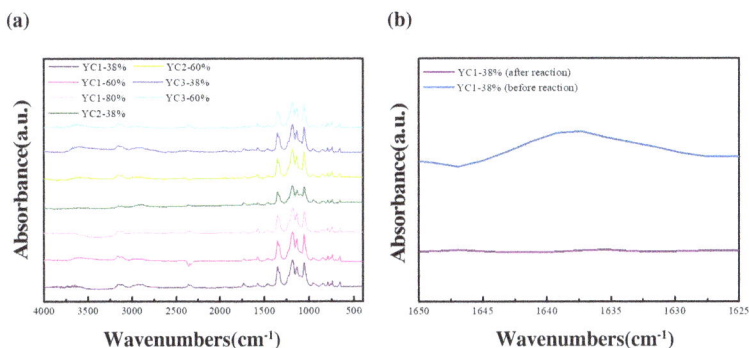

FIGURE 11.2   FT-IR spectra of the samples (a) from 400 to 4000 cm^{-1} and partially enlarged patterns (b) from 1625 to 1650 cm^{-1}.

the entangled polymer matrix. XRD measurements are performed on the LiTFSI, EMIM-TFSI, YC1-60%, YC2-60%, and YC3-60% samples (Figure 11.3). The XRD pattern of LiTFSI exhibits the multiple characteristic peaks indicating its high extent of crystallinity. After undergoing the formation with polymers/IL, most of the peaks corresponding to LiTFSI are disappeared. Compared to pure EMIM-TFSI, the signal intensity of the peak at $2\theta = 13°$ is decreased effectively and the width of the peak at $2\theta = 20°$ becomes broad, indicating the drastic decrease in both grain size and crystallinity. This result gives the suggestion that the prepared samples are mainly in amorphous states and this may attribute that the interactions among the polymers, IL, and lithium salt decrease the crystallinity of LiTFSI.

SEM analysis is used to characterize the surface morphologies. All samples are taken out from the glove box, followed by drying at 80°C in vacuum to ensure

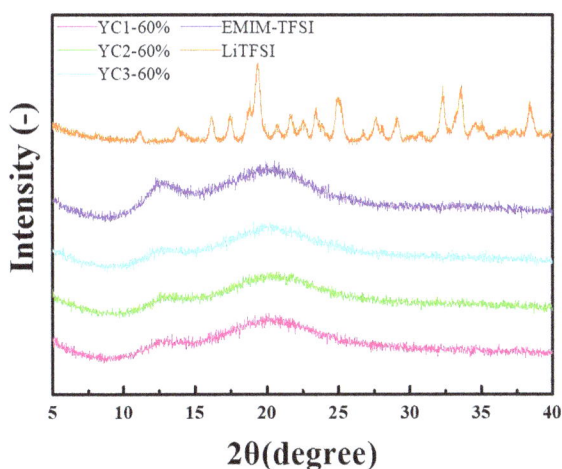

FIGURE 11.3   XRD patterns of LiTFSI, 1E3m-TFSI, and the prepared membranes (YC1-60%, YC2-60%, and YC3-60%).

(a)

(b)

(c)

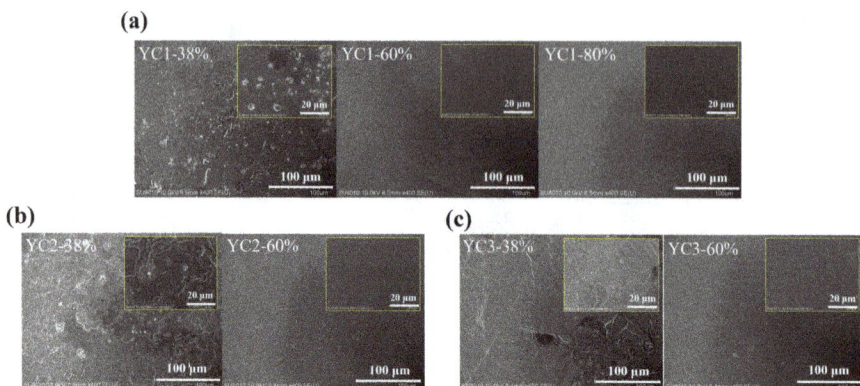

FIGURE 11.4 SEM images of surface of the electrolytes. (a) YC1, (b) YC2, and (c) YC3 series membrane.

that there is no moisture inside the samples before analysis. Figure 11.4a–c exhibits the top surfaces of the YC1, YC2, and YC3 series membranes, respectively. It can be found that these membranes are composed of continuous structures. And the morphology can be referred to the objective evidence of the interactions among the polymer, IL, and LiTFSI domains. For the samples of YC1-38%, YC2-38%, and YC3-38%, some LiTFSI was not dissolved in the matrix completely due to the high crystallinity of the lithium salt, and the corresponding images reveal several particles as well as uneven micropores. With the increase in the IL amount over 60%, their surfaces become homogeneous without any micropores and the LiTFSI particles are well dissolved, which also means most of the added additive can be perfectly trapped inside the polymer matrix. We suggest that the amount of IL influence the miscibility of the composition, and thus, the dispersion of LiTFSI in the polymer matrix can be well improved by adding enough IL. In addition, our previous research claims that the micropores on the membrane surface can be referred to the void spaces that do not facilitate the ion transportation and, instead, increase the interfacial resistance between the electrolytes and the electrodes. Furthermore, the irregular distribution of micropores may result in random deposition of solid electrolyte interphase (SEI) layer during charge–discharge, which is not beneficial to the long-term cycles. Consequently, our SEM images show that the addition of 60 wt% IL not only can improve the solubility of lithium salt but also can afford the formation of membranes without any micropores. It seems that 60 wt% IL can meet a standard and would be regarded as our first choice for further testing.

11.3.2 THERMAL PROPERTIES OF ELECTROLYTES

TGA measurements of the electrolytes are conducted to investigate their thermal properties. The resulting curves exhibit that all of the samples do not have apparent weight loss below 330°C as shown in Figure 11.5a. Their decomposition temperatures ($T_{d5\%}$), which are defined as the temperatures with 5% of weight loss, are found to range between 354°C and 376°C, as listed in Table 11.1. It appears clearly

(a)

(b)

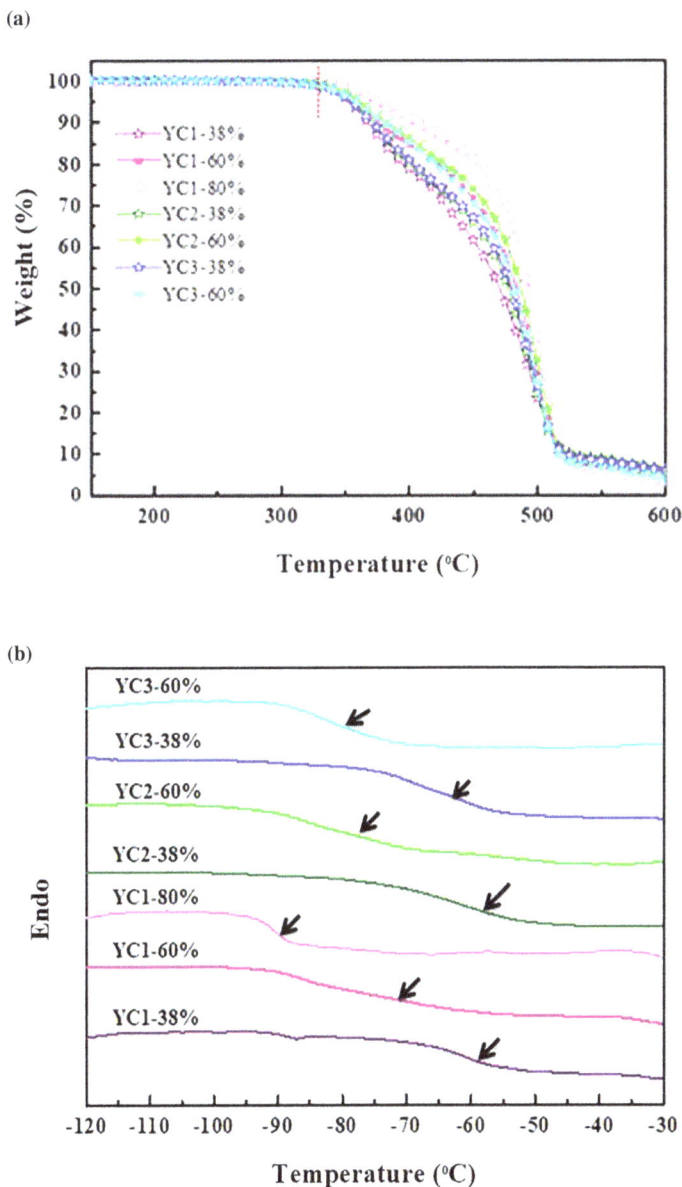

FIGURE 11.5   (a) TGA and (b) DSC thermograms of the prepared membranes.

that these samples exhibit one-step thermal decomposition behavior due to the good mixing of each composition and the whole structures start to decompose at around 330°C. It can be found that their T_d values increase with the increment in the IL amount. It sounds a little subversion from our knowledge. This trend is due to the truth that IL possesses higher thermal stability than that of PEG-based polymers even if it is in liquid phase at room temperature. Thus, their thermal stability can be enhanced with the increase in the IL amount. We can also find that the samples with the same amount of IL exhibit the same profiles, which further confirms that their thermal behaviors are mainly dominated by IL. Consequently, the samples exhibit qualified thermal stabilities with the high T_d, which would render them to exhibit safety once the batteries are overheating. To realize the phase behaviors as a function of temperature of the prepared electrolytes, DSC analysis is used and the resulting curves are shown in Figure 11.5b. As observed from DSC curves, the melting points (T_m) of the samples are lower than −50°C without any endothermic peaks. The T_m and crystallinity of the materials effectively decrease after the addition of IL and LiTFSI. This decrease can be regarded as the truth that the existence of IL domain can restrict the polymer domains to pack neatly and the polymer/IL/LiTFSI domains exhibit good compatibility with each other, which hence results in lower crystallinity. The results also correspond to those from XRD measurements presented in Figure 11.3.

11.3.3 Ionic Conductivity and Electrochemical Windows

In order to realize how the composition and temperature impact the ionic conductivities of the materials, their ionic conductivities are measured and compared from 30°C to 80°C. The result is shown in Figure 11.6a, and the values at 30°C are shown in Table 11.1. All the electrolytes are found to have higher conductivities with increasing temperature. From our experience, the value of the ionic conductivity should be higher than 10^{-4} S cm^{-1} to meet the requirement of charge–discharge and plays an important factor to affect the performance of batteries.[39,40] It is great that our samples demonstrate excellent conductivities over the measured temperature range, which are qualified for further electrical evaluation. Moreover, the increase in IL from 38% to 60% can increase the ionic conductivity to almost one order, given that there are studies showing that IL can serve as the plasticizer to lower

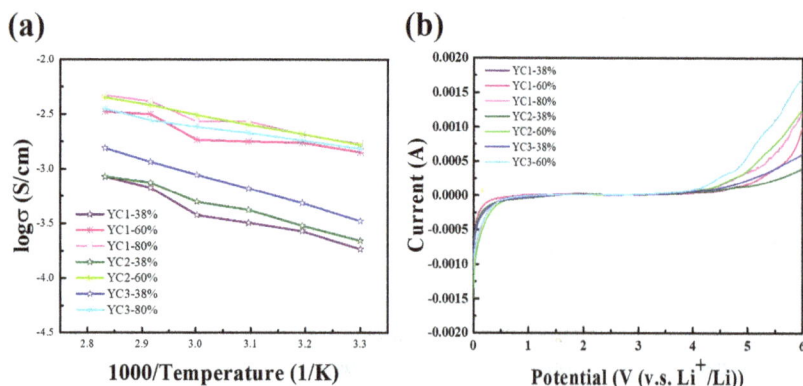

FIGURE 11.6 (a) Ionic conductivity as a function of temperature of the electrolytes. (b) LSV diagram of the electrolytes.

the viscosity and increase the ability of ion transportation. However, the ionic conductivity expected to depend on the contents of IL does not follow the trend in the case of YC1-80%. The continuous addition of IL to 80 wt% only increases the ionic conductivity slightly. Compared to the corresponding SEM images of YC1-60% and YC1-80%, their surfaces have already become homogeneous without obvious difference, and the micro-phase morphologies would not be excessively affected by 60 or 80 wt% of the IL addition. Thus, the ionic conductivity of YC1-80% finally shows no significant enhancement. Referring to the current result, the minimum content of IL, which can construct a homogeneous morphology, dissolve lithium salt completely, and exhibit a relatively high ionic conductivity, is around 60 wt% with respect to the total mass. The further increase in IL would not lead to a significant change in morphology and ionic conductivity. Furthermore, the excess of IL would keep us remain suspicion to the electrochemical stability of the material, which would be explained in the next paragraph

The electrochemical stabilities of the electrolytes are investigated using LSV test (Figure 11.6b). The membranes with higher IL contents result in the decrease in the onset oxidation voltage (YC1-38% > YC1-60% > YC1-80%, YC2-38% > YC2-60%, and YC3-38% > YC3-60%). When the IL addition is adjusted to the same concentration, YC1 series possesses the highest values of the onset oxidation voltage (YC1-38% > YC2-38% > YC3-38%, and YC1-60% > YC2-60% > YC3-60%), among all. In general, the onset oxidation voltage of the electrolyte should be higher than 4.5 V (vs. Li^+/Li) to meet the standard with enough electrochemical stability.[41,42] Our results, however, show the onset oxidation voltage in the range from 3.9–to 4.4 V (vs. Li^+/Li), which may cause structural damage at high voltage. To solve this defect, we have improved the step of polymerization as described in the experiment chapter, and the onset oxidation voltage of YC1-60% can increase to about 4.6 V after we perform the photopolymerization directly on the electrode.

11.3.4 CHARGE–DISCHARGE CAPACITY AND CYCLIC PERFORMANCE

The membranes of YC1-60%, YC2-60%, and YC3-60% are assembled in the cells with $LiFePO_4$ and lithium metal as the cathode and anode, respectively. Figure 11.7a–c shows the charge–discharge curves of the batteries based on YC1-60%, YC2-60%, and YC3-60%. Their corresponding discharge capacity values at specific rates are summarized in Figure 11.8. Note that the charge and discharge rates are set from 0.1 to 1 C. As can be seen, all the cells can deliver around 140 mAh g^{-1} of the capacity at 0.1 and 0.2 C rates without obvious difference. However, the cell with YC1-60% exhibits higher discharge capacities than those with YC2-60% and YC3-60% at high C rates. Considering that they possess similar ionic conductivity, the superior performance of YC1-60% can be referred to its better mechanical stability and electrochemical stability, which has been confirmed by the truth that YC1-60% membrane has the highest degree of cross-linking and the relatively qualified value of onset oxidation voltage shown in Figure 11.6b. As mentioned previously, the low electrochemical stability may cause structural damages at high voltage and that is the fatal flaw our membranes suffer from. Finally, it is difficult to demonstrate long-term charge–discharge without improving the method of polymerization.

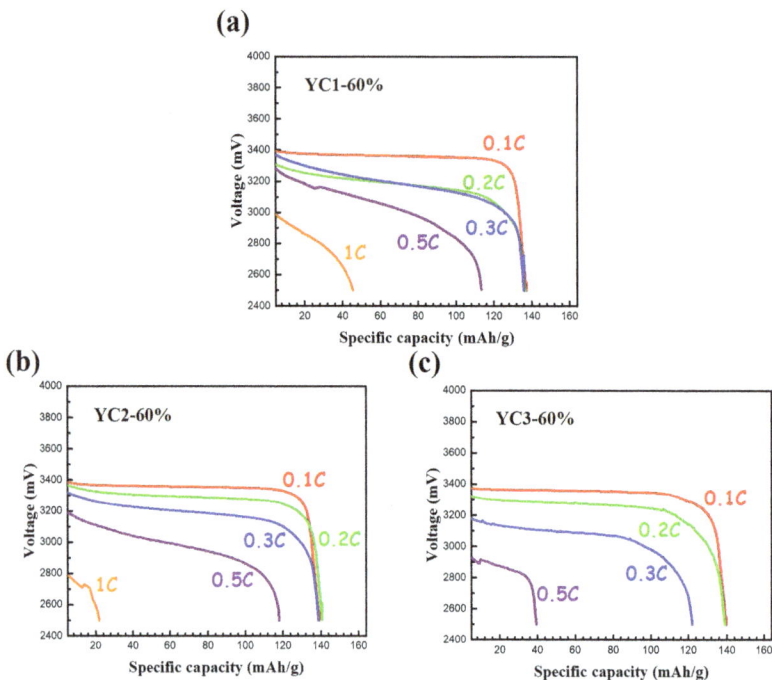

FIGURE 11.7 Discharge curves of the Li/LiFePO$_4$ cells based on (a) YC1-60%, (b) YC2-60%, and (c) YC3-60% at different C rates at 80°C.

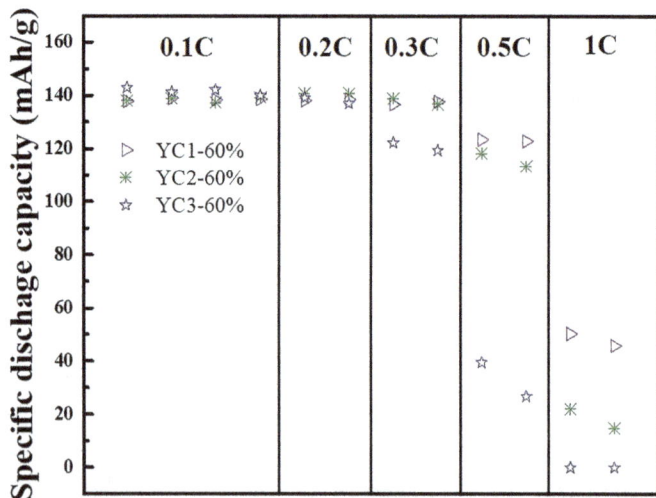

FIGURE 11.8 Discharge capacities of the Li/LiFePO$_4$ cells based on (a) YC1-60%, (b) YC2-60%, and (c) YC3-60% at different C rates at 80°C.

Hence, all membranes were polymerized on electrodes directly, and YC1-60% is chosen to demonstrate the charge–discharge test again at 80°C to check if the capacity can be improved. The result is shown in Figure 11.9. The cell based on YC1-60% prepared by the new method exhibits excellent charge–discharge capacities. The discharge capacities at 0.1, 0.2, 0.3, and 0.5 C are measured to be 161.0, 167.0, 165.0, and 164.0 mAh g^{-1}, respectively; these values are almost as high as 97% of the theoretical value of LiFePO$_4$. Their perfect platform of charge–discharge also indicated the improvement of polarization. The gap between the platform of charge and discharge is only around 150 mV even at 0.5 C rate. In addition, the Coulombic efficiency is determined as the ratio of the discharge capacity to the charge capacity. The values of our batteries are always higher than 97% throughout the cycling process, revealing their high charge transfer reversibility. Note that the lower Coulombic efficiencies during the process of 0.1 C rate may be regarded as the reason of incomplete activation. It seems that the increased capacity, improved stability and the lower voltage polarization can be regarded as decreased interfacial resistances between electrolytes/electrodes, which can possibly be attributed to our new method of photopolymerization. Furthermore, the long-term charge–discharge cycling can provide a comprehensive evaluation of the electrolyte system since the thermal stability, mechanical stability, chemical stability, electrochemical stability, ionic conductivity, and durability of the materials are all factors that would affect the cycling performance. As shown in Figure 11.9, the discharge capacities of the cell based on YC1-60% are maintained at a steady value (higher than 160 mAh g^{-1}), confirming that the material can afford more than 20 cycles without faded performance. In summary, polymerization and membrane formation on the cathode directly can effectively decrease the impedance of interface and improve the stability, contributing to the better performance while charging/discharging.

We compare the current results with some existing IL-based electrolytes, which can be classified into tetraalkylammonium-based electrolytes, guanidinium-based electrolytes, pyrrolidinium-based electrolytes, and imidazolium-based electrolytes. The battery performances of the above researches are concluded in Table 11.2. For example, tetraalkylammonium-based electrolytes were developed by Li et al., and revealed a discharge capacity of 120 mAh g^{-1} at 60°C for the assembled Li/LiFePO$_4$ cell.[43] Guanidinium-based electrolytes were also made by the group of Li et al. and the assembled Li/LiFePO$_4$ cell can deliver a capacity to around 130 mAh g^{-1} at 80°C,[44] while Appetecchi et al. prepared the pyrrolidinium-based electrolyte to export a capacity of around 145 mAh g^{-1} at 40°C.[18] And Yin et al. demonstrated the imidazolium-based electrolytes to show a capacity of 160 mAh g^{-1} at 40°C.[16] All these results were conducted at 0.1 C rate. The ionogels used in this research, by contrast, possess a charge–discharge performance of 164.0 mAh g^{-1} at 0.5 C rate and 60°C, which is superior to the above literatures. Moreover, most of these electrolytes were prepared through thermal polymerization and formed by solvent casting, which is a relatively complicated, costly, and time-consuming process. In this research, the method of photopolymerization is an innovative process of preparation as it combines the steps of polymerization and solvent casting without the use of any organic solvent, which is conducive to shortening the preparation time and increasing the cost-effectiveness.

0.1C: 0, 1~5 cycle

0.1C

0.2C: 6~10 cycles,

0.2C

0.3C: 11~15 cycles,

0.3C

0.5C: 16~20 cycles,

0.5C

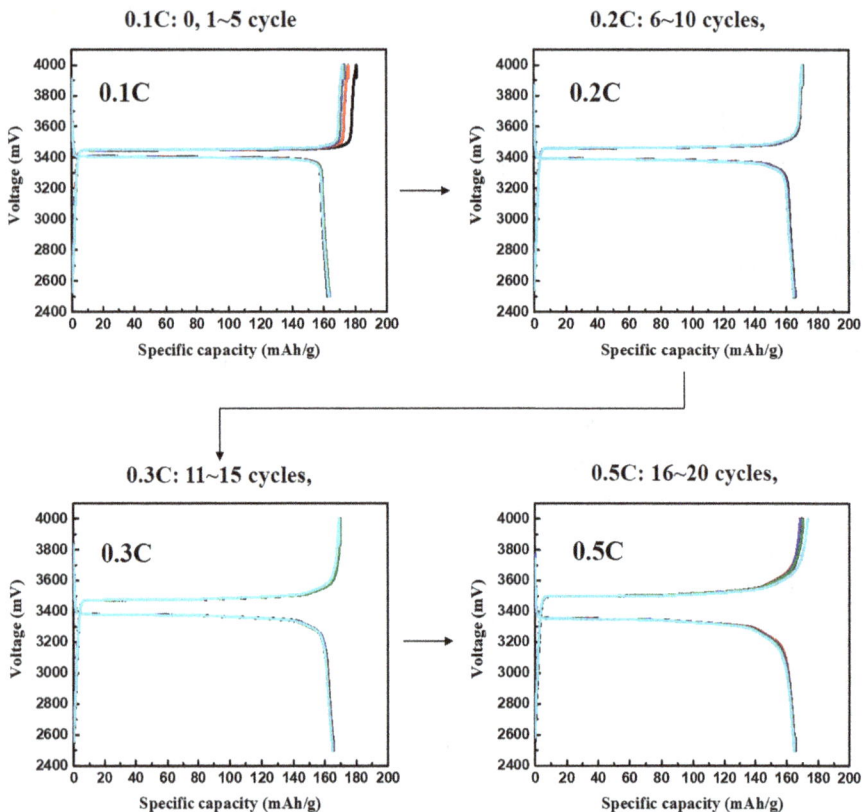

FIGURE 11.9 Charge–discharge curves of the Li/LiFePO₄ cell based on YC1-60% prepared by curing on the cathode directly.

TABLE 11.2

Summary of the Performance of Li/LiFePO₄ Cells Assembled with Various Types of Ionic Liquid–Based Electrolytes

Kinds of Ionic Liquid–Based Electrolytes	Li/LiFePO4 Cell Capacity (mAh g⁻¹)	References
Tetraalkylammonium-based electrolyte	120 (0.1 C, 60°C)	43
Guanidinium-based electrolyte	130 (0.1 C, 80°C)	44
Pyrrolidinium-based electrolyte	145 (0.1 C, 40°C)	18
Imidazolium-based electrolyte	160 (0.1 C, 40°C)	16
Imidazolium-based ionogel	164 (0.5 C, 60°C)	This work

11.4   CONCLUSION

The imidazolium-based ionogels with various amounts of ionic liquid additive and degrees of cross-linking are prepared via facile photopolymerization and evaluated for application in lithium–ion batteries. The membranes with enough addition of ionic liquid exhibit high ionic conductivity and good miscibility between the components, thus resulting in homogeneous surface without any micropores. Some of the membranes exhibit qualified electrochemical windows, evidenced by the relatively high onset oxidation voltages (4.5 V). Moreover, the discharge capacities of the cell based on YC1-60% can reach 165 mAh g^{-1} even at 0.5 C rate. The cell also exhibits excellent capacity retention to 20 cycles. The above properties would render these types of membranes to be a promising candidate as the polyelectrolyte for Li/LiFePO$_4$ batteries.

REFERENCES

1. M. Armand and J. M. Tarascon, *Nature*, 2008, 451, 652–657.
2. B. Kang and G. Ceder, *Nature*, 2009, 458, 190–193.
3. Y. K. Sun, S. T. Myung, B. C. Park, J. Prakash, I. Belharouak and K. Amine, *Nat Mater*, 2009, 8, 320–324.
4. B. Scrosati and J. Garche, *J Power Sources*, 2010, 195, 2419–2430.
5. P. G. Balakrishnan, R. Ramesh and T. P. Kumar, *J Power Sources*, 2006, 155, 401–414.
6. J. B. Goodenough and Y. Kim, *Chem Mater*, 2010, 22, 587–603.
7. S. F. Lux, I. T. Lucas, E. Pollak, S. Passerini, M. Winter and R. Kostecki, *Electrochem Commun*, 2012, 14, 47–50.
8. M. Smiglak, W. M. Reichert, J. D. Holbrey, J. S. Wilkes, L. Y. Sun, J. S. Thrasher, K. Kirichenko, S. Singh, A. R. Katritzky and R. D. Rogers, *Chem Commun*, 2006, 2554–2556. doi: 10.1039/b602086k.
9. M. J. Earle, J. M. S. S. Esperanca, M. A. Gilea, J. N. C. Lopes, L. P. N. Rebelo, J. W. Magee, K. R. Seddon and J. A. Widegren, *Nature*, 2006, 439, 831–834.
10. H. Nakagawa, Y. Fujino, S. Kozono, Y. Katayama, T. Nukuda, H. Sakaebe, H. Matsumoto and K. Tatsumi, *J Power Sources*, 2007, 174, 1021–1026.
11. Y. D. Jin, S. H. Fang, M. Chai, L. Yang, K. Tachibana and S. Hirano, *J Power Sources*, 2013, 226, 210–218.
12. X. G. Sun, C. Liao, N. Shao, J. R. Bell, B. K. Guo, H. M. Luo, D. E. Jiang and S. Dai, *J Power Sources*, 2013, 237, 5–12.
13. H. Sakaebe, H. Matsumoto and K. Tatsumi, *Electrochim Acta*, 2007, 53, 1048–1054.
14. A. Fernicola, F. Croce, B. Scrosati, T. Watanabe and H. Ohno, *J Power Sources*, 2007, 174, 342–348.
15. M. Galinski, A. Lewandowski and I. Stepniak, *Electrochim Acta*, 2006, 51, 5567–5580.
16. K. Yin, Z. X. Zhang, X. W. Li, L. Yang, K. Tachibana and S. I. Hirano, *J Mater Chem A*, 2015, 3, 170–178.
17. K. Yin, Z. X. Zhang, L. Yang and S. I. Hirano, J Power Sources, 2014, 258, 150–154.
18. G. B. Appetecchi, G. T. Kim, M. Montanina, M. Carewska, R. Marcilla, D. Mecerreyes and I. De Meatza, *J Power Sources*, 2010, 195, 3668–3675.
19. X. W. Li, Z. X. Zhang, S. J. Li, L. Yang and S. Hirano, *J Power Sources*, 2016, 307, 678–683.
20. M. Safa, A. Chamaani, N. Chawla and B. El-Zahab, *Electrochim Acta*, 2016, 213, 587–593.

21. P. F. Zhang, M. T. Li, B. L. Yang, Y. X. Fang, X. G. Jiang, G. M. Veith, X. G. Sun and S. Dai, *Adv Mater*, 2015, 27, 8088–8094.
22. A. S. Shaplov, P. S. Vlasov, E. I. Lozinskaya, D. O. Ponkratov, I. A. Malyshkina, F. Vidal, O. V. Okatova, G. M. Pavlov, C. Wandrey, A. Bhide, M. Schonhoff and Y. S. Vygodskii, *Macromolecules*, 2011, 44, 9792–9803.
23. X. J. Chen, J. Zhao, J. Y. Zhang, L. H. Qiu, D. Xu, H. G. Zhang, X. Y. Han, B. Q. Sun, G. H. Fu, Y. Zhang and F. Yan, *J Mater Chem*, 2012, 22, 18018–18024.
24. R. Tejero, D. Lopez, F. Lopez-Fabal, J. L. Gomez-Garces and M. Fernandez-Garcia, *Polym Chem-UK*, 2015, 6, 3449–3459.
25. S. Yi, W. Leon, D. Vezenov and S. L. Regen, *ACS Macro Lett*, 2016, 5, 915–918.
26. M. M. Obadia, A. Jourdain, A. Serghei, T. Ikeda and E. Drockenmuller, *Polym Chem-UK*, 2017, 8, 910–917.
27. M. Yoshizawa, M. Hirao, K. Ito-Akita and H. Ohno, *J Mater Chem*, 2001, 11, 1057–1062.
28. F. Lind, L. Rebollar, P. Bengani-Lutz, A. Asatekin and M. J. Panzer, *Chem Mater*, 2016, 28, 8480–8483.
29. F. Lu, X. P. Gao, A. L. Wu, N. Sun, L. J. Shi and L. Q. Zheng, *J Phys Chem C*, 2017, 121, 17756–17763.
30. N. Sun, X. P. Gao, A. L. Wu, F. Lu and L. Q. Zheng, *J Mol Liq*, 2017, 248, 759–766.
31. M. E. Taylor and M. J. Panzer, J Phys Chem B, 2018, 122, 8469–8476.
32. Y. Yu, F. Lu, N. Sun, A. L. Wu, W. Pan and L. Q. Zheng, *Soft Matter*, 2018, 14, 6313–6319.
33. R. X. He, M. Echeverri, D. Ward, Y. Zhu and T. Kyu, *J Membr Sci*, 2016, 498, 208–217.
34. G. P. Fu, J. Dempsey, K. Izaki, K. Adachi, Y. Tsukahara and T. Kyu, *J Power Sources*, 2017, 359, 441–449.
35. G. P. Fu and T. Kyu, *Langmuir*, 2017, 33, 13973–13981.
36. G. P. Fu, M. D. Soucek and T. Kyu, *Solid State Ionics*, 2018, 320, 310–315.
37. C. Piedrahita, V. Kusuma, H. B. Nulwala and T. Kyu, *Solid State Ionics*, 2018, 322, 61–68.
38. S. Stalin, S. Choudhury, K. H. Zhang and L. A. Archer, *Chem Mater*, 2018, 30, 2058–2066.
39. K. S. Liao, T. E. Sutto, E. Andreoli, P. Ajayan, K. A. McGrady and S. A. Curran, *J Power Sources*, 2010, 195, 867–871.
40. S. Inceoglu, A. A. Rojas, D. Devaux, X. C. Chen, G. M. Stone and N. P. Balsara, *ACS Macro Lett*, 2014, 3, 510–514.
41. K. Karuppasamy, P. A. Reddy, G. Sriniyas, A. Tewari, R. Sharma, X. S. Shajan and D. Gupta, *J Membr Sci*, 2016, 514, 350–357.
42. Y. S. Yun, J. H. Kim, S. Y. Lee, E. G. Shim and D. W. Kim, *J Power Sources*, 2011, 196, 6750–6755.
43. M. T. Li, B. L. Yang, L. Wang, Y. Zhang, Z. Zhang, S. H. Fang and Z. X. Zhang, *J Membr Sci*, 2013, 447, 222–227.
44. M. T. Li, L. Yang, S. H. Fang, S. M. Dong, S. Hirano and K. Tachibana, *J Power Sources*, 2011, 196, 8662–8668.

12 Back-Contact Perovskite Solar Cells

Tai-Fu Lin, Ming-Hsien Li, Pei-Ying Lin,
Itaru Raifuku, Joey Lin, and Peter Chen
National Cheng Kung University

CONTENTS

12.1 INTRODUCTION

When the ion-based batteries are charging, batteries can convert electricity into chemical energy which is stored in batteries. While discharging, ion-based batteries undergo a reverse process to convert chemical energy into electricity for charging other devices. The working principle of solar cells is totally different to the ion-based batteries. The solar cells can convert solar light (photon energy) into electricity (chemical potential of electron) which cannot be stored in solar cells. As a result, solar cells cannot work without solar light illumination. It is worth noting that the solar cells have the potential to integrate with ion-based batteries for constructing a renewable energy system with light harvesting and energy storage. The solar cells can serve as a power supply for charging the ion-based batteries.

In conventional perovskite solar cells (PSCs), perovskite light absorber with several hundred nanometer thickness is sandwiched by two selective contact electrodes, namely electron-selective layer (ESL) and hole-selective layer (HSL), as revealed in Figure 12.1a. Typically, a transparent conducting oxide (TCO) layer, such as indium tin oxide (ITO) or fluorine-doped tin oxide (FTO), serves as the illumination window. The light enters the solar cells from the glass substrate, TCO layer, HSL, or ESL and is finally absorbed by the perovskite absorber. The electron and hole excited in perovskite active layer will diffuse in the opposite direction and be collected by the ESL and HSL, respectively. The optical transmission loss due to the refractive index mismatch, scattering from rough surface, and parasitic absorption of carrier transport material reduces the incident light intensity along with the optical flux. Moreover, deposition of carrier transport layer on the top of perovskite could require high temperature or harsh chemical process that could damage the perovskite layer.

The perovskite solar cells employing back-contact architecture, the so-called back-contact perovskite solar cells (BC-PSCs), can resolve the above issues. In the back-contact architecture, the selective contact of ESL and HSL and metal contacts are patterned and positioned on the same side below the perovskite active layer (refer to Figure 12.1b and c). Such device structure prevents the damage of top electrode deposition on the perovskite active layer and allows the light illumination directly on the perovskite absorber to reduce the optical transmission loss. The bottom TCO layer can be replaced with a metal electrode that provides a lower sheet resistance than TCO and reduces the resistive loss of device.

According to the position of patterned electrodes, the BC-PSCs can be divided into coplanar and non-coplanar types. Figure 12.1b shows the BC-PSCs with coplanar finger electrodes, namely interdigitated electrode (IDE), in which two finger electrodes are patterned and interdigitated to separately collect carriers. The carrier diffusion length of perovskite is usually within few micrometers [1]. The gap between electron- and hole-selective finger electrodes is highly controlled to collect carrier before recombination. Photolithographic techniques are generally applied to fabricate IDE; however, the integrity of finger electrode is critical to the charge collection. To simplify the fabrication process, Udo Bach's group proposed BC-PSCs with non-coplanar finger electrodes, namely quasi-interdigitated electrode (QIDE), in which an insulator is inserted between top patterned electrode and bottom planar electrode as shown in Figure 12.1c [2,3]. Inserting an insulator between them can effectively prevent the short-circuit contact between two selective electrodes. Furthermore, we can increase finger density, compared to the IDE structure, to further enhance charge collection capability.

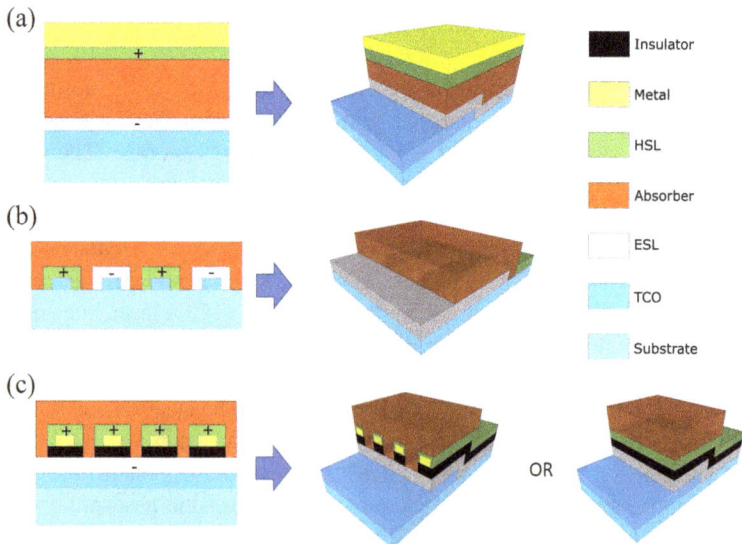

FIGURE 12.1   Device structure of (a) sandwiched PSCs, (b) IDE-based BC-PSCs, and (c) QIDE-based BC-PSCs.

12.2   COPLANAR BACK-CONTACT STRUCTURE

T. Ma et al. used a numerical simulation to investigate the structural parameters in terms of finger electrode width and gap on the photovoltaic performance of the coplanar back-contact-type perovskite solar cell [4]. The coplanar back-contact-type structure is shown in Figure 12.2a, in which the finger electrode of HSL and ESL with the same finger width are placed at the same plane and staggered to form a gap between two finger electrodes. Metal oxide semiconductors of TiO_2 and NiO are respectively used as ESL and HSL in the simulation. Figure 12.2b shows the calculated J-V

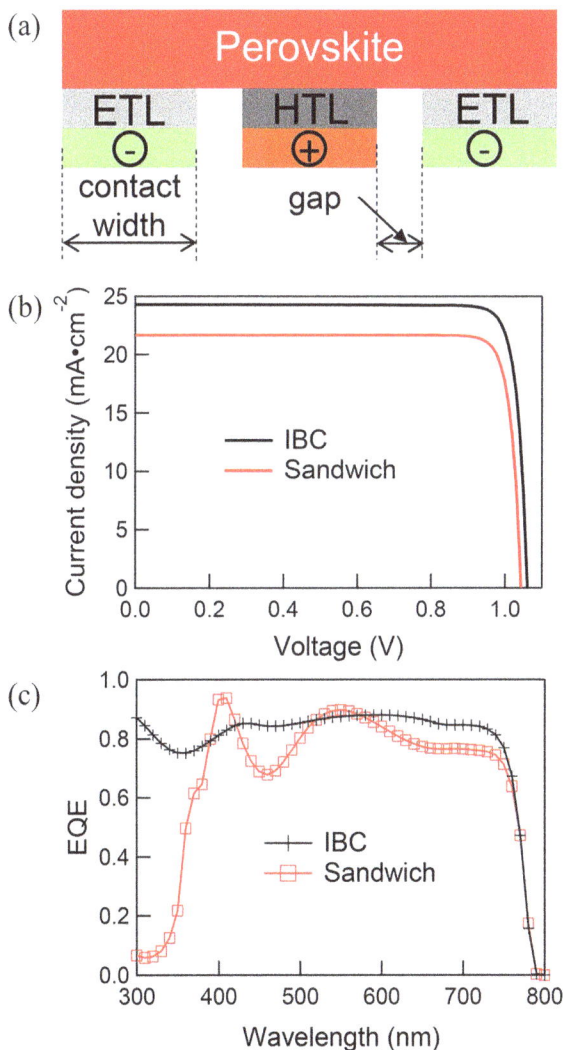

FIGURE 12.2   (a) Structures of the coplanar back-contact-type structure; calculated (b) J-V curves and (c) EQE response of coplanar back-contact-type (black line) and sandwich-type (gray line) perovskite solar cell [4].

curves of PSCs with the sandwich and the coplanar back-contact structures. The short-circuit current density (J_{SC}) of coplanar back-contact structure exhibits an optimized value of 24.3 mA cm^{-2} when the width and gap are 1 and 0.1 µm, respectively, leading to a power conversion efficiency (PCE) of 22.77%. The traditional sandwich structure delivers a lower efficiency mainly due to a reduced J_{SC} of 21.7 mA cm^{-2}. The increased J_{SC} in BC-PSCs is attributed to the reduced light loss at short wavelength region (300–380 nm) as shown in the calculated external quantum efficiency (EQE) spectra (refer to Figure 12.2c). The authors also identify that the default value of finger electrode width ≤5 µm and gap ≤0.5 µm in coplanar BC-PSCs is acceptable for achieving high-efficiency BC-PSCs and simplifying the fabrication process.

It is a common technique to introduce self-assembled monolayers (SAMs) to modify the work function (WF) of metals and semiconductors in organic light-emitting diodes and organic field-effect transistors. The molecular assemblies reorganize into oriented arrange and spontaneously form on surfaces by adsorption. SAMs can affect the energy level by imparting a dipole moment on the surface [5–8]. In order to align the potential difference between two finger electrodes, U. Bach's group apply SAM treatment at the gold–perovskite interfaces to modulate the work function of gold finger electrodes [9,10].

The interdigitated gold electrode is immersed into a 4-methoxythiophenol (OMeTP) solution, resulting in the formation of OMeTP SAMs on two finger electrodes. The OMeTP SAM on anode is subsequently desorbed electrochemically. The desorbed Au electrode is then exposed to a solution of 4-chlorothiophenol (CITP) in order to form the cathode electrode. Kelvin probe force microscopy (KPFM) is used to measure the surface potential changes of the interdigitated gold electrode after surface modification. The formation of the OMeTP SAM on electrode increases contact potential difference (CPD) of 400 mV relative to the bare electrode, and the CPD relative to the bare electrode further increases to 550–580 mV after CITP treatment. The encapsulated device delivers a V_{OC} of 0.55 V, a J_{SC} of 9.73 mA cm^{-2} and a FF of 36.7%, leading to a PCE of 1.96%. The much lower photovoltaic performance of SAM-modified back-contact PSC is mainly attributed to the gap between interdigitated fingers. The gap of SAM-modified back-contact PSC is ~4.3 µm, which is almost tenfold the ideal value of 0.5 µm as demonstrated by simulation. The photogenerated carriers would recombine before they reach interdigitated finger electrodes. Moreover, the SAM-modified interdigitated finger electrodes only deliver less than 600 mV in work function difference, leading to a reduction in open-circuit voltage (V_{OC}) of the resultant device [9].

The same group further optimize the device performance by controlling the perovskite grain size, since the charge-carrier diffusion length is related to perovskite grain size [10]. They grow four grain cluster sizes of perovskite which are approximately 6.0, 2.0, 0.99, and 0.23 µm and denoted as large (L), medium (M), small (S), and extra small (XS), respectively. The maximum theoretical V_{OC} from BC-PSCs is determined by the work function difference induced by the self-assembled dipole monolayers coated on the gold electrodes. The V_{OC} increases with the perovskite grain size, showing an average value of 0.46, 0.4, 0.31, and 0.09 V for L, M, S, and XS cluster sizes, respectively. The average J_{SC} values for devices with L, M, S, and XS cluster sizes are 9.44, 7.64, 5.88, and 0.69 mA cm^{-2}, respectively. However, there is no clear trend in FF for devices with different grain size. It is noted that perovskite films with large cluster sizes exhibit longer charge recombination lifetimes, leading to a higher V_{OC}. Bach's group provided an approach via SAM treatment and optimized the device performance by improving the quality of the perovskite film and/or reducing the gap between the interdigitated electrodes. These pioneering works give a new insight into further advancing the photovoltaic performance of back-contact perovskite solar cells.

12.3   NON-COPLANAR BACK-CONTACT STRUCTURE

A. N. Jumabekov et al. first reported the QIDE-based BC-PSCs in 2016 [2]. Figure 12.3a shows the fabrication process of QIDE-based BC-PSCs. ZnO is sputtered on the ITO substrate to serve as ESL. Photoresist layer is spin-coated onto the substrate and exposed to UV light through a chrome metal photomask. The UV-exposed regions are removed using developer solution, and then, Al_2O_3, Al, and Ni layers are sequentially evaporated onto the substrate using electron-beam evaporator. Then, residual photoresist was removed by washing the substrates in acetone. HSL finger electrode is formed by oxidizing the surface of Ni and Al layer with an annealing temperature of 300°C. After that, the device was completed by depositing perovskite layer onto the QIDE. Figure 12.3b shows the band diagram of QIDE-based BC-PSCs. Electron and hole generated in perovskite layer are respectively collected by ZnO/ITO and NiO_x/

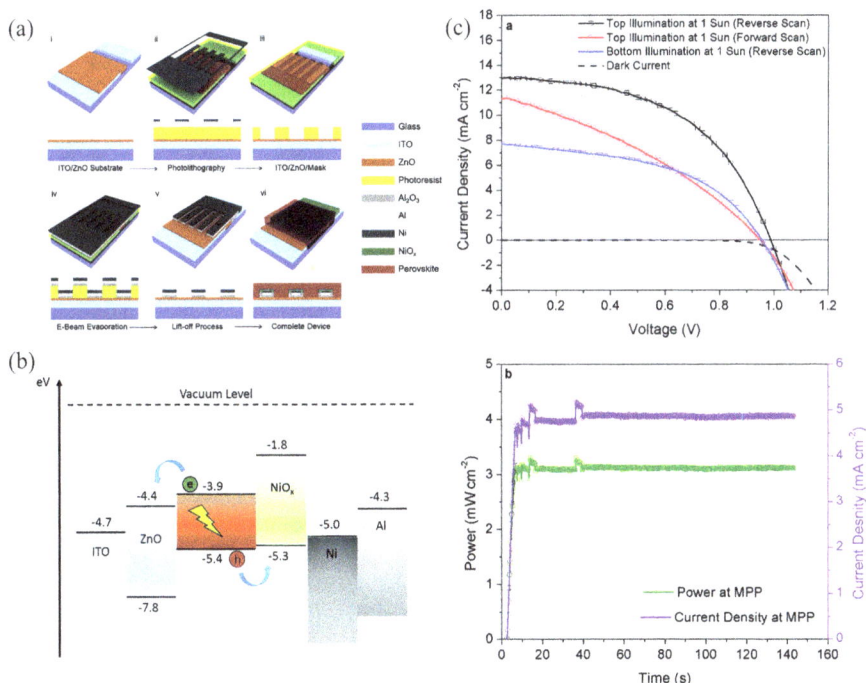

FIGURE 12.3   (a) Fabrication process, (b) band diagram and maximum power point tracking, and (c) *J-V* characteristic curves of QIDE-based BC-PSCs [2].

Ni/Al electrodes. Al_2O_3 layer acts as an insulator to prevent direct contact between cathode and anode. Figure 12.3c shows the *J-V* curves of the QIDE-based BC-PSCs. For the device illuminated from the perovskite layer (top illumination), the BC-PSCs under reverse scan obtain a PCE of 6.54% with a J_{SC} of 12.95 mA cm^{-2}, a V_{OC} of 0.98 V, and a FF of 51.5%. The device under forward scan shows a PCE of 3.68%, indicating the device exhibited a large hysteresis. They also carried out maximum power point tracking to determine the steady-state PCE. The devices show a stable power output with a PCE of around 3.2%, which is similar to the PCE value under forward scan. The origin of hysteresis in common perovskite solar cells has been discussed, and some reasons such as ferroelectric effect, ion migration, and capacitance effect have been suggested [11–13]. However, there is no discussion about the hysteresis in BC-PSCs. Therefore, investigating the origin of hysteresis in BC-PSCs would be one of the interesting topics. When the device is illuminated from the substrate (bottom illumination), the device shows a PCE of 3.52% with a J_{SC} of 7.7 mA cm^{-2}, a V_{OC} of 0.95 V, and a FF of 48.11%.

As shown in Figure 12.4a, metal electrode causes the optical loss due to its light absorption and reflection. To overcome the optical losses, G. DeLuca et al. proposed BC-PSCs with semitransparent QIDE by replacing the metal electrodes with the transparent ITO electrodes [3]. The cross-sectional SEM image of the semitransparent QIDE-based BC-PSCs is shown in Figure 12.4b. First, they optimized the deposition process of ITO transparent electrode to deliver a sheet resistance of 25–35 Ω sq^{-1}. CuSCN is applied as HSL to match the work function of ITO. The device illuminated from the perovskite layer achieves a PCE of 1.37% with a J_{SC} of 5.52 mA cm^{-2}, a V_{OC} of 0.84 V, and a FF of 0.30. By contrast, when the device is illuminated from the back side, a higher PCE of 1.66% is obtained with a J_{SC} of 5.60 mA cm^{-2}, a V_{OC} of 0.88 V, and a FF of 0.34. A similar J_{SC} value indicates that the absorption loss by conventional QIDE is successfully improved by substituting the metal electrode with the ITO electrode. They also performed simulations that reveal that the PCE of semitransparent QIDE-based BC-PSCs can be improved up to 11.5% and 13.3% for the front and back illumination, respectively.

Recently, G. D. Tainter et al. reported QIDE-based BC-PSCs employing SnO_2 and NiO_x as ESL and HSL, respectively, whose device architecture is illustrated in Figure 12.5a [14]. Figure 12.5b shows the *J-V* curve of QIDE-based BC-PSCs with a gap of 1 μm for NiO_x finger electrode. The device illuminated from the perovskite

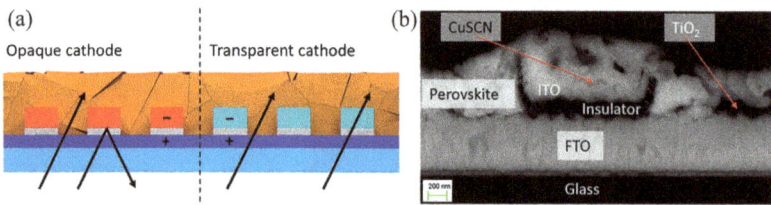

FIGURE 12.4 (a) Light transmission, absorption, and reflection paths in QIDE with metal electrodes (left) and transparent electrodes (right). (b) Cross-sectional SEM of the semitransparent QIDE-based BC-PSCs [3].

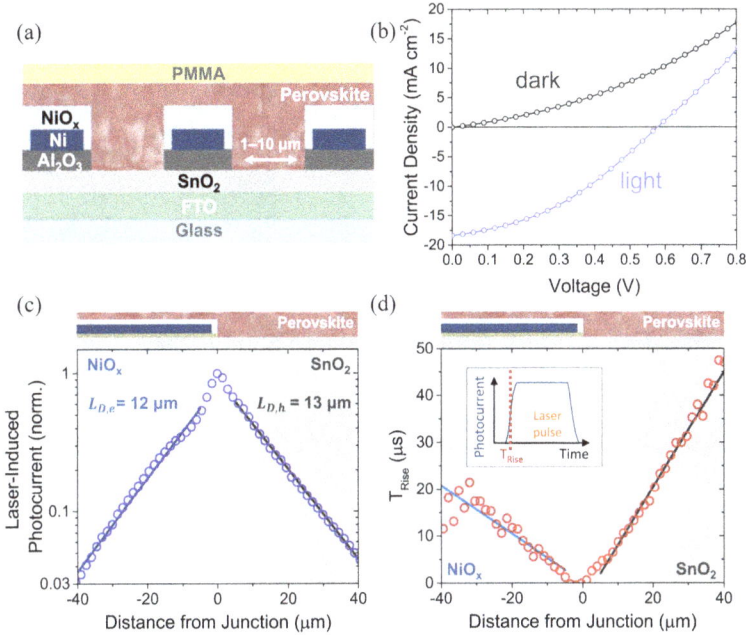

FIGURE 12.5 (a) Schematic illustration of QIDE-based BC-PSCs. (b) *J-V* curves of QIDE-based BC-PSCs with NiO_x finger gap of 1 µm. Dependence of (c) laser-induced photocurrent and (d) rise time with distance from the electrode junction [14].

layer shows a PCE of 4.1% with a J_{SC} of 18.4 mA cm^{-2}, a V_{OC} of 0.57 V, and a FF of 38%. Figure 12.5c and d represent the laser-induced photocurrent and rise time as a function of electrode position. The calculated surface recombination velocity is 1.7 and 1.1 cm s^{-1} for the perovskite covered on the NiO_x and SnO_2 surface, respectively. The hole and electron carrier lifetime for the region of perovskite film coverage on the SnO_2 and NiO_x electrode are calculated to be 27 and 17 µs, respectively. Low surface recombination velocity along with long carrier lifetime supports high J_{SC} value of QIDE-based BC-PSCs. It is noted that QIDE-based BC-PSCs suffer from lower V_{OC} and FF than the conventional perovskite solar cells due their poor shunt resistance. Further optimization on the device architecture or fabrication process is required to improve the photovoltaic performance.

Although the top-down approach for the fabrication of interdigitated electrodes via the photolithography or electron-beam lithography is well established, the finger electrode could be easily broken or even disconnected during the photolithography process. Such damage in finger electrode hinders the charge extraction and collection. Udo Bach's group replaced the finger electrode with the porous electrode by applying a sacrificial lithographic mask that provides facile, large-area, and cost-effective fabrication for the QIDE-type BC-PSCs [15]. The fabrication process of porous QIDE is shown in Figure 12.6. The authors introduced self-assembled monolayers of (3-aminopropyl)triethoxysilane (APTES) to functionalize the TiO_2/FTO substrate

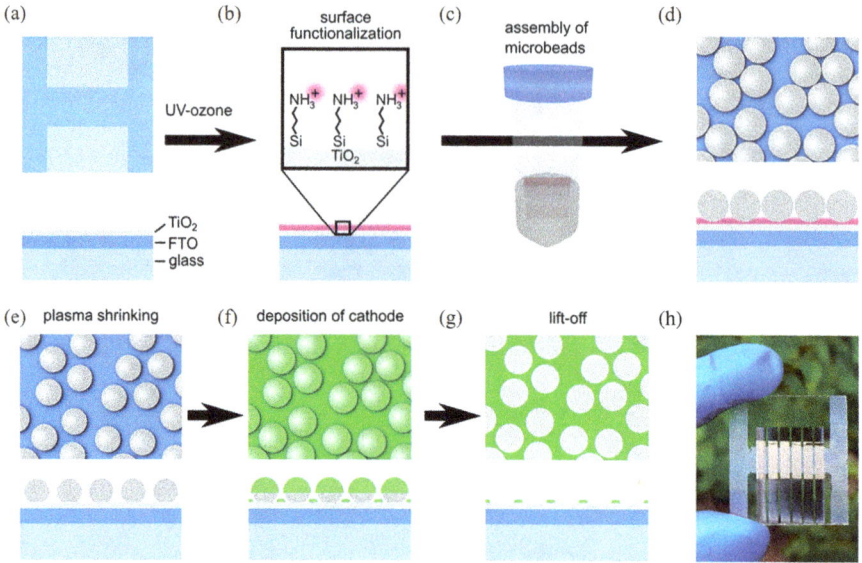

FIGURE 12.6 Fabrication process of porous QIDE-based BC-PSCs: (a) patterned FTO/TiO$_2$ anode substrate; (b) APTES SAM treatment on the top of anode; (c) immersion of substrate in the colloidal suspension to assemble the polystyrene microbeads; (d) monolayer of polystyrene microbeads assembled on the surface of TiO$_2$ anode; (e) plasma etching for reducing the size of polystyrene microbeads; (f) deposition of counter electrode and cathode; (g) removal of polystyrene microbeads; (h) photographic image of porous QIDE-based BC-PSCs [15].

(refer to Figure 12.6a), which facilitates the formation of dense monolayer of microbeads on the substrate (refer to Figure 12.6b). The coverage of microbeads on the substrate surface can be controlled by the pH value of the colloidal solution which adjusts the electrostatic attraction between negatively charged microbeads and positively charged substrate. The colloidal suspension with a pH value of 5.5 is prepared, which contains 1-μm-sized polystyrene microbeads with carboxyl groups. The substrate is dipping into colloidal suspension (refer to Figure 12.6c), on which the microbeads are self-assembled to form a monolayer as presented in Figure 12.6d. The plasma etching for the polystyrene microbeads was employed to define the width of porous electrode (refer to Figure 12.6e). An Al$_2$O$_3$ insulating layer (120 nm), an Al wetting layer (30 nm), and NiCo alloy (50 nm) are sequentially evaporated onto the microbead-coated substrate using electron-beam evaporation (refer to Figure 12.6f). The porous electrode is thus formed by removing the microbeads with toluene or tetrahydrofuran (refer to Figure 12.6g). After the annealing treatment at 300°C for 15 min, the porous electrode was completed by converting NiCo into NiCoO$_x$ cathode. The perovskite light absorber is finally deposited on the top of porous electrode to complete the BC-PSCs with an architecture of FTO/TiO$_2$/Al$_2$O$_3$/Al/NiCo/NiCoO$_x$/perovskite.

The deprotonation of carboxyl groups on microbeads surface leads to negatively charged microbeads, while the protonation of amine groups of APTES gives rise to the TiO$_2$-coated substrate with positive charge. As a result, the deposition rate of microbeads on the APTES-treated substrate is higher than that without APTES treatment, resulting in a denser coverage of microbeads as demonstrated in Figure 12.7a (with APTES surface treatment) and b (without APTES surface treatment). After the microbead shrinkage by plasma etching, deposition of $Al_2O_3/Al/NiCo$, and removal of microbeads, the surface morphology of porous electrode is thus created as revealed in Figure 12.7c (with APTES surface treatment) and d (without APTES surface treatment). It indicates that APTES-treated substrate exhibits a higher porosity than that without APTES treatment. The enhanced porosity for APTES-treated substrate also increases the transmission across the 300–1100 nm wavelength range (refer to Figure 12.7e). Eventually, the back-contact perovskite solar cells integrated with porous electrode are demonstrated, among which the APES-treated devices deliver a J_{SC} of 7.2 mA cm^{-2}, a V_{OC} of 0.9 V, a FF of 27%, and a PCE of 1.75%.

The same group further fabricated a honeycomb QIDE (H-QIDE) in BC-PSC, in which the honeycomb array was well arranged by photolithography [16]. The BC-PSCs integrated with HQIDE are illustrated in Figure 12.8a. The fabrication process and device structure are similar to the previous work except the fabrication of honeycomb QIDE by photolithography. The bottom TiO_2 compact layer works as ESL, and the conversion of annealed NiCo into $NiCoO_x$ serves as HSL. An Al_2O_3 insulator is inserted to separate these two electrodes. The electrode dimensions have a dramatic impact on the charge collection and device performance due to the limitation of carrier diffusion length in the perovskite active layer. Two-sized HQIDEs with different diameter and gap are designed and denoted as S-HQIDE (diameter of 2.7 μm and gap of 1.3 μm in Figure 12.8b and c) and L-HQIDE (diameter of 8 μm and gap of 1.5 μm in Figure 12.8d and e). The BC-PSCs integrated with S-HQIDE achieve a better photovoltaic performance (PCE = 5.39%) than the L-HQIDE counterpart (PCE = 1.46%), mainly on the enhanced photocurrent. The result is reasonable since a small-sized HQIDE can enhance the charge collection and thus improve the J_{SC} for the S-HQIDE device.

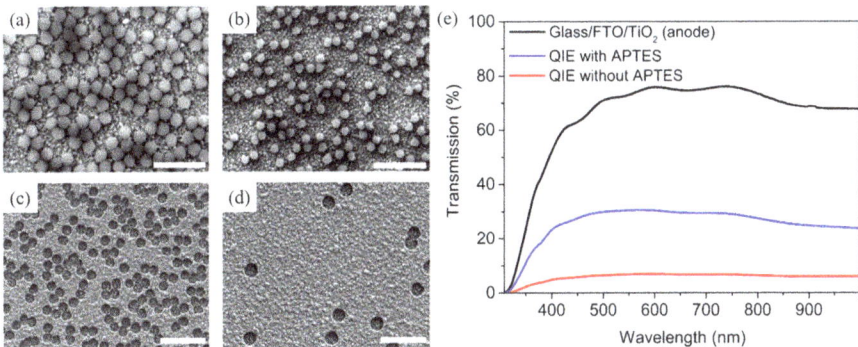

FIGURE 12.7 SEM images of monolayer polystyrene microbeads on (a) the APTES-functionalized glass/FTO/TiO$_2$ anode substrate and (b) the glass/FTO/TiO$_2$ anode substrate. SEM images of porous QIDE (c) with and (d) without APTES treatment. The scale bar is 4 μm. (e) UV–Vis–NIR transmission spectra of glass/FTO/TiO$_2$ anode substrate, QIDE without and with APTES treatment [15].

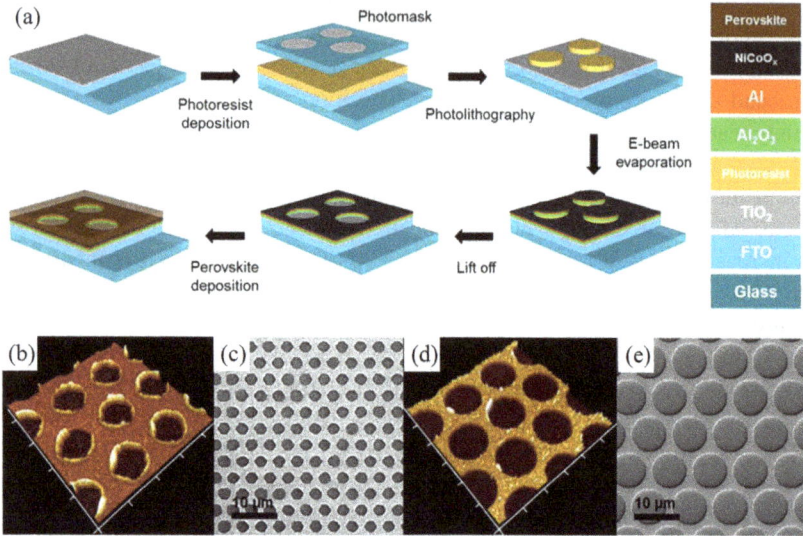

FIGURE 12.8 (a) Fabrication process of HQIDE-based BC-PSCs. (b) AFM and (c) SEM images of S-HQIDE; (d) AFM and (e) SEM images of L-HQIDE [16].

The same group further demonstrated HQIDE-based BC-PSCs via dipole-field modification on two gold finger electrodes [17]. Two self-assembled benzenethiol derivatives with opposite dipole moments are adsorbed onto the gold electrodes that replace the charge-transporting layers to selectively extract charge due to the work function asymmetry. The fabrication flow is presented in Figure 12.9a. Au/SiO$_2$/Au stacking layer is first deposited on top of the glass substrate. Photo-lithography is employed to define the HQIDE composed of Au/SiO$_2$. Thus, the bottom planar gold layer (denoted as "a", colored in blue in Figure 12.9b) and top Au/SiO$_2$ mesh (denoted as "b", colored in red in Figure 12.9b) are separated by SiO$_2$ insulator. The bottom planar gold layer and top gold mesh are surface-modified with monolayer of OMeTP. The honeycomb-shaped top Au/SiO$_2$ electrode is then exposed to CITP solution to undergo oxidative electrochemical reaction and replace OMeTP with CITP SAM. The formation of OMeTP and CITP on bottom planar electrode and top HQIDE, respectively, creates a surface potential difference of 360 mV which is capable of opposite charge extraction. Three feature sizes of HQIDE in terms of honeycomb diagonal length and pitch (denoted as "d" and "p" in Figure 12.9b) are controlled to correlate the device photovoltaic performance. The results indicate that HQIDE-based BC-PSCs having the smallest diagonal length and pitch ($d = 5.0\,\mu m$ and $p = 6.2\,\mu m$) display the best PCE of 2.61%. Larger feature size limits the charge collection efficiency and the resultant J_{SC}.

12.4 CONCLUSION

In conclusion, we briefly introduce the development of the back-contact perovskite solar cells. The efficiency trend and photovoltaic parameters of back-contact perovskite solar cells are summarized in Figure 12.10 and Table 12.1, respectively. Compared with

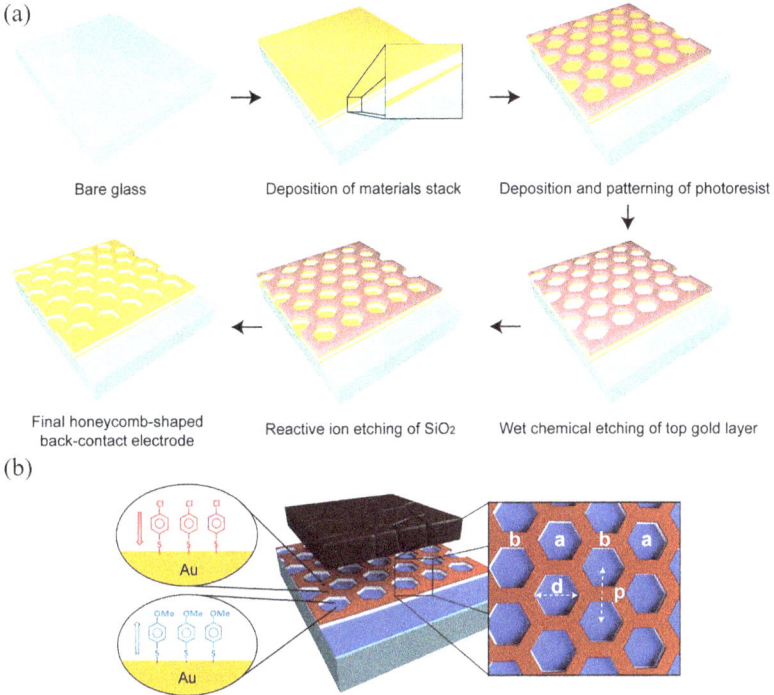

FIGURE 12.9   (a) Fabrication flow of HQIDE for BC-PSCs. (b) The work function modification of bottom planar Au electrode (marked by blue) by OMeTP and top HQIDE by CITP (marked by red) [17].

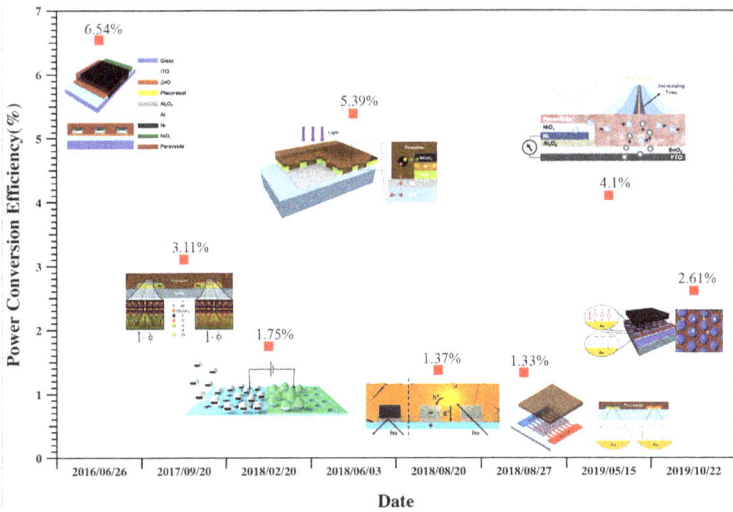

FIGURE 12.10   Efficiency evolution of BC-PSCs.

TABLE 12.1

Photovoltaic Parameters of Different Types of BC-PSCs. The Device Illuminated from the Perovskite Layer Is Denoted as "Front," While that Illuminated from the Substrate Is Denoted as "Back"

Type		Illumination Side	V_{OC} (V)	J_{SC} (mA cm^{-2})	FF (%)	PCE (%)	References
Coplanar	IDE	Front	0.64	9.87	37.32	2.36	[9]
		Back	0.64	8.77	36.69	2.23	
		Front	0.46	9.44	30.58	1.33	[10]
Non-coplanar	Porous	Back	0.83	7.1	66	3.88	[18]
	QIDE	Front	0.98	12.95	51.5	6.54	[2]
		Back	0.95	7.7	48.4	3.52	
	QIDE	Front	0.84	5.52	30	1.37	[3]
		Back	0.88	5.6	34	1.66	
	QIDE	Front	0.57	18.4	38	4.1	[14]
	QIDE	Front	0.9	7.2	27	1.75	[15]
	HQIDE	Front	0.7	16.4	46	5.39	[16]
	HQIDE	Front	0.52	12.52	40.16	2.61	[17]

the conventional sandwiched PSCs, back-contact perovskite solar cells still have much room for improvement in photovoltaic performance. Some key issues are highlighted to improve the performance of back-contact devices. First, in order to increase the charge collection efficiency, the gap of IDE or QIDE should be comparable with the carrier diffusion length of perovskite. Second, the series resistance should be reduced to improve the FF of devices. For instance, ESL and HSL should be well separated by insulator and the metal electrodes are required to be passivated to prevent the carrier recombination at metal electrode/perovskite interface.

REFERENCES

1. Brenner, T.M., et al., Hybrid organic – inorganic perovskites: low-cost semiconductors with intriguing charge-transport properties. *Nature Reviews Materials*, 2016. **1**(1): p. 15007.
2. Jumabekov, A.N., et al., Back-contacted hybrid organic–inorganic perovskite solar cells. *Journal of Materials Chemistry C*, 2016. **4**(15): pp. 3125–3130.
3. DeLuca, G., et al., Transparent Quasi-interdigitated electrodes for semitransparent perovskite back-contact solar cells. *ACS Applied Energy Materials*, 2018. **1**(9): pp. 4473–4478.
4. Ma, T., et al., Unveil the full potential of integrated-back-contact perovskite solar cells using numerical simulation. *ACS Applied Energy Materials*, 2018. **1**(3): pp. 970–975.
5. Goh, C., S.R. Scully, and M.D. McGehee, Effects of molecular interface modification in hybrid organic-inorganic photovoltaic cells. *Journal of Applied Physics*, 2007. **101**(11): p. 114503.
6. Bruening, M., et al., Simultaneous control of surface potential and wetting of solids with chemisorbed multifunctional ligands. *Journal of the American Chemical Society*, 1997. **119**(24): pp. 5720–5728.

7. Liscio, A., et al., Bottom-up fabricated asymmetric electrodes for organic electronics. *Advanced Materials*, 2010. **22**(44): pp. 5018–5023.

8. Kobayashi, S., et al., Control of carrier density by self-assembled monolayers in organic field-effect transistors. *Nature Materials*, 2004. **3**(5): pp. 317–322.

9. Lin, X., et al., Dipole-field-assisted charge extraction in metal-perovskite-metal back-contact solar cells. *Nature Communications*, 2017. **8**(1): p. 613.

10. Lin, X., et al., Effect of grain cluster size on back-contact perovskite solar cells. *Advanced Functional Materials*, 2018. **28**(45): p. 1805098.

11. Wei, J., et al., Hysteresis analysis based on the ferroelectric effect in hybrid perovskite solar cells. *The Journal of Physical Chemistry Letters*, 2014. **5**(21): pp. 3937–3945.

12. Zhang, T., et al., Understanding the relationship between ion migration and the anomalous hysteresis in high-efficiency perovskite solar cells: a fresh perspective from halide substitution. *Nano Energy*, 2016. **26**: pp. 620–630.

13. Cojocaru, L., et al., Origin of the hysteresis in I–V curves for planar structure perovskite solar cells rationalized with a surface boundary-induced capacitance model. *Chemistry Letters*, 2015. **44**(12): pp. 1750–1752.

14. Tainter, G.D., et al., Long-range charge extraction in back-contact perovskite architectures via suppressed recombination. *Joule*, 2019. **3**(5): pp. 1301–1313.

15. Jumabekov, A.N., et al., Fabrication of back-contact electrodes using modified natural lithography. *ACS Applied Energy Materials*, 2018. **1**(3): pp. 1077–1082.

16. Hou, Q., et al., Back-contact perovskite solar cells with honeycomb-like charge collecting electrodes. *Nano Energy*, 2018. **50**: pp. 710–716.

17. Lin, X., et al., Honeycomb-shaped charge collecting electrodes for dipole-assisted back-contact perovskite solar cells. *Nano Energy*, 2019: p. 104223.

18. Hu, Z., et al., Transparent conductive oxide layer and hole selective layer free back-contacted hybrid perovskite solar cell. *The Journal of Physical Chemistry C*, 2017. **121**(8): pp. 4214–4219.

13 Engineering of Conductive Polymer Using Simple Chemical Treatment in Silicon Nanowire-Based Hybrid Solar Cells

Po-Hsuan Hsiao, Ilham Ramadhan Putra, and Chia-Yun Chen
National Cheng Kung University

CONTENTS

13.1 INTRODUCTION

The demand for clean energy continues to be followed up by the development of renewable energy coping with the environmentally friendly, cheap, and efficient routes. Solar energy is one of the energy sources offering clean, sustainable, and stable energy. In 2018, the worldwide solar photovoltaic (PV) power was expected to generate approximately 2.8% of the worldwide electricity [1]. In recent years, silicon (Si)-based solar cells have been considered the most widely used type of PV devices. Unfortunately, the manufacturing of this type of solar cells requires a fairly complex process. In this regard, the various types of solar cells followed by the development of alternative materials essentially turn out to be a prospective field on both academic research and industrial applications. The emerging PV structures such as dye-sensitized, organic-based,

233

and inorganic thin film-based solar cells are widely being escalated and utilizing various materials. Moreover, one more type of "layer-by-layer" solar cells that possess the potential impact on energy-related applications is hybrid-solar cells.

The "hybrid" term is typically known as the sequence of organic–inorganic materials applied for the device with active junction carrying on the charge separation of photogenerated electrons and holes. Further completed device or cell system development utilizing current collector layers as the top and bottom electrodes becomes the main point in various developments of solar cells. The main focus on the device engineering is to build up remarkable performance and efficiency by considering several factors especially the charge transport and separation. These include materials selection, p- and n-type materials, nanostructural approaches, passivation layer, interface contact improvement, and physical engineering techniques such as junction thickness determination. Moreover, several studies spotlighted to other parts, for example, are the electrode materials selection as current collectors and hole (h^+) or electron (e^-) selective layer applied between the active layer and current collectors. In the organic–inorganic junction, the structure consists of transparent conductive polymer in combination with the inorganic semiconductor, contributing to an adequate transport and separation of photogenerated carriers. From various materials acquired for the cell construction, the arrangement of PEDOT:PSS polymer and n-type crystalline Si was proposed by considering both superiorities. This type of hybrid solar cells has attracted worldwide attention since it was proposed in 2010 and further developed as the potential solar cells at present. Compared with ion-based batteries as the energy-storage system, hybrid solar cells turn out be the efficient way for supplying the electric energy in a stable and reliable way. By combing hybrid solar cells with battery system, this can be an ecofriendly and renewable platform for supplying energy required in use. Thus, this book can be served as the bridge for the readers to fully understand the related technology of energy generation and storage.

Poly(3,4-ethylenedioxythiophene):poly(styrenesulfonate) (PEDOT:PSS) is a conductive polymer that offers numerous superiorities such as tunable electrical conductivity by uncomplicated treatments, facile fabrication to form thin film, acceptable thermal stability, material flexibility, and superior transparency in the visible range. In contrast, crystalline Si as the n-type material was considered to offer several advantages such as higher minority carrier lifetime, adjustable doping type, and concentrations, thus making it a preferable candidate for creating p–n junction with p-type organic materials. In this architecture of hybrid solar cells, PEDOT:PSS/planar n-type Si junctions have led to achieving cells efficiency by up to 12%. Later, many approaches to improve the performance of these hybrid solar cells were discovered including nanostructural technique [2], passivation layer at the interface between PEDOT:PSS/n-type Si, and additional charge selective layer configuration [3].

Interestingly, the PEDOT:PSS material as the p-type layer in this type of cell arrangement surely has an important impact in the junction section, and the study of this organic is substantial to be conducted. Specifically, this polymer in its commercial product has a low applicable electrical conductivity around 0.2–10 S cm^{-1}. This value, of course, becomes a problem when applied it to hybrid solar cells that require a preferable high electrical conductivity. Many studies on the improvement of electrical conductivity in PEDOT:PSS have been widely carried out, such as applying physical treatments by heat or light and chemical treatment including cosolvents, surfactants, acids, and ionic liquids, known to increase its conductivity up to 10^3 orders for many applications. Furthermore, these chemical utilizations conducted by different combinations of pretreatment, direct addition, and post-treatment may benefit the desired applications based on treated PEDOT:PSS. Specifically, the underlying mechanism regarding the conductivity improvement correlates with the structural and morphological changes. However, in some cases of the post-treatment, the process may influence the thin-film uniformity or even damage the prepared thin film [4]. For these reasons, an easy-and-efficient approach to overcome some of these issues needs to be studied.

The use of cosolvents and surfactants is certainly an interesting consideration because it offers simplicity and reliability in the fabrication process. Of the various uses of chemicals, the addition of cosolvents using ethylene glycol (EG) is known to increase the conductivity of PEDOT:PSS to around 620 S cm^{-1} [5]. In contrast, the addition of surfactants such as fluorosurfactant provides superior wetting properties of the PEDOT:PSS solution, which enables the facile deposition on a variety of hydrophobic substrates [6]. By combining both organic solutions and additional annealing processes, the improvement of both conductivity and uniformity in PEDOT:PSS thin film can actually benefit the deposition of functional PEDOT:PSS on the nanostructured n-type Si. For example, the application of emerging PEDOT:PSS/planar Si for the cell design has been carried out and proven to improve the photovoltaic performance [7,8]. However, the research for combining PEDOT:PSS thin film with nanowires (SiNWs) is still limited. In fact, the SiNWs have various advantages, and one of the well-known applications in solar cells is being able to provide a light-trapping effect so as to enable the optimization of incoming light transmitted from the transparent PEDOT:PSS layer [9]. Therefore, PEDOT:PSS/SiNW has been considered an efficient way for the construction of high-performance hybrid solar cells.

13.2   PEDOT:PSS WITH TUNABLE ELECTRICAL CONDUCTIVITY

13.2.1   PEDOT:PSS Fabricated by "Baytron P" Routes

PEDOT chains were generally synthesized by the polymerization process of its monomer known as EDOT (3,4-ethylenedioxythiophene) or its derivatives through electrochemical oxidation, transition metal-mediated coupling, or oxidative chemical process. The electrochemical oxidation process of EDOT derivatives is one of the advantageous methods since it needs a short synthetic period, small amounts of EDOT monomer, and can generate free-standing thin films. Several electrolytes, such as polyelectrolytes, or aqueous micellar can be utilized for the electrochemical process. Moreover, the polymerization of substituted EDOTs (alkyl sulfonate, alkylated, etc.) can also be processed by electrochemical routes, and further modification of this process is called electrochemical desilylation [10,11]. Another method termed "transition metal-catalyzed coupling" was utilized to fabricate neutral PEDOT based on the direct formation. The oxidative chemical polymerization has mostly applied in synthesizing PEDOT. The chemical polymerization of EDOT derivatives can be processed using various oxidants. Generally, the oxidizing agents are Fe(OTs)$_3$ or FeCl$_3$, which results in the insoluble and infusible compounds in a film type with black color.

Furthermore, the polymerization process of alkoxylated or alkylated EDOT derivatives (the $R \geq C_{10}H_{21}$) produced the random-soluble PEDOT derivatives in various organic solvents including THF, CH_2Cl_2, and $CHCl_3$. Aforementioned, using several fabrication processes above, PEDOT intrinsically not only has insoluble properties but also possesses several sound properties including high conductivity with value around 100 S cm^{-1}, stable in the oxidized state, and relatively high transparency compared with other organic conductors. Since the insolubility of PEDOT becoming a critical issue for the fabrication process, the uncomplicated oxidation chemical routes of Baytron P (P stands for polymer) process were discovered by Bayer AG and further broadened numerous researches related to this conductive polymer [12].

Baytron P synthesis process is known as the most fundamentally convenient PEDOT-based compound fabrication procedure by polymerization method of EDOT in the water-soluble polyelectrolyte solution (generally using template polymers such as polystyrene sulfonic acid or PSSa) by utilizing sodium persulfate (also known as sodium peroxodisulfate – $Na_2S_2O_8$) as the oxidizing agent. The polymerization can produce aqueous PEDOT/PSS dispersion with dark-blue color and soluble in the water allowing facile practicable conductive polymer. The reaction of Baytron P process is briefly shown in Figure 13.1. It exhibits several interesting properties such as insoluble in many common solvents, transparent, and sound mechanical durability. However, the PEDOT:PSS complex usually possesses reduced conductivity around 10 S cm^{-1}, which is becoming a critical issue for particular applications.

13.2.2 PSS Functions in Commercial PEDOT:PSS Complex

Pristine PEDOT possesses an electronic bandgap in the range of 1.6–1.7 eV owing to the onset π–π^* absorption and has the maximum absorption located at 2.0 eV. The maximum absorption wavelength is 610 nm that results in a dark-blue color [12]. However, this pristine polymer owns a lower redox potential that

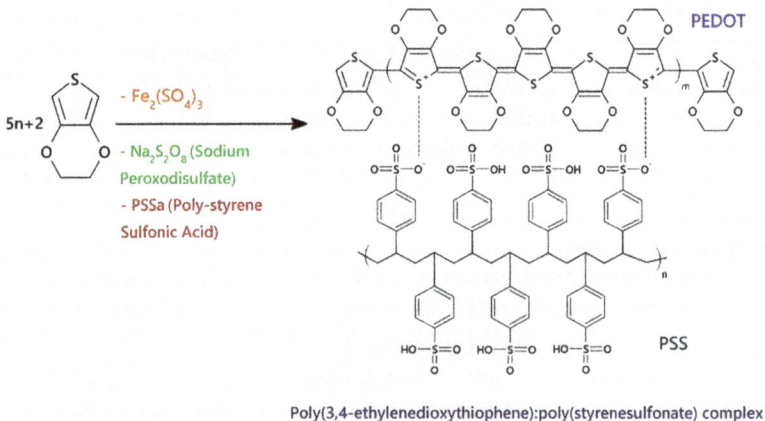

Poly(3,4-ethylenedioxythiophene):poly(styrenesulfonate) complex

FIGURE 13.1 Baytron P synthesis routes applying sodium peroxodisulfate as the oxidizing agent.

intended to be oxidized in air and hard to be processed. The aforementioned Bayer P polymerization routes are offering the stable complex PEDOT:PSS in the aqueous dispersion to clear this issue up. In PEDOT:PSS-based aqueous solutions, PSS brings two functions: the first function is to assist the PEDOT components being dispersed in the aqueous medium, such as water, yields the dark-blue microdispersion, which is facile to process and can be stably utilized. It is certainly beneficial in applications for various ease solutions-based process as the example is thin-film fabrication utilizing simple spin coating, drop casting, or advance process needed.

In contrast, by applying Baytron P polymerization procedures, the product of PEDOT chains is known as a positively doped conjugated polymer (PEDOT$^+$). The positive charge or the radical cation was gained through the oxidation process of EDOT monomers. The term p-doping in conducting polymers studies can be described as the process that involves the oxidation treatment of the polymer backbone, and immediately the electronic structure is modified [13]. By this finding, the second function of PSS component is as the anion or charge balancing dopant of cation (PEDOT$^+$). Since PSS is an anionic polyelectrolyte, it results in high stability complex with PEDOT owing to the fact that this ionic set cannot be separated by capillary electrophoresis [14]. These two important functions of PSS attribute to many aspects of PEDOT:PSS behaviors. Specifically, it affects the structural morphology and the electrical conductivity of PEDOT:PSS complex.

13.2.3   PSS Investigations of Electrical Conductivity in PEDOT:PSS

Although the Baytron P process can achieve facile processing of PEDOT:PSS aqueous solution, the decreasing value of its conductivity becomes the main issue for practical applications. Several reasons are being outlined generally caused by the influence of PEDOT structure and distribution, as well as the PSS existence in the complex. As mentioned earlier, the PEDOT segments as the metallic part inside the complex play an important role in PEDOT:PSS conductivity. As the polythiophene derivative, it consists of main thiophene rings that bring the charges from the conjugated electrons. In the PEDOT:PSS complex, the p-type PEDOT is interpreted as the initial electronic state as the yield of oxidation processes in the polymerization process. As discussed before, the PEDOT segment has its own intermediate electronic structure between quinoid and benzoid that assumed to allow the conjugated arrangement in PEDOT and providing the different distribution of charge carriers on the polymer chains [15]. The benzoid structure is known to give a coiled structure, and the quinoid structure is assumed to promote extend-coiled or linear conformation. The coiled structure of PEDOT in PEDOT:PSS is deliberated to lower the conductivity of the complex because of incompletely distribution of π-electrons over the whole PEDOT chains [16]. In contrast, the prior explanation in PSS structural-morphological investigation is notably that PSS has not only two main functions on the stability of PEDOT:PSS complex but also significantly influences the structural behavior of PEDOT:PSS solution. Therefore, the structural-morphological behavior of conjugated polymer claimed correlates to electrical conductivity properties of PEDOT:PSS.

General treatment of pristine PEDOT:PSS can be divided into three procedures, namely *pretreatment* such as the engineering on polymerization process by chemical oxidation; *direct treatment*, for example, is the mixing process with various chemical additives; and *post-treatment* route after the formation of dried-solid PEDOT:PSS. Moreover, in many studies, the combination of these procedures surely exhibits higher and better conductivity values. Interestingly, in several investigations, some grades of PEDOT:PSS mentioned above can produce different final conductivities after the same treatment using certain additives, for example, the study of conductivity improvement of Clevios P and PH1000 with a close initial conductivity in the range of 0.2–0.35 S cm^{-1}. Hydrofluoroacetone trihydrate (HFA) treatments applied in both grades revealed significant difference in final conductivity by the value of 171 S cm^{-1} for Clevios P and 1164 S cm^{-1} for Clevios PH1000 [17]. This finding deduced the improvement is because of a longer PEDOT chain arrangement of Clevios PH1000 compared to Clevios P. However, in some studies, other grades of commercial PEDOT:PSS obviously are still be used for preferred applications.

The physical source and chemical additive treatments are used to tune the conductivity of pristine PEDOT:PSS. Physical treatments, such as employment of UV irradiation, can manipulate the work function and tune conductivity by reducing the number of charge-trapping defects followed by the conformational change to linear or expanded coil arrangement of PEDOT segments [18]. Moreover, heat-treatment methods can elevate the electrical conductivity due to the change of oxidation state of PEDOT:PSS films [19]. In contrast, chemical treatments, such as polar solvents, surfactants, ionic liquids, and acids, promise a simple and inexpensive treatment that can improve the conductivity value to 10^3 orders of magnitude higher than the pristine sample [20]. Dimethylsulfoxide (DMSO) and ethylene glycol (EG) are normally used through the direct mixing or the post-treatment procedure that can achieve superior values more than 500 S cm^{-1}. Furthermore, by addition, the mixing treatment using surfactants such as fluorosurfactant or triton-X100 into mixed DMSO/EG-PEDOT:PSS can offer excellent conductivity performance with the sound film uniformity. The summary of both polar solvents and several other additive treatments in diverse studies is collected in Table 13.1.

The utilization of DMSO and EG in diverse routes and the combination can clearly enhance the electrical conductivity of pristine PEDOT:PSS films and applied for many emerging applications such as organic solar cells, thermoelectric devices, and hybrid solar cells in recent years. The enhancement of conductivity value caused by these additives is demonstrated with regard to PEDOT segment's structural changes from coiled to more linear conformation, removal of excess PSS, and hopping actions among metallic particles identified by Raman spectroscopy, XPS-ESCA, and AFM instruments. Moreover, the current studies claimed that sulfuric acid (H_2SO_4) can enhance the PEDOT:PSS conductivity achieving around 4300 S cm^{-1} through post-treatment routes [25]. However, the direct mixing procedures using this acid can lead to the agglomeration of PEDOT:PSS and thus hinder the facile fabrication capability of PEDOT:PSS solution due to dehydrating reaction. Moreover, not all post-treatment processes by chemicals can be applied in PEDOT:PSS-based device fabrication routes because for some cases such as post-treatment using alcohol that degrades the stability of thin film deposited on substrates [4].

TABLE 13.1

Several DMSO and EG Treatments for Commercial Pristine PEDOT:PSS with Regard to Tune Electrical Conductivity in Several Emerging Devices Applications

Initial Conductivity (S cm⁻¹)	Additives	Treatment Processes	Final Conductivity (S cm⁻¹)	Emerging Applications	References
0.3 (Clevios PH1000)	DMSO, Triton X-100, alcohols	Direct addition (DMSO and TX mixing) + posttreatment (dropped alcohols)	650 (DMSO-TX); 826 (DMSO-TX-IPA); 908 (DMSO-TX-ethanol); 1105 (DMSO-TX-methanol)	Planar silicon/organic hybrid solar cells	[7]
$R_s = 1.3 \times 10^6\ \Omega\ sq^{-1}$ (Clevios PH1000)	EG	Direct addition (mixing)	$R_s = 190\ \Omega\ sq^{-1}$	Planar silicon/organic hybrid solar cells	[8]
0.3 (Clevios PH1000)	EG, DMSO	Direct addition (mixing) + post treatment (immerse in EG bath)	639 (EG); 620 (DMSO)	Thermoelectric	[21]
0.3 (Clevios PH1000)	EG, PEG	Direct addition (mixing)	640 (EG); 805 (PEG)	Organic solar cells	[22]
0.7 (Clevios PH1000)	DMSO	Post-treatment (drop method)	1185	Liquid crystal display	[23]
0.1–10 (Clevios PH1000)	EG	Direct addition + post-treatment (immerse method)	1418	Organic solar cells	[24]

In a brief description, the following discussion is several points of view regarding the mechanism how the conductivity of pristine PEDOT:PSS can be improved. The first mechanism is proposed by the identification using X-ray photoelectron spectroscopy (XPS) by the mixing of DEG into PEDOT:PSS complex. These findings confirmed that the removal of PSS from PEDOT:PSS complex noticeable by XPS spectra recognition and area calculation of pointed out peaks presenting the sulfur elements in the energy range of 160–170 eV. The sulfur peaks used on the identification are the S2p bands located at 163–167 eV later confirmed as the doublet peaks of sulfur atoms in PEDOT. The intensities of PEDOT peaks identified with respect to S2p bands from PSS constituent showed to increase by the addition of DEG later suggested as the removal of partially PSS in the film [26]. Another study also confirmed the excess PSS washing on the film surface during the additive post-treatment applying post-spin-rinsing method (PSRM) using the DMSO solvent [27].

A further approach based on the identification of PEDOT chain conformational changes before and after EG treatment was obtained by Raman spectroscopy. As-prepared PEDOT:PSS was proclaimed consisting of intermediate structures, there are benzoid and quinoid structures in its PEDOT segments as aforecited above. The resonant structure of benzoid, which is confirming the coiled structure, will change to the quinoid arrangement as the linear or expanded coiled structure results in the conductivity enhancement mechanism. The changes of this conformational arrangement can be observed by the changes of molecules and chemical structural alteration validated by peak shifting of the spectra. The same study also revealed that the conformational change corresponds to the delocalization process of polaron to bipolaron charges over several units in PEDOT chains recognized by the electron spin resonance (ESR). A polaron is conformed as a positive charge on a unit, and a bipolaron is two positive charges delocalized over several units later; both mechanisms are known to enhance the conductivity of pristine PEDOT:PSS [28].

Other assumptions are correlated to phase separation and screening effect constructed after the addition of additives. The phase separation is attributed to the phase separation of PEDOT and PSS by the formation of hydrogen bonding among cosolvents and both PSS and PSSH, resulting in the linearly oriented PEDOT chains [22]. Another study revealed the screening effect occurred due to dielectric properties of solvents that reduce the Coulomb interaction between PEDOT and PSS constituents. Further study in polar solvents including EG and DMSO addition based on THz-TDS measurements confirms the charge screening responsible for the increase of conductivity [27,29]. The conductivity enhancement by hopping, doping effects, and morphological changes was proposed by other investigators. The resistance (R)–temperature (T) relationship of conducting polymers at $T < 220$ K was particularly analyzed by one-dimensional variable range hopping (VRH) model [30].

13.3 TREATED PEDOT:PSS FOR SILICON NANOWIRES-BASED HYBRID SOLAR CELLS

This study was conducted for functioning PEDOT:PSS as the hole transport layer as well as p-type component in hybrid solar cells. A simple route utilizing ethylene glycol (EG-99%) as solvent has been proposed in enhancing the conductivity of pristine

PEDOT:PSS thin film. EG was chosen for cosolvent and secondary doping agent since this solvent has two polar groups and superior dielectric constant leading to the phase separation of PEDOT:PSS constituents and further results in the better conductivity. The incorporation of EG studied in different weight percents (wt%) from 1 to 11 wt% with PEDOT:PSS was performed. Moreover, to obtain preferable thin-film uniformity on the substrate, the surfactant as the wettability improver is needed. Fluorosurfactant zonyl (FS) is not only claimed to provide better adhesive of thin film but also can reduce the sheet resistance providing additional conductivity enhancement of PEDOT:PSS in the film shape [6]. The fabrication route of thin films is proposed by detail in Figure 13.2. The pretreatment of pristine PEDOT:PSS solution was conducted by the filtering process. And then, EG cosolvent was added into the filtered PEDOT:PSS in several amounts from 3, 7, and 11 wt%, and the samples are named as EG3, EG7, and EG11, respectively. As the reference, pristine PEDOT:PSS was used and noted as EG0. Moreover, the small amount of FS incorporated into the PEDOT:PSS/EG followed by the mixing process under atmosphere for several hours. Thin-film fabrication was utilized by applying spin-coating methods on the substrates. Later, the drying process was performed at 140°C for several minutes in creating solidified dispersion. Electrical properties of the treated PEDOT:PSS/EG thin films were evaluated by two approaches using four-point probe (FPP) and Hall-effect analyzer for examining the charge carrier concentrations and mobility. The film thickness was estimated by ellipsometry, and its morphology was observed by scanning electron microscopy (SEM).

Figure 13.3 presents the transmittance spectra, SEM observation, and thin-film-coated results of the pristine and treated samples on the glass substrates, respectively. The spectra of the bare glass substrate indicate a quite high transmittance up to 95% and stable in the whole wavelength. Furthermore, the samples from 0 to 11 wt% EG

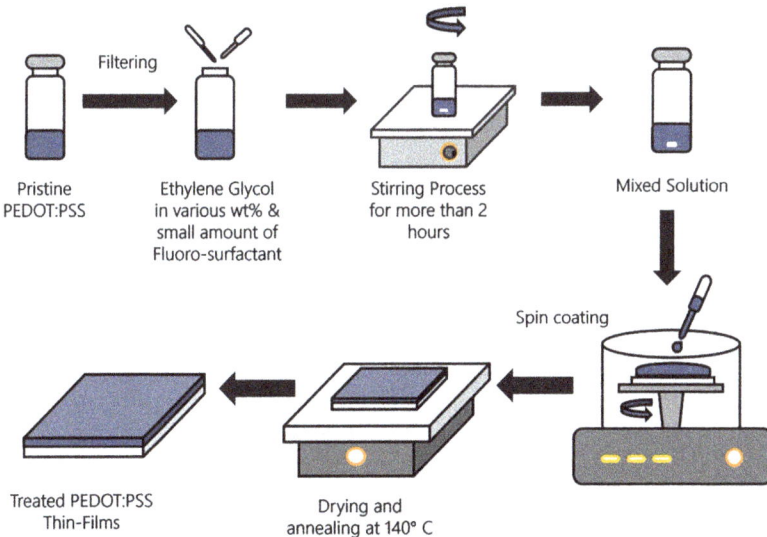

FIGURE 13.2   Thin-film fabrication routes by incorporation of EG and FS.

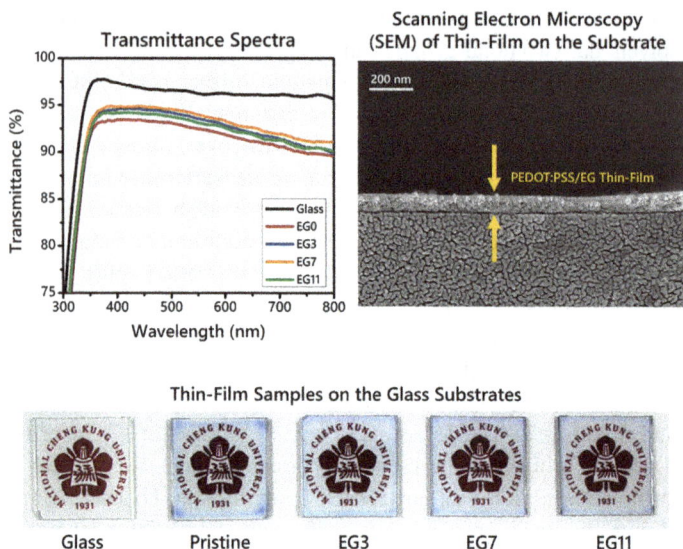

Transmittance Spectra

Scanning Electron Microscopy (SEM) of Thin-Film on the Substrate

Thin-Film Samples on the Glass Substrates

Glass   Pristine   EG3   EG7   EG11

FIGURE 13.3 Optical properties of PEDOT:PSS thin films. The topsides consist of the transmittance percentage of thin films on the glass substrates (a) and the cross-sectional scanning electron microscopy (SEM) figures of EG7 uniform film (b). The bottom side is the figure of fabricated thin films by different wt% EG incorporation.

addition are observed to provide the optical transmittance more than 90% in the whole applied wavelengths. The 300–800 nm wavelength average for each sample is found to be 90.72%, 91.77%, 92.15%, and 91.44% for EG0, EG3, EG7, and EG11, respectively. In detail, it can be noted the pristine PEDOT:PSS has lower average value in comparison with the treated films. Moreover, the thickness confirmation with ellipsometry showed the values of 150.34, 142.93, 129.11, and 137.65 nm, for EG0 to EG11, respectively.

Both analyses are confirmed the macro-observation of thin film showed at the bottom side of Figure 13.3. It seems the EG0 (pristine) possesses the light dark-blue color with a darker part at the corner of film in comparison with the treated PEDOT:PSS. The results indicate that all the EG treated samples turn to be more transparent than the pristine one. However, by adding 11 wt% of EG results in the threshold of excess EG where the thickness is significantly increased and thus reduces the transmittance of films. Further electrical analysis confirms the high-elevated improvement values after EG treatment by utilizing two approaches employing surface-sheet resistivity and charge concentration/mobility approaches.

Furthermore, Figure 13.4a evidences that the conductivity values of treated PEDOT:PSS are significantly improved. The FPP and Hall-effect approaches bring out supports for each other. The sheet resistance were found to be 160.29×10^3, 91.57, 87.52, and 103.75 Ω sq^{-1} for EG0, EG3, EG7, and EG11, respectively. Further investigation of Hall-effect analysis clearly described the secondary doping mechanism of EG in PEDOT:PSS. The measurement of pristine PEDOT:PSS shows the high mobility ($7.7912 \, cm^2 V^{-1} s^{-1}$) with the low carrier concentration ($5.53 \times 10^{17} cm^{-3}$) that results in poor conductivity with the value of 0.69 S cm^{-1}. Furthermore, by addition of 3 and

(a) Thin-Film Conductivity Value

(b) Raman Spectroscopy Spectra

(c) Sulfur Peaks (S2p)

PEDOT : PSS Ratio

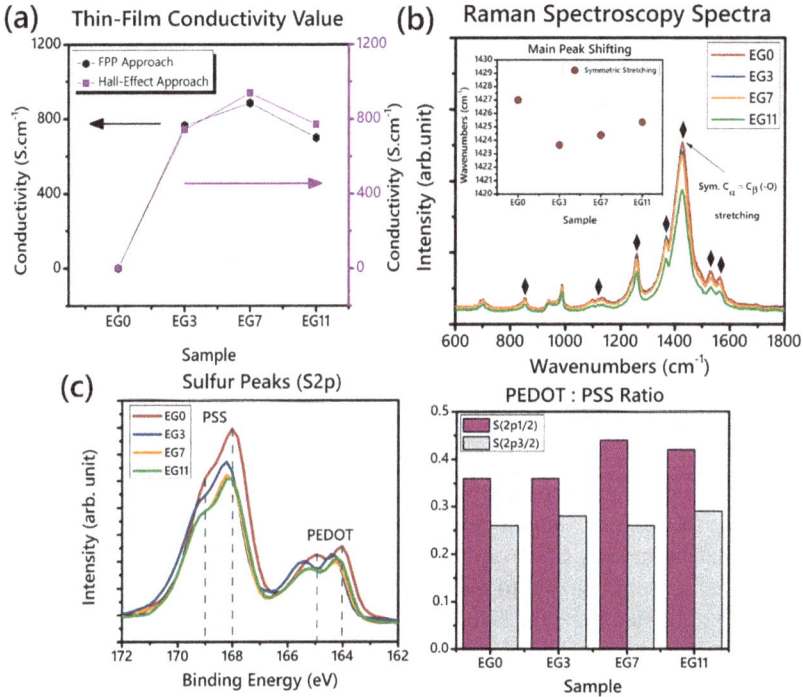

FIGURE 13.4 Analysis for the conductivity enhancement by the incorporation of EG including (a) conductivity by four-point probe and Hall-effect approaches, (b) Raman spectroscopy in identifying the structural change in PEDOT chains, and (c) ESCA-XPS analysis for PEDOT:PSS ratio identification.

7 wt%, the mobility experienced to decrease with 0.60251 and 0.061306 cm^{2}V^{-1}s^{-1} that is followed by increasing in charge carrier density to 7.72×10^{21} cm^{-3} and 9.56×10^{22} cm^{-3}, respectively. Further addition of EG to 11 wt% is found to increase the mobility to 0.32 cm^{2}V^{-1}s^{-1} and decrease the carrier concentration to 1.51×10^{22} cm^{-3}. This results in the film conductivity to 743.75, 938.22, and 770.79 S cm^{-1} by mixing 3, 7, and 11 wt% of EG, respectively, as shown in Table 13.2.

TABLE 13.2

Several Parameters from Hall-Effect Measurement and the Conductivity Values

Sample	Mobility (cm²V⁻¹·s⁻¹)	Bulk Carrier Concentration (cm⁻³)	Conductivity (S cm⁻¹)
EG0 (pristine)	7.79	5.53×10^{17}	0.6898
EG3	0.60	7.72×10^{21}	743.75
EG7	0.061	9.56×10^{22}	938.22
EG11	0.32	1.51×10^{22}	770.79

Further investigation of conductivity improvement in PEDOT:PSS is traced by Raman spectroscopy analysis, which is a useful assessment to observe structural change of conducting polymers chains specifically at the molecular level. The study conducted using 632.8 nm excitation laser with wavenumbers range of 900–1630 cm^{-1}. Figure 13.4b provides the Raman spectra by the comparison of sample including EG0 (pristine), EG3, EG7, and EG11. The intense and sharp particular peaks of PEDOT chains were observed at 987, 1126, 1255, 1363, 1427, 1533, and 1565 cm^{-1}, which are in accordance with previous reports [8,28,31,32]. The peaks centered at 987 and 1126 cm^{-1} are the oxyethylene ring deformation and C–O–C deformation vibrational modes, respectively. Moreover, peaks at 1255 and 1363 cm^{-1} are attributed to the $C_\alpha=C'_\alpha$ (inter-ring) stretching and $C_\beta=C_\beta$ stretching, respectively. The notable peak of symmetric $C_\alpha=C_\beta$ stretching is the mode at 1427 cm^{-1}, and a pair peaks at 1533 and 1565 cm^{-1} can be assigned to be the asymmetric $C_\alpha=C_\beta$ stretching modes. In the figure, the intensity of EG11 is decreased compared with all other spectra, which is due to the high concentration of excess EG in PEDOT:PSS film. The shifting of symmetric $C_\alpha=C_\beta$ stretching modes toward lower wavenumbers is found in comparison with that of the pristine sample, as shown in the inserted figure of Figure 13.4b. It suggests that by mixing EG from 3, 7, and 11 wt%, it attributes to the changes of PEDOT chains from benzoid to quinoid resonant structures. The benzoid resonant structure is suggested to be the favored structure for the coiled conformation; in contrast, the quinoid is the favored structure for linear or expanded-coil conformation of PEDOT chains. In pristine PEDOT:PSS film, both conformations were indicated to exist, and after EG treatment, the benzoid structures are expected to transform to quinoid states. Furthermore, the quinoid arrangement leads the extended coil or linear conformation, which can generate a lower energy barrier for interchain interaction and further result in decreasing the inter-domain charge hopping region [22].

Further analysis to figure out the possible effect of EG incorporation in PEDOT:PSS regarding the conductivity enhancement is by thin-film surface investigation utilizing ESCA-XPS instrument. Figure 13.4c is representing the ESCA spectra of sulfur signals and at the estimated PEDOT to PSS ratio. By the results of curve fitting, the spectra confirmed four main peaks of S(2p) consisting of S(2p$_{3/2}$) and S(2p$_{1/2}$) representing the sulfur atoms from PEDOT and PSS. The PEDOT indication peaks are at the low energy at 164 eV for S(2p$_{3/2}$) and 165 eV for S(2p$_{1/2}$) [8]. In contrast, the PSS peaks recognized at 168 eV for S(2p$_{3/2}$) and 169 eV for S(2p$_{1/2}$) that is detected at the higher binding energy spectra as the influence of the electronegative oxygen connection in the sulfonate moiety. Based on the spectra observation, the shifting of spectra in EG addition samples is noticed in comparison with the EG0 sample. The shifting value is estimated at around 0.3 eV and corresponding to the chemical shift at the surface of the thin films as the consequences of the chemical environment changes promoted by the addition of EG in different concentrations that are correlated with aforementioned discussions from Raman analysis. Advanced investigation of PEDOT:PSS/EG thin films indicates the PEDOT to PSS ratio by calculating the relative areas of PEDOT and PSS peaks. Figure 13.4c at the right side shows the ratio at surfaces indicate that the ratio is slightly increased by adding 3 wt% of EG with the value of 0.36 and then reaches 0.44 by adding 7 wt% of EG. Subsequently, the more addition of EG reduces the PEDOT:PSS ratio to 0.42. From both observations, the addition of EG increases the existence of PEDOT:PSS chains at the surface of the deposited film. The EG solvent is believed to remove the excess PSS and thus modifies the PEDOT:PSS ratio [33].

The proposed mechanism representing the influence of 3–7 wt% EG addition is demonstrated as shown in Figure 13.5. In the pristine condition, the PEDOT:PSS arrangement is mostly attributed to coulombic interactions that result in coiled structures. Moreover, through EG cosolvent addition, the conductivity improvement is found because of structural change in PEDOT chains followed by the phase separation between the PEDOT and PSS chains. The driving force regarding these conformational changes is caused by the polar groups of cosolvents that interact with the PEDOT segment. Since the EG has two or more polar groups, the interaction between the polar groups (–OH) and positive charges on PEDOT chains is occurred. Moreover, the size of PEDOT nanocrystals is increased because of more linear arrangement in the nanocrystals. This linear orientation of PEDOT chains allows the interaction of PEDOT interchains that can lower the energy barrier for charge hopping and in turn facilitates the charge transfer. The p–n heterojunction is formed by combining the treated p-type PEDOT:PSS and n-type Si nanowires (SiNWs). The arrangement of fabricated cells is shown in Figure 13.6a consisting of treated PEDOT:PSS/EG as the p-junction layer, n-type SiNWs as the n-junction layer, aluminum as the bottom electrode, and ITO as the top electrode. The $2 \times 2\,cm^2$ cells were measured using the solar simulator, and J–V assessment is presented in Figure 13.6b. The solar cells were made by using different contents of EG from 3 to 11 wt% since this range of the samples possesses the improved electrical conductivity compared with untreated PEDOT:PSS.

The involved working mechanism is presented in Figure 13.6c. Since the PEDOT:PSS/EG represents a high transparent hole-transport layer, the solar light is mostly transmitted into the SiNW regions, where the significant light absorption is

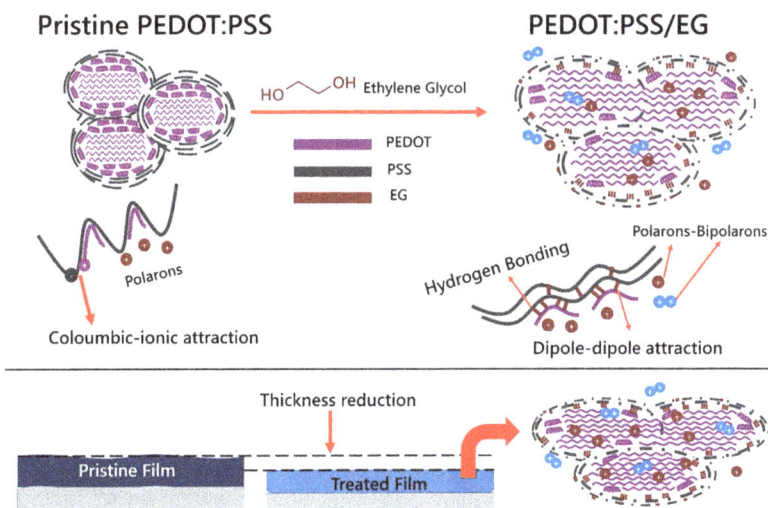

FIGURE 13.5   The mechanism of conductivity enhancement influenced by EG cosolvent addition thin films.

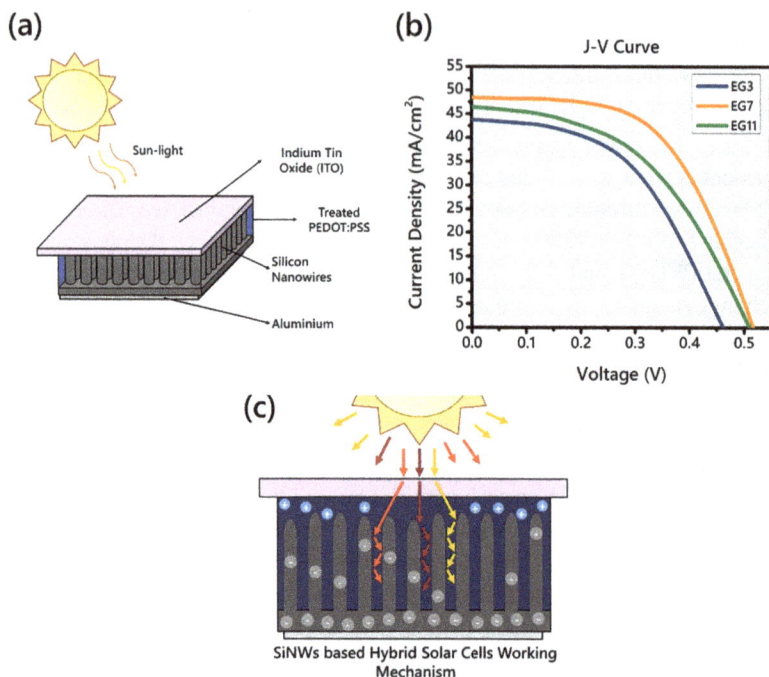

FIGURE 13.6   (a) Silicon nanowires-based hybrid solar cell arrangement, (b) J–V curve of the cells for different wt% of EG addition, and (c) the working mechanism of hybrid solar cells.

encountered due to the light-trapping effect from the oriented nanowires. It causes the generation of photoexcited carriers, and the holes are transported toward p-type electrodes and collected by the top electrode; in contrast, the electrons are separated from the heterojunction and eventually collected by the aluminum electrode. Since the PEDOT:PSS/EG7 layer possesses the superior light transmittance, which benefits the generation of more photoexcited carriers, while the underlying SiNWs can efficiently absorb the visible light. The theoretical study to unveil the light-absorption capability of bare SiNWs and perovskite-coated SiNWs was reported, which indicated that the SiNWs possess the significant absorptance of lights covering the wavelengths from 400 to 800 nm [34]. Such preferred structures therefore yield the improved PCE up to 14.12% by well construction of large-area heterojunction and sound light management, as shown in Figure 13.6b.

13.4   CONCLUSION

In conclusion, the synthesis, structure, and electrical properties of PEDOT:PSS layer were studied. The electrical conductivity of pristine PEDOT:PSS is found to be rather low, which is merely under 10 S cm^{-1}. However, it can be adjusted into higher values using diverse methods including physical treatment or chemical process. The enhancement of electrical conductivity can be up to 10^3 orders of magnitude by utilizing polar solvents, EG and DMSO mostly. The underlying mechanism for conductivity improvement has been studied using XPS, Raman spectroscopy, AFM, Hall-effect, GIWAXS, and many more. Furthermore, the clarification of optimization in treated PEDOT:PSS polymer for constructing hybrid solar cells was achieved by adding 7 wt% of EG and 0.1 wt% of FS additives. The optical transmittance can be achieved more than 90% by testing the addition of various amounts of EG and has an advantage in optimizing the cell performance. The conductivity of thin film can be achieved up to 938 S cm^{-1} with high uniformity without employing post-treatment after thin-film fabrication. The Raman analysis and XPS results give more clear evidence about the conductivity enhancement that is attributed to the structural conformation. Also, the reduction of excess PSS chains at the surface is encountered, which can therefore increase the PEDOT/PSS ratio, initiate the phase separation of PEDOT and PSS chains inside the thin film, and positively influence the carrier doping level in the forms of polarons and bipolarons. Based on the optimized design of treated PEDOT:PSS films, the conversion efficiency of hybrid solar cells can reach 14.12% through such facile, inexpensive, and reliable treatment. This can pave ways particularly for energy-based devices such as emerging solar cells, thermoelectric, and electrochromic applications.

ACKNOWLEDGMENT

This study was financially supported by the Ministry of Science and Technology of Taiwan (MOST 106-2221-E-006-240) and (MOST 107-2221-E-006-013-MY3), and the Hierarchical Green-Energy Materials (Hi-GEM) Research Center, from the Featured Areas Research Center Program within the framework of the Higher Education Sprout Project by the Ministry of Education (MOE) and the Ministry of Science and Technology (MOST 107-3017-F-006-003) in Taiwan. The authors greatly thank Center for Micro/Nano Science and Technology, National Cheng Kung University with the facilities provided for conducting material characterizations.

REFERENCES

1. Joint Research Centre. European Comission, "PV Status Report 2018," 2018.
2. W. Lu, C. Wang, W. Yue, and L. Chen, "Si/PEDOT:PSS core/shell nanowire arrays for efficient hybrid solar cells," *Nanoscale*, vol. 3, no. 9, pp. 3631–3634, 2011.
3. C.-Y. Chen, T.-C. Wei, P.-H. Hsiao, and C.-H. Hung, "Vanadium oxide as transparent carrier-selective layer in silicon hybrid solar cells promoting photovoltaic performances," *ACS Appl. Energy Mater.*, 2019. doi: 10.1021/acsaem.9b00565.
4. D. Alemu, H.-Y. Wei, K.-C. Ho, and C.-W. Chu, "Highly conductive PEDOT:PSS electrode by simple film treatment with methanol for ITO-free polymer solar cells," *Energy Environ. Sci.*, vol. 5, no. 11, p. 9662, 2012.
5. W. Wichiansee and A. Sirivat, "Electrorheological properties of poly(dimethylsiloxane) and poly(3,4-ethylenedioxy thiophene)/poly(stylene sulfonic acid)/ethylene glycol blends," *Mater. Sci. Eng. C*, vol. 29, pp. 78–84, 2008.
6. M. Vosgueritchian, D. J. Lipomi, and Z. Bao, "Highly conductive and transparent PEDOT:PSS films with a fluorosurfactant for stretchable and flexible transparent electrodes," *Adv. Funct. Mater.*, vol. 22, no. 2, pp. 421–428, 2012.
7. Q. Li, J. Yang, S. Chen, J. Zou, W. Xie, and X. Zeng, "Highly conductive PEDOT:PSS transparent hole transporting layer with solvent treatment for high performance silicon/ organic hybrid solar cells," *Nanoscale Res. Lett.*, vol. 12, pp. 1–8, 2017.

8. J. P. Thomas, L. Zhao, D. McGillivray, and K. T. Leung, "High-efficiency hybrid solar cells by nanostructural modification in PEDOT:PSS with co-solvent addition," *J. Mater. Chem. A*, vol. 2, no. 7, p. 2383, 2014.

9. C.-Y. Chen and C.-P. Wong, "Unveiling the shape-diversified silicon nanowires made by HF/HNO 3 isotropic etching with the assistance of silver," *Nanoscale*, vol. 7, no. 3, pp. 1216–1223, 2015.

10. M. Łapkowski and A. Proń, "Electrochemical oxidation of poly(3,4-ethylenedioxythiophene) – 'in situ' conductivity and spectroscopic investigations," *Synth. Met.*, vol. 110, pp. 79–83, 2000.

11. G. Steiner and C. Zimmerer, "Poly(3,4-ethylenedioxythiophene) (PEDOT)." 2013. doi: 10.1007/978-3-642-32072-9_128.

12. L. Groenendaal, F. Jonas, D. Freitag, H. Pielartzik, and J. R. Reynolds, "Poly(3,4-ethylenedioxythiophene) and its derivatives: past, present, and future," *Adv. Mater.*, vol. 12, no. 7, pp. 481–494, 2000.

13. H. J. Ahonen, J. Lukkari, and J. Kankare, "N- and p-doped poly(3,4-ethylenedioxythiophene): two electronically conducting states of the polymer," *Macromolecules*, vol. 33, no. 18, pp. 6787–6793, 2000.

14. S. Ghosh and O. Inganäs, "Self-assembly of a conducting polymer nanostructure by physical crosslinking: applications to conducting blends and modified electrodes," *Synth. Met.*, vol. 101, no. 1, pp. 413–416, 1999.

15. S. Garreau, G. Louarn, J. P. Buisson, G. Froyer, and S. Lefrant, "In situ spectro-electrochemical Raman studies of poly(3,4-ethylenedioxythiophene) (PEDT)," *Macromolecules*, vol. 32, no. 20, pp. 6807–6812, 1999.

16. J. Ouyang, C. W. Chu, F. C. Chen, Q. Xu, and Y. Yang, "High-conductivity poly(3,4-ethylenedioxythiophene):poly(styrene sulfonate) film and its application in polymer optoelectronic devices," *Adv. Funct. Mater.*, vol. 15, no. 2, pp. 203–208, 2005.

17. Y. Xia and J. Ouyang, "Significant different conductivities of the two grades of poly(3,4-ethylenedioxythiophene):poly(styrenesulfonate), Clevios P and clevios PH1000, arising from different molecular weights," *ACS Appl. Mater. Interfaces*, vol. 4, no. 8, pp. 4131–4140, 2012.

18. A. Moujoud, S. H. Oh, H. S. Shin, and H. J. Kim, "On the mechanism of conductivity enhancement and work function control in PEDOT:PSS film through UV-light treatment," *Phys. Status Solidi Appl. Mater. Sci.*, vol. 207, no. 7, pp. 1704–1707, 2010.

19. Y. Kim, A. M. Ballantyne, J. Nelson, and D. D. C. Bradley, "Effects of thickness and thermal annealing of the PEDOT:PSS layer on the performance of polymer solar cells," *Org. Electron. Phys. Mater. Appl.*, vol. 10, no. 1, pp. 205–209, 2009.

20. K. Sun, S. Zhang, P. Li, Y. Xia, X. Zhang, D. Du, F. H. Isikgor, and J. Ouyang, "Review on application of PEDOTs and PEDOT: PSS in energy conversion and storage devices," *J. Mater. Sci.: Mater. Electron.*, vol. 26, no. 7, pp. 4438–4462, 2015.

21. G. H. Kim, L. Shao, K. Zhang, and K. P. Pipe, "Engineered doping of organic semiconductors for enhanced thermoelectric efficiency," *Nat. Mater.*, vol. 12, no. 8, pp. 719–723, 2013.

22. D. Alemu Mengistie, P. C. Wang, and C. W. Chu, "Effect of molecular weight of additives on the conductivity of PEDOT:PSS and efficiency for ITO-free organic solar cells," *J. Mater. Chem. A*, vol. 1, no. 34, pp. 9907–9915, 2013.

23. T. R. Chou, S. H. Chen, Y. Te Chiang, Y. T. Lin, and C. Y. Chao, "Highly conductive PEDOT:PSS films by post-treatment with dimethyl sulfoxide for ITO-free liquid crystal display," *J. Mater. Chem. C*, vol. 3, no. 15, pp. 3760–3766, 2015.

24. Y. H. Kim, C. Sachse, M. L. MacHala, C. May, L. Müller-Meskamp, and K. Leo, "Highly conductive PEDOT:PSS electrode with optimized solvent and thermal posttreatment for ITO-free organic solar cells," *Adv. Funct. Mater.*, vol. 21, no. 6, pp. 1076–1081, 2011.

25. N. Kim, S. Kee, S. H. Lee, B. H. Lee, Y. H. Kahng, Y. R. Jo, B. J. Kim, and K. Lee, "Highly conductive PEDOT:PSS nanofibrils induced by solution-processed crystallization," *Adv. Mater.*, vol. 26, no. 14, pp. 2268–2272, 2014.

26. X. Crispin, S. Marciniak, W. Osikowicz, G. Zotti, A. D. Van Der Gon, F. Louwet, M. Fahlman, L. Groenendaal, F. De Schryver, and W. R. Salaneck, "Stability of poly (3,4-ethylene dioxythiophene)-poly(styrene sulfonate): a photoelectron spectroscopy study," *Polymer (Guildf).*, vol. 41, no. 21, pp. 2561–2583, 2003.

27. X. Zhang, J. Wu, J. Wang, J. Zhang, Q. Yang, Y. Fu, and Z. Xie, "Highly conductive PEDOT:PSS transparent electrode prepared by a post-spin-rinsing method for efficient ITO-free polymer solar cells," *Sol. Energy Mater. Sol. Cells,* vol. 144, pp. 143–149, 2016.

28. J. Ouyang, Q. Xu, C. W. Chu, Y. Yang, G. Li, and J. Shinar, "On the mechanism of conductivity enhancement in poly(3,4-ethylenedioxythiophene):poly(styrene sulfonate) film through solvent treatment," *Polymer (Guildf).*, vol. 45, no. 25, pp. 8443–8450, 2004.

29. J. Y. Kim, J. H. Jung, D. E. Lee, and J. Joo, "Enhancement of electrical conductivity of poly(3,4-ethylenedioxythiophene)/poly(4-styrenesulfonate) by a change of solvents," *Synth. Met.*, vol. 126, nos. 2–3, pp. 311–316, 2002.

30. S. Zhang, Z. Fan, X. Wang, Z. Zhang, and J. Ouyang, "Enhancement of the thermoelectric properties of PEDOT:PSS: via one-step treatment with cosolvents or their solutions of organic salts," *J. Mater. Chem. A*, vol. 6, no. 16, pp. 7080–7087, 2018.

31. W. W. Chiu, J. Travaš-Sejdić, R. P. Cooney, and G. A. Bowmaker, "Studies of dopant effects in poly(3,4-ethylenedioxythiophene) using Raman spectroscopy," *J. Raman Spectrosc.*, vol. 37, no. 12, pp. 1354–1361, 2006.

32. S. Garreau, J. L. Duvail, and G. Louarn, "Spectroelectrochemical studies of poly(3,4-ethylenedioxythiophene) in aqueous medium," *Synth. Met.*, vol. 125, pp. 325–329, 2001.

33. S. Funda, T. Ohki, Q. Liu, J. Hossain, Y. Ishimaru, K. Ueno, and H. Shirai, "Correlation between the fine structure of spin-coated PEDOT:PSS and the photovoltaic performance of organic/crystalline-silicon heterojunction solar cells," *J. Appl. Phys.*, vol. 120, no. 3, p. 33103, 2016.

34. C. Y. Chen, T. C. Wei, Y. C. Lai, and T. C. Lee, "Passivating silicon-based hybrid solar cells through tuning PbI2 content of perovskite coatings," *Appl. Surf. Sci.*, vol. 511, p. 145541, 2020.

14 Concluding Remarks

Ngoc Thanh Thuy Tran, Sanjaya Brahma,
Wen-Dung Hsu, Shih-Yang Lin, Jow-Lay Huang,
Masahiro Yoshimura, Jeng-Shiung Jan,
Chia-Yun Chen, Peter Chen, and Ming-Fa Lin
National Cheng Kung University

Chin-Lung Kuo
National Taiwan University

Phung My Loan Le
University of Science, Vietnam National University

CONTENT

In general, this book covers the development of the theoretical frameworks, the high-resolution measurements, and the excellent performance of batteries. It can significantly promote the full understanding in basic sciences, application engineerings, and commercial products. However, the close combinations of cathode, electrolyte, and anode materials are required to create the best ion transport in batteries [1–3], become an open issue, and are worthy of the systematic investigations. That is, the critical mechanisms, which determine the stationary currents of ions in three core components, are not thoroughly identified from the up-to-date theoretical and experimental results, such as the non-well-verified roles of charged impurities [4,5], vacancies [6], grain boundaries [7], interfaces between distinct systems [8], significantly modulated bond lengths [9], electronic spin configurations (ferromagnetism or antiferromagnetism) [10], Coulomb screening behaviors [11,12], active chemical reactions [13], controllable temperatures [14], and quantum Bose–Einstein/Fermi–Dirac boson/fermion statistics temperatures [15]. As to lithium-ion-related batteries, the first-principles method [16–22] and the machine learning method [23] are efficient in exploring the lattice symmetries [17], energy bands/discrete electronic states [18], charge transfers [19], free carrier densities [20], energy gaps [21], spatial charge/spin density distributions [10], van Hove singularities of energy spectra [20], formation energies of distinct vacancies [22], ionization energies [23]. The multi-orbital hybridizations in chemical bonds [24], atom- and orbital-dependent spin configurations [25], and the intrinsic geometric symmetries [26] are deduced to be responsible for the diversified essential properties. The similar simulations would be further developed for the other emergent green energy materials [27–29]. As to the experimental syntheses, measurements, and designs, they have been successfully done for the Li-ion batteries, silicon-nanowire-based hybrid solar cells, and perovskite solar cells. The core materials in the aforementioned devices, with the save, cheap, light, rapid-ion, and long-lifetime characteristics, are the studying focuses in significantly enhancing the operation performances. This can be achieved through the chemical modifications (absorptions, dopings, and substitutions); the physical methods; the optimal changes among the solid-, liquid-, and gel-state electrolytes; and the delicate treatments of junctions/contacts. More high-resolution experiments would be finished in the near-future researches.

As the demand for LIBs is increasing, it is critical to increase its specific energy. Archiving high-voltage LIBs has recently gained much attraction as one way to deal with this issue. Theoretical and experimental studies have been carried out to find out the candidate for high-voltage cathode materials [30–32]. Searching for electrolytes that compatible with a low-voltage anode such as graphite and high-voltage cathode about 4.5–5 V plays an important role. In this work, machine learning based on the generation model has been established for screening electrolyte molecules from key factors-electron affinity and ionization energy values. Introducing machine learning into material science is trying to replace expert-based knowledge with data in the material design; however, it is worth to know that even the advancement of machine learning model cannot achieve excellent performance without the assistance of expert-based knowledge. The proposed method developed in this work can significantly reduce the development time and cost for new materials design.

The results from first-principles calculations indicate that introducing N-dopants can increase the amount of defects in graphene, which can provide more energetically favorable sites to bind Li or Na atoms. The migration energy barrier of vacancy defects was found to become higher after doping with N, which can effectively avoid the capacity loss due to the suppression of vacancy aggregation. However, introducing many N-dopants near defect sites may lower down the Li/Na reversible capacity. Our results thus suggest that the concentration of N-doping should be carefully controlled in order to obtain optimal Li/Na storage capacity of graphene-based nanomaterials.

The 3D ternary $Li_4Ti_5O_{12}$, the Li^+-based battery anode, presents the unusual lattice symmetry (a triclinic crystal), band structure, charge density, and density of states under the first-principles calculations. It belongs to a large direct-gap semiconductor of $E_g^d \sim 2.98$ eV. The atom-dominated valence and conduction bands, the spatial charge distribution, and the atom- and orbital-decomposed van Hove singularities are available in the delicate identifications of multi-orbital hybridizations in Li–O and Ti–O bonds. The extremely nonuniform chemical environment, which induces the very complicated hopping integrals, directly arises from the large bonding fluctuations and the highly anisotropic configurations. Also, the developed theoretical framework is very useful for fully understanding the cathodes and electrolytes of oxide compounds.

The rich and unique fundamental properties of the emergent 3D Li_2SiO_3 compound are thoroughly investigated from the first-principles calculations. The critical multi-orbital hybridizations, which survive in Li–O and Si–O bonds, are accurately identified from the atom-dominated band structure, the charge density distributions in the significantly modulated chemical bonds, and the atom- and orbital-projected van Hove singularities. The theoretical framework could be further developed for the other electrolyte, anode, and cathode materials of Li$^+$-based batteries [33], for example, the important differences among the various LiXO-related compounds, and the diversified phenomena driven by the various components [34]. Interestingly, the highly anisotropic and nonuniform environments need to be included in the phenomenological models [35]. For example, the suitable tight-binding model is expected to have the position-dependent hopping integrals and the orbital-create on-site Coulomb potentials in simulating the VASP energy bands and density of states. Whether the intrinsic atomic interactions of Hamiltonian could be expressed in the analytic form is the near-future studying focus; furthermore, they are very useful in exploring the other diverse phenomena, such as the magnetic quantization in a uniform perpendicular magnetic field [36]. The solid-state anode material of Li$_2$SiO$_3$ with 24 atoms in a primitive unit cell is an orthorhombic structure. There are 32 Li–O and 16 O–Si–O chemical bonds, in which each atom has four neighboring other atoms. Their bond lengths, respectively, present the large modulations over 10% and 6% S. Most importantly, the very strong orbital hybridizations create a very wide indirect gap of $E_g \approx 5.077$ eV, being close to the largest one in diamond ≈ 5.5 eV. Its magnitude is even lower than the optical threshold absorption frequency.

The high-resolution optical reflection/absorption/transmission spectroscopies are available in detecting the obvious insulating property of the transparent material. The occupied valence and the unoccupied conduction bands are highly asymmetric to each other about the Fermi level. Furthermore, there exist the wide energy range of (-8.0 eV, E^v, 8 eV), the strong/various dispersion relations with the wave vector, the high anisotropy, and the frequently noncrossing/crossing/anticrossing behaviors. Moreover, the van Hove singularities, which arise from the band-edge states, appear as the six/three dominating/minor special structures in density of states of the valence/conduction energy spectrum ($E > 0$, $E < 0$). They only belong to the broadening asymmetric/symmetric peaks and shoulders and play a critical role in examining the multi-orbital hybridizations of Li–O and Si–O bonds, $2s$-($2s$, $2p_x$, $2p_y$, $2p_z$), and ($3s$, $3p_x$, $3p_y$, $3p_z$) – ($2s$, $2p_x$, $2p_y$, $2p_z$). The diverse covalent bondings are partially supported by the atom-dominated band structure and charge density distributions in modulated chemical bonds. The theoretical predictions on the optimal geometry, occupied wave-vector-dependent valence bands, and valence and conduction density of states could be verified from X-ray elastic scatterings LEED [35,37] and ARPES, respectively. The calculated results clearly illustrate that LiSiO-based compounds have certain metastable configurations and even an infinite intermediate ones during the charging and discharging processes for the Li$^+$ ion transport in batteries. The similar structural transformations between two metastable structures are expected to occur at any time. The optimal evolution paths, which might become an emergent issue, are under the current investigations.

The 3D ternary Li$_x$Co/NiO$_2$ compounds, being the well-known cathode materials of Li$^+$-based batteries, are predicted to exhibit the rich and unique lattice symmetries (a trigonal lattice), band structures, charge/spin density distributions, and van Hove singularities. The accurate analyses, which are made from the exact first-principles calculations, are available in identifying the significant multi-orbital hybridizations/ spin configurations in Li–O and Co/Ni–O bonds. Specifically, the spin-dependent many-particle Coulomb interactions need to be included in the numerical evaluations simultaneously, while they are fully absent in the anode and electrolyte materials (e.g., LiTiO and LiSiO). The essential electronic and magnetic properties are very sensitive to the lithium concentration. All the Li_xCo/NiO_2 systems belong to the ferromagnetic metals except that $LiCoO_2$ is a spin-degenerate semiconductor with the energy gap of ~1.05 eV. Three kinds of ferromagnetic configurations are revealed during the variation of x. There are a lot valence and conduction bands with the range of -8.0 eV $< E^{c,v} < 3.0$ eV, in which most of them are dominate by transition and oxygen atoms. The intrinsic interactions mainly come from the chemical bondings and the spin-dependent on-site electron–electron Coulomb ones. The theoretical predictions on the optimal geometries, the rich occupied bands, and many van Hove singularities could be verified from the high-resolution measurements of the X-ray diffraction, angle-resolved photoemission spectroscopy, and scanning tunneling spectroscopy, respectively.

The stage-n graphite alkali-intercalation compounds present the rich and unique geometric and electronic properties, being directly identified from the first-principles calculations. The diversified phenomena strongly depend on the type, concentration, and distribution of intercalants. By the delicate calculation and analyses on the optimal geometries, band structures, and densities of states, the π and σ bands could be well defined through the interlayer carbon-alkali bondings/the significant van der Waals, or their chemical bondings are almost orthogonal to each other. Most importantly, the interlayer single-orbital atomic interactions, $2p_z$-$3s$ and $2p_z$-$2p_z$, play critical roles in modifying the main features of electronic energy spectra. Interestingly, the dramatic transformation of E_F is clearly identified from a prominent peak there and a minimum valley of the left-hand neighbor at about $E < -1.0$ eV. Apparently, the modified π and σ bonds create the diverse fundamental properties. The theoretical predictions on band structures and van Hove singularities could be verified from the high-resolution ARPES [38] and STS [39] measurements, respectively.

Graphene nanosheets (GNs) prepared by variety of methods including exfoliation and chemical modification are used extensively as anode (negative electrode) in LIBs due to excellent Li storage capability [40] and have shown significant potential as energy storage material. GNs could achieve significantly high first charge–discharge capacity (1233 and 672 mAh g^{-1}). The high irreversible capacity is due to the reaction of Li ion with oxygen-containing functional groups and formation of the solid electrolyte interface (SEI). Microstructure modification [41] is also achieved by the chemical modification, where flower-like GNS could be synthesized and the corresponding anode material has maintained high capacity (460 mAh g^{-1}) after 100 cycles and that shows good electrochemical performance. Introduction of the porosity and the doping has significantly enhanced the electrochemical properties.

We recently proposed a novel ionic liquid (i.e., 1-ethyl-3-methylimidazolium bis(trifluoromethanesulfonyl)imide) containing a suitable lithium salt (i.e., lithium bis(trifluoromethanesulfonyl)imide, LiTFSI) as advanced electrolyte for lithium batteries. This IL-based electrolyte offers a good compatibility with the lithium metal electrode while maintaining high thermal stability and acceptable conductivity. Excellent performances were also shown in terms of interface stability, cycle life, and rate capability when the EMITFSI + LiTFSI solution was tested in lithium cells. We also prepared IL-containing polymer membranes by immobilizing a solution of LiTFSI in EMI-TFSI mixed with other organic solvents, such as ethylene carbonate, propylene carbonate, and dimethyl carbonate (EC, PC, and DMC), into a poly(vinylidene fluoride-co-hexafluoropropylene) (PVDF-HFP) matrix. These gel electrolytes showed promising electrochemical performances.

Some interesting examples of mixing ionic liquids with aprotic organic solvents to form hybrid electrolytes have been investigated. Our researches observed that by mixing LiTFSI with a reasonable concentration in ethylene carbonate and propylene carbonate with 1-ethyl-3-methylimidazolium bis(trifluoromethanesulfonyl)imide ionic liquid is possible to improve the safety without compromising performances. They observed that electrolytes containing 20% of IL offer the best compromise between viscosity and ionic conductivity, even cycling performance.

The stability of electrode-electrolyte interphase (solid electrolyte interphase) as well as ion-diffusion mechanism to improve continuously actual problems in rechargeable batteries should be carried out. Moreover, ionic liquids are also promising candidates for lithium-sulfur or lithium-oxygen.

To date, the development of secondary batteries has focused on small electric appliances and portable IT devices. Lithium secondary batteries are expected to build upon these achievements and create new applications, such as green energy, wireless charging, self-development, recycle, from portable to wearable, and flexible. It is an important task to make advance preparations for future batteries by considering the functions of these applications. Among the future applications of secondary batteries, medium- and large-sized cells show significant promise. Energy storage systems, in the form of batteries for electric vehicles and robots, or high-performance lithium secondary batteries capable of storing alternative energy such as solar, wind, and marine energy, are viewed as a key component of next-generation smart grid technology.

Microcells and flexible batteries are other types of future secondary batteries. Microcells can be applied to RFID/USN, MEMS/NEMS, and embedded medical devices, while flexible batteries are mainly used in wearable computers and flexible displays. The structural control and manufacturing process of these batteries are different from today's methods. The development of all-solid-state lithium secondary batteries is also highly anticipated. With massive recalls caused by frequent battery explosions, an important challenge that lies ahead is to resolve the instability problem of existing liquid electrolytes through the application of electrolytes consisting of polymers or organic/inorganic composites, along with the development of suitable electrode materials and processes.

A series of polymer electrolytes are prepared through polymerization of 1-ethyl-3-vinylimidazolium bis(trifluoromethanesulfonylimide) with poly(ethylene glycol) diacrylate (PEGDA) ($Mn = 700$ g mol^{-1}), poly(ethylene glycol) methyl ether methacrylate (PEGME) ($Mn = 500$ g mol^{-1}), and lithium bis(trifluoromethanesulfonyl)imide (LiTFSI) in an ionic liquid to obtain mechanically robust membranes. The prepared samples exhibit high thermal stabilities and display advantageous ionic conductivities

256 Lithium-Ion Batteries and Solar Cells

[42–47]. It appears that their electrochemical stabilities can be improved after we adjust the process by polymerizing the mixture on the $LiFePO_4$ cathode directly, which is also responsible for the remarkable discharge capacity and improved capacity retention of the $Li/LiFePO_4$ cells based on the prepared electrolytes. In summary, this research demonstrates a simple preparation through the facile photopolymerization method, and the current findings indicate that this type of electrolytes is promising for lithium ion batteries.

Interestingly, this chapter covers the synthesis, structure, and electrical properties of pristine and treated PEDOT:PSS layer as well its application in silicon-nanowires-based solar cells. It can significantly promote the full understanding of fundamental sciences, engineering deliberation, and emerging products. By the delicate facile method through the incorporation of EG as the cosolvents, the direct mixing of 7 wt% EG in precursors results in the optimum conductivity up to 10^3 orders of magnitude for the reduction of sheet resistance and remarkably improved up to 938 S cm^{-1} with high uniformity without employing post-treatment after thin-film fabrication. The Raman analysis and XPS results give more clear evidence about the conductivity enhancement that is attributed to the structural conformation. Also, the reduction of excess PSS chains at the surface is encountered, which can, therefore, increase the PEDOT/PSS ratio, initiate the phase separation of PEDOT and PSS chains inside the thin film, and positively influence the carrier doping level in the forms of polarons and bipolarons. Furthermore, the transmittance of deposited PEDOT:PSS films achieves more than 90% for the improvement of corresponding transparency in the visible-frequency region. The successful syntheses, detailed analyses, and well-behaved designs realized the explicit efficiency improvement of photovoltaic cells achieving >14% by the well management of electrical and optical properties of PEDOT:PSS layer. This can pave ways particularly for energy-based devices such as emerging solar cells, thermoelectric, and electrochromic applications [48,49]. However, the further chemical modifications (dopings and substitutions) [50,51], the physical methods, and delicate treatments of junctions/contacts with more comprehensive approaches would be finished in the near-future researches.

REFERENCES

1. Tu, Z., Choudhury, S., Zachman, M. J., Wei, S., Zhang, K., Kourkoutis, L. F., & Archer, L. A. (2017). Designing artificial solid-electrolyte interphases for single-ion and high-efficiency transport in batteries. *Joule*, 1(2), 394–406.
2. Jiang, F., & Peng, P. (2016). Elucidating the performance limitations of lithium-ion batteries due to species and charge transport through five characteristic parameters. *Scientific Reports*, 6, 32639.
3. Lin, C. E., Zhang, H., Song, Y. Z., Zhang, Y., Yuan, J. J., & Zhu, B. K. (2018). Carboxylated polyimide separator with excellent lithium ion transport properties for a high-power density lithium-ion battery. *Journal of Materials Chemistry A*, 6(3), 991–998.
4. Shan, X., Chen, S., Wang, H., Chen, Z., Guan, Y., Wang, Y.,... & Tao, N. (2015). Mapping local quantum capacitance and charged impurities in graphene via plasmonic impedance imaging. *Advanced Materials*, 27(40), 6213–6219.

5. Walsh, A., Scanlon, D. O., Chen, S., Gong, X. G., & Wei, S. H. (2015). Self-regulation mechanism for charged point defects in hybrid halide perovskites. *Angewandte Chemie International Edition*, 54(6), 1791–1794.

6. Wang, J., Ma, A., Li, Z., Jiang, J., Feng, J., & Zou, Z. (2015). Effects of oxygen impurities and nitrogen vacancies on the surface properties of the Ta_3N_5 photocatalyst: A DFT study. *Physical Chemistry Chemical Physics*, 17(35), 23265–23272.

7. Huang, S. Z., Zhang, Q., Yu, W., Yang, X. Y., Wang, C., Li, Y., & Su, B. L. (2016). Grain boundaries enriched hierarchically mesoporous MnO/carbon microspheres for superior lithium ion battery anode. *Electrochimica Acta*, 222, 561–569.

8. Yang, F., Wang, D., Zhao, Y., Tsui, K. L., & Bae, S. J. (2018). A study of the relationship between coulombic efficiency and capacity degradation of commercial lithium-ion batteries. *Energy*, 145, 486–495.

9. Li, Z. Y., Wang, H., Yang, W., Yang, J., Zheng, L., Chen, D.,... & Liu, X. (2018). Modulating the electrochemical performances of layered cathode materials for sodium ion batteries through tuning coulombic repulsion between negatively charged TMO_2 slabs. *ACS Applied Materials & Interfaces*, 10(2), 1707–1718.

10. Tran, N. T. T., Nguyen, D. K., Lin, S. Y., Gumbs, G., & Fa-Lin, M. (2019). Fundamental properties of transition-metals-adsorbed graphene. *ChemPhysChem*, 20(19), 2473–2481.

11. Kjellander, R. (2016). Decay behavior of screened electrostatic surface forces in ionic liquids: The vital role of non-local electrostatics. *Physical Chemistry Chemical Physics*, 18(28), 18985–19000.

12. Ji, H., Joo, M. K., Yi, H., Choi, H., Gul, H. Z., Ghimire, M. K., & Lim, S. C. (2017). Tunable mobility in double-gated $MoTe_2$ field-effect transistor: Effect of coulomb screening and trap sites. *ACS Applied Materials & Interfaces*, 9(34), 29185–29192.

13. Winsberg, J., Hagemann, T., Janoschka, T., Hager, M. D., & Schubert, U. S. (2017). Redox-flow batteries: From metals to organic redox-active materials. *Angewandte Chemie International Edition*, 56(3), 686–711.

14. Zhang, Q., Wang, W., Wang, Y., Feng, P., Wang, K., Cheng, S., & Jiang, K. (2016). Controllable construction of 3D-skeleton-carbon coated $Na_3V_2(PO_4)_3$ for high-performance sodium ion battery cathode. *Nano Energy*, 20, 11–19.

15. Wächtler, F., & Santos, L. (2016). Quantum filaments in dipolar Bose-Einstein condensates. *Physical Review A*, 93(6), 061603.

16. Park, J. H., Liu, T., Kim, K. C., Lee, S. W., & Jang, S. S. (2017). Systematic molecular design of ketone derivatives of aromatic molecules for lithium-ion batteries: First-principles DFT modeling. *ChemSusChem*, 10(7), 1584–1591.

17. Zhang, C., Jiao, Y., He, T., Ma, F., Kou, L., Liao, T.,... & Du, A. (2017). Two-dimensional GeP_3 as a high capacity electrode material for Li-ion batteries. *Physical Chemistry Chemical Physics*, 19(38), 25886–25890.

18. Lu, H., & Sun, S. (2018). Polyimide electrode materials for Li-ion batteries via dispersion-corrected density functional theory. *Computational Materials Science*, 146, 119–125.

19. Ullah, S., Denis, P. A., & Sato, F. (2017). Beryllium doped graphene as an efficient anode material for lithium-ion batteries with significantly huge capacity: A DFT study. *Applied Materials Today*, 9, 333–340.

20. Tran, N. T. T., Nguyen, D. K., Glukhova, O. E., & Lin, M. F. (2017). Coverage-dependent essential properties of halogenated graphene: A DFT study. *Scientific Reports*, 7(1), 17858.

21. Barik, G., & Pal, S. (2019). Energy gap-modulated blue phosphorene as flexible anodes for lithium-and sodium-ion batteries. *The Journal of Physical Chemistry C*, 123(5), 2808–2819.

22. Shiiba, H., Zettsu, N., Nakayama, M., Oishi, S., & Teshima, K. (2015). Defect formation energy in spinel $LiNi_0$. $5Mn_1$. $5O_4-\delta$ using ab initio DFT calculations. *The Journal of Physical Chemistry C*, 119(17), 9117–9124.

23. Lin, M. F., & Hsu, W. D. (Eds.). (2019). *Green Energy Materials Handbook*. CRC Press, Boca Raton, FL. ISBN 9781138605916, chapter 7.

24. Tran, N. T. T., Lin, S. Y., Lin, C. Y., & Lin, M. F. (2017). *Geometric and Electronic Properties of Graphene-Related Systems: Chemical Bonding Schemes*. CRC Press, Boca Raton, FL. ISBN 9781138556522.

25. Niu, Z., Wu, H., Lu, Y., Xiong, S., Zhu, X., Zhao, Y., & Zhang, X. (2019). Orbital-dependent redox potential regulation of quinone derivatives for electrical energy storage. *RSC Advances*, 9(9), 5164–5173.

26. Liu, J., Zhao, T., Zhang, S., & Wang, Q. (2017). A new metallic carbon allotrope with high stability and potential for lithium ion battery anode material. *Nano Energy*, 38, 263–270.

27. Eriksson, S. K., Josefsson, I., Ellis, H., Amat, A., Pastore, M., Oscarsson, J.,... & Hagfeldt, A. (2016). Geometrical and energetical structural changes in organic dyes for dye-sensitized solar cells probed using photoelectron spectroscopy and DFT. *Physical Chemistry Chemical Physics*, 18(1), 252–260.

28. Mohr, T., Aroulmoji, V., Ravindran, R. S., Müller, M., Ranjitha, S., Rajarajan, G., & Anbarasan, P. M. (2015). DFT and TD-DFT study on geometries, electronic structures and electronic absorption of some metal free dye sensitizers for dye sensitized solar cells. *Spectrochimica Acta Part A: Molecular and Biomolecular Spectroscopy*, 135, 1066–1073.

29. Zhang, L., Liu, X., Rao, W., & Li, J. (2016). Multilayer dye aggregation at dye/TiO_2 interface via π...π stacking and hydrogen bond and its impact on solar cell performance: A DFT analysis. *Scientific Reports*, 6, 35893.

30. Li, J., Klöpsch, R., Stan, M. C., Nowak, S., Kunze, M., Winter, M., & Passerini, S. (2011). Synthesis and electrochemical performance of the high voltage cathode material Li [Li_0. $2Mn_0$. $56Ni_0$. $16Co_0$. $O_8]O_2$ with improved rate capability. *Journal of Power Sources*, 196(10), 4821–4825.

31. Zhu, L., Yan, T. F., Jia, D., Wang, Y., Wu, Q., Gu, H. T.,... & Tang, W. P. (2019). $LiFePO_4$-coated $LiNi_0$. $5Co_0$. $2Mn_0$. $3O_2$ cathode materials with improved high voltage electrochemical performance and enhanced safety for lithium ion pouch cells. *Journal of the Electrochemical Society*, 166(3), A5437–A5444.

32. Zhang, H., Gong, Y., Li, J., Du, K., Cao, Y., & Li, J. (2019). Selecting substituent elements for $LiMnPO_4$ cathode materials combined with density functional theory (DFT) calculations and experiments. *Journal of Alloys and Compounds*, 793, 360–368.

33. Zhang, S. S. (2006). A review on electrolyte additives for lithium-ion batteries. *Journal of Power Sources*, 162, 1379–1394.

34. Lin, Y.-T., Lin, S.-Y., Chiu, Y.-H., & Lin, M.-F. (2016). Alkali-induced rich properties in graphene nanoribbons: Chemical bonding. arXiv:1609.05562v1(cond-mat.mes-hall).

35. Du Plessis, A., Yadroitsev, I., Yadroitsava, I., & Le Roux, S. G. (2018). X-ray microcomputed tomography in additive manufacturing: A review of the current technology and applications. doi: 10.1089/3dp.2018.0060.

36. Tang, T., & Luo, D. L. (2010). Density functional theory study of electronic structures in lithium silicates: Li_2SiO_3 and Li_4SiO_4. *Journal of Atomic and Molecular Sciences*, 1, 185.

37. Kuganathan, N., Kordatos, A., & Chroneos, A. (2018). Li_2SnO_3 as a cathode material for lithium-ion batteries: Defects, lithium ion diffusion and dopants. doi: 10.1038/s41598-018-30554-y.

38. Rufieux, P., Cai, J., Plumb, N. C., Patthey, L., Prezzi, D., Ferretti, A.,... & Pignedoli, C. A. (2012). Electronic structure of atomically precise graphene nanoribbons. *ACS Nano*, 6, 6930–6935.

39. Kano, S., Tadaa, T., & Majima, Y. (2015). Nanoparticle characterization based on STM and STS. *Chemical Society Reviews*, 44, 970.

40. Guo, P., Song, H., & Chen, X. (2009). Electrochemical performance of graphene nanosheets as anode material for lithium-ion batteries. *Electrochemistry Communications*, 11, 1320–1324.

41. Wang, G., Shen, X., Yao, J., & Park, J. (2009). Graphene nanosheets for enhanced lithium storage in lithium ion batteries. *Carbon*, 47, 2049–2053.

42. Yin, K., Zhang, Z. X., Li, X. W., Yang, L., Tachibana, K., & Hirano, S. I. (2015). Polymer electrolytes based on dicationic polymeric ionic liquids: Application in lithium metal batteries. *Journal of Materials Chemistry A*, 3, 170–178.

43. Yin, K., Zhang, Z. X., Yang, L., & Hirano, S. I. (2014). An imidazolium-based polymerized ionic liquid via novel synthetic strategy as polymer electrolytes for lithium ion batteries. *Journal of Power Sources*, 258, 150–154.

44. Appetecchi, G. B., Kim, G. T., Montanina, M., Carewska, M., Marcilla, R., Mecerreyes, D., & De Meatza, I. (2010). Ternary polymer electrolytes containing pyrrolidinium-based polymeric ionic liquids for lithium batteries. *Journal of Power Sources*, 195, 3668–3675.

45. Li, X. W., Zhang, Z. X., Li, S. J., Yang, L., & Hirano, S. (2016). Polymeric ionic liquid-plastic crystal composite electrolytes for lithium ion batteries. *Journal of Power Sources*, 307, 678–683.

46. Safa, M., Chamaani, A., Chawla, N., & El-Zahab, B. (2016). Polymeric ionic liquid gel electrolyte for room temperature lithium battery applications. *Electrochimica Acta*, 213, 587–593.

47. Zhang, P. F., Li, M. T., Yang, B. L., Fang, Y. X., Jiang, X. G., Veith, G. M.,... & Dai, S. (2015). Polymerized ionic networks with high charge density: Quasi-solid electrolytes in lithium-metal batteries. *Advanced Materials*, 27, 8088–8094.

48. Kim, G. H., Shao, L., Zhang, K., & Pipe, K. P. (2013). Engineered doping of organic semiconductors for enhanced thermoelectric efficiency. *Nature Materials*, 12(8), 719–723.

49. Alemu Mengistie, D., Wang, P. C., & Chu, C. W. (2013). Effect of molecular weight of additives on the conductivity of PEDOT:PSS and efficiency for ITO-free organic solar cells. *Journal of Materials Chemistry A*, 1(34), 9907–9915.

50. Kim, N., Kee, S., Lee, S. H., Lee, B. H., Kahng, Y. H., Jo, Y. R.,... & Lee, K. (2014). Highly conductive PEDOT:PSS nanofibrils induced by solution-processed crystallization. *Advanced Materials*, 26(14), 2268–2272.

51. Zhang, S., Fan, Z., Wang, X., Zhang, Z., & Ouyang, J. (2018). Enhancement of the thermoelectric properties of PEDOT:PSS: Via one-step treatment with cosolvents or their solutions of organic salts. *Journal of Materials Chemistry A*, 6(16), 7080–7087.

15 Open Issues and Potential Applications

Thi Dieu Hien Nguyen,
Ngoc Thanh Thuy Tran, and Ming-Fa Lin
National Cheng Kung University

CONTENT

The theoretical predictions, which are conducted by the numerical simulations and the phenomenological models, are available in understanding the essential properties of green energy materials, especially for the geometric, electronic, optical, and transport properties [1,2] of the Li^+-ion-based batteries (LIBs) (Figure 15.1). The former covers the first-principles calculations (e.g., Vienna ab initio simulation package (VASP); [3,4]), molecular dynamics with the empirical intrinsic interactions [5], and quantum Monte Carlo within the Pauli principle [6,7]. The numerical simulation methods have been frequently utilized in fully exploring the optimal geometries with various chemical bonds [8]; atom- and spin-dominated energy spectra and wave functions [9]; spatial charge/spin density distributions; and atom-, orbital-, and spin-decomposed density of states [10], as done for cathode, electrolyte, and anode of Li^+-related batteries (Figure 15.1a; details in Chapters 4–7). According to the development of the theoretical framework, the critical multi-/single-orbital hybridizations of the chemical bonds could be delicately identified from the above-mentioned physical quantities. The similar analyses could be generalized to other emergent materials in batteries, e.g., the popular four-component LiFePO cathode [11–13] and silicon/germanium nanotube anode [14]. They will be completed in the near-future studies.

It is well known that the electronic properties, absorption spectra, electrical conductivities, and Coulomb excitations could be understood from the tight-binding model/the effective-mass approximation [15], the dynamic Kubo formula [16], the static one (discussed later [15]), and the random-phase approximation [17]. However, these methods are seldom utilized in the fundamental properties of cathode, electrolyte, and anode materials, mainly owing to a lot of atoms with the various chemical-bonding strengths in a primitive unit cell. For example, there are many oxygen-related chemical bonds under a specific crystal structure (details in Chapters 4–7; [18,19]). The parameters, which are required in the tight-binding model, cover the various orbital-hybridization-dependent hopping integrals [20] and the different ionization energies (the distinct on-site Coulomb potentials; [21]). Specifically, the former is very sensitive to the chemical bond lengths [22]. In general, they are obtained from the well-fitting between this model and the first-principles method in the band

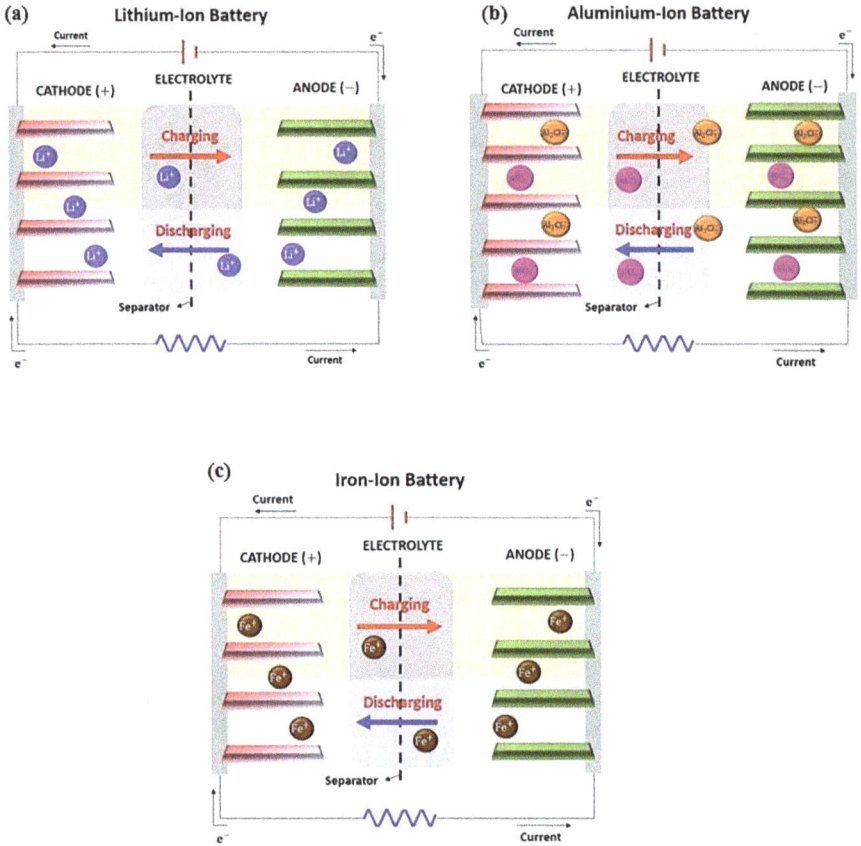

FIGURE 15.1   (a) Lithium-, (b) aluminum-, and (c) iron-ion-based batteries

structure [23]. However, it might be very difficult to achieve this object because of the very complicated results of VASP calculations [3,4,23]. The generalized tight-binding model, which is developed for the magnetic quantization phenomenon [24], will become meaningless, since the magnetically quantized Landau sub-bands are too complicated to identify from the magnetic-field-dependent energy spectra and localized wave functions [24].

The development of the tight-binding model will be very helpful in providing concise physical properties about the optical properties. The up-to-date optical reflection/absorption/transmission spectroscopies [25] are available in examining the diversified valence and conduction energy spectra [25]. When any material is perturbed by the external electromagnetic fields, the occupied electronic states are vertically excited to the unoccupied ones under the conservations of momentum and energy. The initial and final states almost have the same wave vectors because of the negligible photon momentum. Such high-resolution optical measurements could provide the threshold absorption frequency (the optical gap; [25]) and the prominent excitation structures, in which the latter are created by the van Hove singularities of the initial and/or final states. From the theoretical viewpoints, one needs to calculate the number of excitation channels and the intensity of vertical transitions. According to the Fermi golden rule [26] or the linear Kubo formula [27], the optical absorption spectrum is due to the perturbation of an electric dipole moment. That is, it is proportional to the joint density of states through the initial and final states and each channel intensity, where the former and the latter are, respectively, associated with energy bands and wave functions. How to evaluate the optical transition intensity, or efficiently link the close relations of the tight-binding model and the dynamic Kubo formula, is just the studying focus. Moreover, the systematic studies have been successfully conducted on the layered graphene-related systems for their optical [28] and magneto-optical properties [28]. More heavy works need to be done to fully explore the diversified absorption behaviors of the multicomponent cathode, electrolyte, and anode materials [29–31].

The dielectric Coulomb screening (details in [32]) is one of the mainstream topics in any condensed matter systems. This phenomenon, which directly reflects the intrinsic response of macroscopic charge fluctuations, is well characterized by the dimensionless energy loss function (a universal quantity; [27]). In general, the electronic excitation behaviors, the single-particle electron–hole pairs (the Landau damping; [33]) and collective excitations (plasmon modes; [34,35]), could be examined/verified from the momentum- and frequency-enriched inelastic scatterings of the incident high-energy electron beam (electron energy loss spectroscopy [EELS]; [36]). Most important, the high-resolution EELS is capable of identifying the diverse (momentum, frequency)-dependent excitation-phase diagrams with the simultaneous e–h regions and plasmon dispersion relations [37]. The rich and unique Coulomb excitations in layered graphene systems have been presented under the efficient couplings between the generalized tight-binding model and the modified random-phase approximation, where the intralayer and interlayer hopping integrals and Coulomb interactions are included in the calculations simultaneously [38]. The critical factors and the concise physical properties are clearly illustrated in this book [38]. For example, the low-frequency plasmons, which, respectively, arise from the quantization modes of the collective free-carrier and free-electron oscillations, are very sensitive to the direction and magnitude of the transferred momentum [39]. They are greatly diversified by the temperature [39], carrier doping [40], layer number [40], stacking configuration [40,41], dimension [41], planar or cylindrical symmetry [41], gate voltage [41], and magnetic field [34,42]. Very interesting, the dimensionless screened response function is well defined for the coupled 2D structures. The developed theoretical framework is also useful in understanding the Coulomb decay rates of the excited carriers [43], e.g., the quasiparticle lifetimes due to the Coulomb scatterings with the strong energy and wave-vector dependences for monolayer graphene [44], silicene [45,46], and germanene [47].

The small first Brillouin zone would lead too much difficulty of the random-phase approximation in the momentum- and frequency-dependent bare and screened response functions, because most of the transferred momenta during the electron–electron interactions are frequently beyond the smallest reciprocal lattice vector [47]. The up-to-date experimental growths show that many emergent materials have large unit cells under the unequal lattice symmetries. For example, the very popular

Li$^+$-based batteries are purely made up of the lithium oxide ternary-/four-component compounds in terms of cathode, electrolyte, and anode materials, e.g., corresponding to the 3D LiFe/Co/NiPO [48,49–54], LiSi/Ge/SnO [47,55,56], and LiSc/Ti/VO systems, respectively [57–60]. The chemical bonds could exist only through the multi-orbital hybridizations with oxygen atoms or only the X–O bonds are revealed in the stable 3D optimal geometric structures, such as the 48 Li–O and 52 Ti–O bonds with the great modulated bond lengths in the VASP calculations (Chapter 4; [58]). Moreover, the complicated multi-orbital mixings are delicately identified from the atom- and orbital-decomposed van Hove singularities, the spatial charge density distribution, and the atom-dominated energy spectrum. Such mechanisms have generated significant energy gaps in various battery materials. Apparently, the orbital-dependent on-site Coulomb energies and a lot of orbital- and bond-length-related hopping integrals need to be covered in the Hamiltonian of the tight-binding model. The Coulomb matrix elements, which come to exist in the dielectric screening effects, might be closely related to $(2s, 2p_x, 2p_y, 2p_z)$ and $(4s, 3d_{x^2-y^2}, 3d_{xy}, 3d_{yz}, 3d_{zx}, 3d_{z^2})$ orbitals. Most important, how to efficiently solve the strong local field effects for large unit cells and well characterize the dimensionless energy loss function is the studying focus. Maybe, the development of the dielectric function tensor [59,60] could provide an effective way of exploring the complicated Coulomb screening phenomena. In short, the semiconducting/insulating behaviors with important band gaps and the complex Coulomb scattering processes are expected to become the negative factors in creating the plasmon modes of valence electrons even under the long-wavelength limit.

Very interesting, the ion-based batteries present the rich and unique phenomena after the electrical-chemical reactions. During the charging processes, the Li$^+$-/Al$^+$-/Fe$^+$-ions [61–67] are released from cathode [64–66], transport through electrolyte [64–66], and enter into anode [64–66]. And then, the opposite is true for the discharge processes [64–66]. Apparently, the ion transports have created drastic changes in terms of geometric structures [68], chemical environments [68], and scattering events [68]. For example, the reduced Li$^+$-ions and electrons in 3D ternary LiFe/Co/NiO cathode would lead to the dramatic transformation of lattice symmetry. It might exist a lot of intermediate geometric configurations between two metastable ones, in which they possess very large unit cells. How to link the complicated relations of the former and the latter is an important issue. For the well-characterized cathodes, the VASP calculations can investigate the main features of geometric properties, e.g., the various Li–O and Fe–/Co/Ni–O bond lengths under a specific crystal structure with the atom number below 100–200 [69–71]. However, the other simulation methods [70,71] are required to solve the optima geometries for the short-lifetime configurations beyond the critical atom number in a unit cell. The similar geometric transformations are expected to be revealed in electrolyte and anode materials. Specifically, there exist the negatively charged anode systems because of the electron current through the external lead circuit. Such light-free carriers might display outstanding screening abilities, as observed for conduction electrons in metals [72]. Their important roles on the intercalation and de-intercalation phenomena and the fundamental properties, which come to exist in graphite Li$^+$-intercalated compounds/LiTiO [72], are worthy of the systematic investigations, such as the dielectric effects on the Li$^+$-concentration-dependent ground-state energies, band structures, wave functions, and metallic/semimetallic/ semiconducting behaviors. Furthermore, whether the Li$^+$-ions or the neutral Li-atoms survive in the intercalated/de-intercalated anodes is one of the studying focuses [73,74]. Even for the first-principles simulations, the simultaneous consideration of the electron cloud, anode material, and Li$^+$-intercalation/ Li$^+$-de-intercalation will become a serious challenge. It should be noticed that the up-to-date various experimental measurements could not clarify this issue. The experimental examinations might require the techniques and equipment [73,74].

From the theoretical and experimental viewpoints, the ion transports play critical roles in developing lithium-/aluminum-/iron-based batteries [67,75–77] with the outstanding high-rate capability and high power. Their mechanisms and physical, chemical, and material properties need to be clarified in the near-future investigations. For example, the Li$^+$-related batteries consist of the nanostructured materials: (I) cathode of LiFe/Co/NiO [48–54], (II) electrolyte of LiSi/Ge/SnO [47,55,56], and (III) anode of LiSc/Ti/VO [57–60], belonging to the ternary lithium oxide compounds. However, it is very difficult to predict/examine/identify the nature of Li-ion transport behaviors within the core components. Very interesting, recently the high-resolution experimental techniques are developed to study the local Li-ion diffusion [78], such as the rapid transport through grain boundaries in Li$_x$CoO$_2$ using conductive atomic force microscopy with a spatial resolution of 10 nm. They can provide important information of optimizing cathode, electrolyte, and anode materials [23,29–31] and developing nanoionics [79]. Most important, the significant factors, which are closely related to the nanostructures materials, might cover the interlayer spacing or the grain boundary [80], the concentration of Li$^+$ vacancies [80], the extremely nonuniform chemical environments (e.g., the great modulation of bond lengths; [81]), and the Coulomb screening effects due to the valence electrons. How to link them simultaneously in the theoretical framework of the ion transports will become an emergent challenge, such as the prediction of the steady current density in cathode, electrolyte, and anode materials under various scattering events [82]. As to the trivalent and bivalent cases (aluminum and iron ions discussed later; [83]), the ion transports are getting more complicated because of the large ionic structures [84,85] and the non-negligible chemical reactions at the interfaces between electrolyte and anode or cathode materials [86]. The development of nanoionics is urgently required in the experimental [87] and theoretical [88] explorations.

Silicon/germanium/carbon nanotubes and nanowires might have high performances or potential applications in Li$^+$-based batteries [89]. The hollow carbon nanotubes were first discovered by I. Iijima in 1991 through an electric arc with specific catalysts [90]. Each single-walled nanotube possesses a perfect cylindrical surface as a result of carbon hexagons (Figure 15.2a; [91]), leading to the unusual periodical boundary condition [91,92]. All electronic states are described by the longitudinal wave vector and the transverse angular momentum, as measured from the axial direction [93]. The achiral and chiral carbon nanotubes have been clearly examined by the high-resolution scanning tunneling microscopy (STM; [94,95]). Furthermore, the extension of STM and scanning tunneling spectroscopy (STS; [95]) is capable of identifying the low-energy electronic properties simultaneously. That is, the STS measurements can distinguish the semiconducting and metallic behaviors [96].

(a) ● C

(b) ● Si

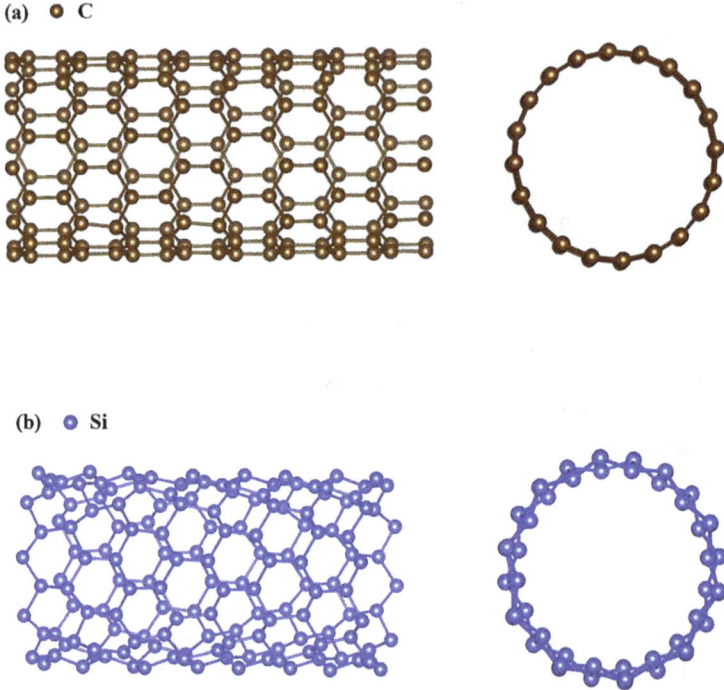

FIGURE 15.2   Geometric structures of (a) carbon nanotubes and (b) silicon nanotubes

According to the up-to-date experiment examinations [96] and theoretical predictions [97,98] about the low-lying electronic energy spectra near the Fermi level, the radius- and chiral-angle-dependent single-walled carbon nanotubes could be classified into three kinds [99,100]: (I) metals, (II) narrow-, and (III) middle-gap semiconductors, in which they, respectively, have 1D linear conduction and valence bands intersecting at E_F, and energy gaps inversely proportional to radius and the square of radius. When carbon nanotubes are further synthesized together, the multiwalled [101,102] or bundle systems are almost uniformly grown in experimental laboratories under the large-scale quantity [101,102]. This phenomenon will provide a lot of available positions during the Li^+-ion intercalations and de-intercalations. From the viewpoints of mathematics, physics, chemistry, and materials, carbon nanotubes are worthy of the systematic investigations on the basic [101,102] and applied [104] sciences, e.g., the many-particle Coulomb excitations and decay rates in the current studies [105,106]. The chemical modifications on them are expected to greatly diversify the application functionalities, e.g., the enhanced storage of atoms or molecules or ions by the intercalation [107], chemisorption [108], and substitution [109].

It should have certain important differences and similarities between 1D silicon or germanium [110,111] and carbon nanotubes [112]. Silicon nanotubes and nanowires have been successfully synthesized through the chemical vapor deposition by using an electric in the absence of any catalysts [113,114]. Each silicon nanotube could be regarded as a cylindrically rolled-up monolayer silicene (Figure 15.2b), as clearly illustrated between carbon nanotube and graphene [114,115]. Geometric structures of any nanotubes are well characterized by the radius and chirality; therefore, the essential physical, chemical, and material properties are very sensitive to these two critical factors. However, a buckling structure (a nonplanar honeycomb lattice; [45–47]), corresponding to different heights of the A and B sublattices, is induced by significant competition of the sp^2 and sp^3 bondings on a monolayer silicene [116]; the absence of the latter on carbon nanotubes). This will be directly reflected on a cylindrical surface, further leading to the enhancement of the bonding misorientation and the Si-four-orbital hybridization. Moreover, the spin-orbital couplings (the quantum effect under the relative theory; [116,117]) need to be taken into consideration during the calculations of electronic properties. As a result, how many kinds of silicon nanotubes in terms of band gaps might require the delicate calculations and analyses, as done for carbon nanotubes [118]. The first-principles method [119] and the tight-binding model [120] would be suitable in fully exploring the low-energy fundamental properties. For example, they are reliable in investigating whether the metallic silicon nanotubes [121] could survive under the specific geometries. Most important, the potential applications on ion-based batteries are closely related to the intercalations/de-intercalations of the metallic ions or atoms. Only the VASP calculations can deal with the chemical modifications on the essential properties and thus provide the full information about the useful transport functionalities of silicon nanotubes, e.g., the optimal geometric structure, the highest ion concentration, the spatial charge density distributions, and the semiconducting or metallic property for the ion-intercalated silicon nanotube compounds. The detailed comparisons with the carbon nanotube intercalation compounds would be completed under the accurately numerical simulations [103,112] and the concise physical, chemical, and material properties [105,108]. Obviously, the storage of atoms and molecules inside silicon nanotubes or silicon nanotube bundles [122,123] becomes very meaningful for the potential applications of green energies, such as the capture of hydrogen gases [124,125] and carbon oxides [125–127] and the intercalation of Li^+-ions/Li-atoms. The active chemical environments [128] and the important orbital hybridizations [128] of various chemical bonds are predicted to play critical roles in the basic [128] and applied [128] sciences. The theoretical framework under the numerical simulations [129] and the phenomenological modes [130] will be developed or proposed in the near-future researches.

One big drawback of currently used LIBs is its safety and fire hazards due to the volatility of conventional organic liquid electrolytes inside the battery. To address this issue, solid electrolytes have been developed to eliminate the use of flammable liquids. All-solid-state battery is considered as the next-generation batteries with high power, high energy density, and high intrinsic safety storage system, in which the solid electrolyte is one of the key components that determine its performance [131–133]. Solid electrolytes can be divided into two categories, namely solid inorganic electrolyte and solid polymer electrolyte. The former consists of various types of structures, including crystalline electrolytes (perovskite [134], NASICON [135], garnet [136], and LISICON types [137]), amorphous electrolytes (oxide [138] and sulfide glassy [139] types), thio-LISICON-type [140], LiPON-type [141], and glass-ceramic [142] electrolytes. For the case of solid polymer electrolytes, polyethylene oxide (PEO) is the most commonly used. Despite the advantages of solid electrolytes exhibit over liquid electrolytes, there is still intrinsic challenges such as low ionic conductivity, grain boundary resistance, and dendrite formation with Li metal anodes. The search for stable solid electrolytes with high conductivity is currently an important issue to compete the liquid ones in commercial batteries. Among all the solid electrolytes, NASICON (natrium superionic conductor), garnet, and LISICON (lithium superionic conductor) exhibit the high ionic conductivities at the room temperature [143]. Specifically, the garnet-type $Li_7La_3Zr_2O_{12}$ (LLZO) solid electrolyte is commonly observed in two phases (tetragonal and cubic), in which its ionic conductivity is higher than that of tetragonal LLZO by two orders of magnitude [144]. Cubic LLZO is considered as one of the best solid electrolytes due to its remarkable chemical and thermal stability. Thermodynamic simulations in terms of the phase diagram and phase stability are expected to provide a better understanding of the solid electrolyte interface problems with electrode materials, including Li-dendrite penetration on anode side (Li-metal) and bad interfacial connection on the cathode side. Further improvements to select the compatibility combination of electrode and electrolyte materials for solid-state LIBs are necessary to satisfy the rapidly increasing demand, especially electric vehicles and mass energy storage systems.

In addition to lithium-ion-related batteries, aluminum-ion-based ones are successfully applied in the commercial products, such as hybrid electric vehicles (HEVs) [132,133], plug-in hybrid electric vehicles (PHEVs) [145], and the electrical grid [146–148]. In general, they principally consist of the anode, electrolyte, and cathode materials, respectively, corresponding to the metallic aluminum [145], the ionic liquid, and the semimetallic/semiconducting graphite/MoS_2 [65,149–155]. Specifically, the first condensed matter system is a thin Al foil with a thickness of $\sim20^{-6}$m. Very interesting, an ionic liquid electrolyte (an ionically conductive medium) comes to exist by mixing 1-ethyl-3-methylimidazolium chloride ((EMIm)Cl) and aluminum chloride ($AlCl_3$), being able to produce redox active chloroaluminate anions ($AlCl_4^-$ and $Al_2Cl_7^-$). These two kinds of anions are responsible for the ion transport; furthermore, the experimental processes need to prevent the electrolyte from being modified by chemical reactions. A thin-film pristine graphite of thickness ~50–100 10^{-6}m provides the enough and at spaces for the intercalations/de-intercalations of the anion ions, since the interlayer $3p_Z$ orbital interactions of the graphitic sheets belong to the weak but significant van der Waals forms. The intercalation effects on the optimal geometric structures and the fundamental electronic properties could be investigated through the VASP calculations, e.g., the combined distribution configuration of $AlCl_4^-$ and $Al_2Cl_7^-$ anions [151,152,156,157], and the dramatic changes in the spatial charge density distributions [65,158,159] and the low-lying valence and conduction bands near the Fermi level [158]. Most important, the aluminum-ion batteries belong to a class of rechargeable ones, where Al-related ions can provide energy by transporting from the anode to cathode materials. In the recharging processes, the aluminum-related ions come back to the negative electrode and can exchange three electrons per ion. This clearly illustrates that the insertion of one Al^{3+} is equivalent to three Li^+-ions in conventional intercalation cathodes [160]. The trivalent charge carrier in Al^{3+} is just merit and drawback, where the transfer of three-unit charges by one ion greatly enhances energy storage capacity [156–161] but creates too strong electrostatic intercalation for the host materials under the well-defined electrochemical behavior.

Recently, the third kind of secondary batteries, the iron-ion-based one (Figure 15.1c), has been successfully developed for the first time under a specific capacity of 207 mAh g^{-1} at 30 mA g^{-1}. The most important merit is quite cheap, and the other characteristics require high-resolution experimental examinations [67]. Very interesting, this new kind of battery mainly consists of mild steel, layered vanadium pentoxide (V_2O_5; [67]), and tetraethylene glycol dimethylether (TEGDME; [68]), respectively, corresponding to anode, cathode, and electrolyte solvent. The first material is mostly made up of iron (99.33% in weight), while few components mainly originate from carbon (0.15%) and manganese (0.52%). Apparently, the 3D ternary iron compound possesses a superlarge unit cell, since the concentrations of the substitution atoms are very dilute. This will result in a rather high barrier in the first-principles calculations of the optimal geometric structures [67], and thus, the other essential physical, chemical, and material properties [68], for example, how many chemical bonds and the great bond-length modulations have to be determined in the numerical simulations, mainly owing to the creation of the extremely non-uniform chemical environment in a primitive unit cell [68]. As to the binary V_2O_5 cathode, the pristine system has a well-behaved crystal structure [68] and belongs to an interesting ferroelectric semiconductor [68]. However, the multi-orbital hybridizations in oxygen-related chemical bonds are worthy of the thoroughly theoretical investigations through the spatial charge density distributions around vanadium and oxygen atoms and atom- and orbital-decomposed density of states [162], as done earlier in Chapters 4–7 for Li$^+$-ion-battery cathode or electrolyte or anode materials using the VASP calculations. The intercalations/de-intercalations of Fe^{2+} ions into/from the cathode material would be very interesting topics, e.g., the existence of metastable and intermediate configurations and the diversified electronic properties. Moreover, the replacement of cathode and electrolyte materials in Fe^{2+}-related batteries might greatly enhance the performance of ion transport, i.e., the optimal combinations of the cathode, electrolyte, and anode materials need a series of experimental tests [67].

REFERENCES

1. Tarascon, J.-M.; Armand, M. Issues and challenges facing rechargeable lithium batteries. *Nature* **2001**, 414, 359–367.
2. Salvadori, A.; Grazioli, D.; Geers, M. G. D.; Danilov, D.; Notten, P. H. L. A multiscale-compatible approach in modeling ionic transport in the electrolyte of Lithium-ion batteries. *Journal of Power Sources* **2015**, 293, 892–911.
3. Vaderbilt, D. Soft self-consistent pseudopotential in a generalized eigenvalue formalism. *Physics Review B* **1990**, 41, 7892–7895.
4. Krese, G.; Furthmuller, J. Efficient iterative schemes for ab ratio total-energy calculations using a plane-wave basis set. *Physics Review B* **1996**, 54, 11169–11186.
5. Kashammer, P.; Sinno, T. Interactions of twin boundaries with intrinsic point defects and carbon in silicon. *Journal of Applied Physics* **2013**, 114, 083505.
6. Austin, B. M.; Zubarev, D. Y.; Lester, W. A. Quantum Monte-Carlo and related approaches, *Chemical Reviews* **2012**, 112(1), 263–288.
7. Needs, R. J.; Towler, M. D.; Drummond, N. D. Continuum variational and diffusion quantum Monte-Carlo calculations. *Journal of Physics: Condensed Matter* **2010**, 22(2), 023201.

8. Kitta, M.; Akita, T.; Maeda, Y.; Kohyama, M. Preparation of a spinel $Li_4Ti_5O_{12}(1\ 1\ 1)$ surface from a rutile TiO_2 single crystal. *Applied Surface Science* **2012**, 258, 3147–3151.

9. Shih, P.-H.; Do, T.-N.; Gumbs, G.; Huang, D.; Pham, H. D.; Lin, M. F. Rich magnetic quantization phenomena in AA bilayer silicene. *Scientific Reports* **2019**, 9, 14799.

10. Kitta, M.; Matsuda, T.; Maeda, Y.; Akita, T.; Tanaka, S.; Kido, Y.; Kohyama, M. Atomistic structure of a spinel $Li_4Ti_5O_{12}(1\ 1\ 1)$ surface elucidated by scanning tunneling microscopy and medium energy ion scattering spectrometry. *Surface Science* **2014**, 619, 5–9.

11. Yi, T.-F.; Li, X.-Y.; Liu, H.; Shu, J.; Zhu, Y.-R.; Zhu, R.-S. Recent developments in the doping and surface modification of $LiFePO_4$ as cathode material for power lithium ion battery. *Ionics* **2012**, 18(6), 529–539.

12. Zaghib, K.; Goodenough, J. B.; Mauger, A.; Julien, C. Unsupported claims of ultra-fast charging of $LiFePO_4$ Li-ion batteries. *Journal of Power Sources* **2009**, 194(2), 1021–1023.

13. Sun, C.; Rajasekhara, S.; Goodenough, J. B.; Zhou, F. Monodisperse porous $LiFePO_4$ microspheres for a high-power Li-ion battery cathode. *Journal of the American Chemical Society* **2011**, 133, 2132–2135.

14. Bensalah, N.; Kamand, F. Z.; Mustafa, N.; Matalqeh, M. Silicon-germanium bilayer sputtered onto a carbon nanotube sheet as anode material for lithium-ion batteries. *Journal of Alloys and Compounds* **2019**, 152088.

15. Kittel, C. *Introduction to Solid State Physics*, 8th Edition, Wiley, New York, **2004**.

16. Nakayama, T.; Shima, H. Computing the Kubo formula for large systems. *Physical Review E* **1998**, 58(3), 3984–3992.

17. Chen, G. P.; Voora, V. K.; Agee, M. M., Balasubramani, S. G.; Furche, F. Random-phase approximation methods. *Annual Review of Physical Chemistry* **2017**, 68, 421–445.

18. Ouyang, C. Y.; Zhong, Z. Y.; Lei, M. S. Ab initio studies of structural and electronic properties of $Li_4Ti_5O_{12}$ spinel. *Electrochemistry Communications* **2007**, 9, 1107–1112.

19. Sun, X.; Radovanovic, P. V.; Cui, B. Advances in spinel $Li_4Ti_5O_{12}$ anode materials for lithium-ion batteries. *New Journal of Chemistry* **2015**, 39, 38–63.

20. Silva-Guillen, J.; San-Jose, P.; Roldan, R. Electronic band structure of transition metal dichalcogenides from ab initio and Slater-Koster tight-binding model. *Applied Sciences* **2016**, 6(10), 284.

21. Lin, C.-Y.; Wu, J.-Y; Chiu, C.-W; Lin, M.-F. *Coulomb Excitations and Decays in Graphene-Related Systems*, 1st Edition, CRC Press, Boca Raton, FL, **2019**.

22. Veerapandian, M.; Lee, M. H.; Krishnamoorthy, K.; Yun, K. Synthesis, characterization and electrochemical properties of functionalized graphene oxide. *Carbon* **2012**, 50, 4228–4228.

23. Ullah, A.; Majid, A.; Rani, N. A review on first principles-based studies for improvement of cathode material of lithium ion batteries. *Journal of Energy Chemistry* **2018**, 27(1), 219–237.

24. Lin, C.-Y.; Wu, J.-Y.; Ou, Y.-J.; Chiu, Y.-H.; Lin, M.-F. Magneto-electronic properties of multilayer graphenes. *Physical Chemistry Chemical Physics* **2015**, 17(39), 26008–26035.

25. Suzuki, S.; Takamura, M.; Yamamoto, H. Transmission, reflection, and absorption spectroscopy of graphene microribbons in the terahertz region. *Japanese Journal of Applied Physics* **2016**, 55, 06GF08.

26. Zhang, J. M.; Liu, Y. Fermi's golden rule: Its derivation and breakdown by an ideal model. *European Journal of Physics* **2016**, 37, 6.

27. Oughstun, K. E. *Electromagnetic and Optical Pulse Propagation*. Springer Series in Optical Sciences, 1st Edition, Springer, New York, **2019**.

28. Tran, N. T. T.; Lin, S.-Y.; Lin, C.-Y.; Lin, M.-F. *Geometric and Electronic Properties of Graphene-Related Systems: Chemical Bonding Schemes*, 1st Edition, CRC Press, Boca Raton, FL, **2017**.
29. Stanley Whittingham, M. Lithium batteries and cathode materials. *Chemical Reviews* **2004**, 104(10), 4271–4302.
30. Mekonnen, Y.; Sundararajan, A.; Sarwat, A. I. A Review of Cathode and Anode Materials for Lithium-Ion Batteries. Southeast Conference, IEEE, Norfolk, VA, **2016**, pp. 1–6.
31. Khan, S. A.; Ali, S.; Saeed, K.; Usman, M.; Khan, I. Advanced cathode materials and efficient electrolytes for rechargeable batteries: Practical challenges and future perspectives. *Journal of Materials Chemistry A* **2019**, 7, 10159–10173.
32. Friedberg, R.; Zhao, H. S. Short-range Coulomb screening in a dielectric crystal. *Physical Review B* **1991**, 44(5), 2297–2305.
33. Ou, Y. C.; Sheu, J. K.; Chiu, Y. H.; Chen, R. B.; Lin M. F. Influence of modulated fields on the Landau level properties of graphene. *Physics Review B* **2011**, 83, 195405.
34. Chen, R. B.; Chiu, C. W.; Lin, M. F. Magneto plasmons in simple hexagonal graphite. *RSC Advances* **2015**, 5, 53736–53740.
35. Chuang, Y. C.; Wu, J. Y.; Lin, M. F. Electric field induced plasmon in AA stacked bilayer graphene. *Annals of Physics* **2013**, 339, 198.
36. Hart, J. L.; Lang, A. C.; Le, A. C.; Longo, P.; Trevor, C.; Twesten, R. D.; Taheri, M. L. Direct detection electron energy-loss spectroscopy: A method to push the limits of resolution and sensitivity. *Scientific Reports* **2017**, 7(1), 8243.
37. Hwang, E. H.; Sensarma, R.; Sarma, D. S. Plasmon-phonon coupling in graphene. *Physics Review B* **2010**, 82(19), 195406.
38. Chiu, Y. H.; Lai, Y. H.; Ho, J. H; Chuu, D. S.; Lin, M. F. Electronic structure of a two-dimensional graphene monolayer in a spatially modulated magnetic field: Peierls tight-binding model. *Physical Review B* **2008**, 77, 045407(6).
39. Lin, M. F.; Shyu, F. L. Temperature-induced plasmons in a graphite sheet. *Journal of Physical Society Japan* **2000**, 69, 607–610.
40. Lin, M. F.; Shyu, F. L. Plasmons and optical properties of semimetal graphite. *Journal of Physical Society Japan* **2000**, 69, 3781–3784.
41. Shyu, F. L.; Lin, M. F.; Lu, Y. T. Electronic excitations in cylinder superlattices. *Journal of Physical Society Japan* **1999**, 68, 3352–3359.
42. Wu, J. Y.; Gumbs, G.; Lin, M.-F. Combined effect of stacking and magnetic field on plasmon excitations in bilayer graphene. *Physics Review B* **2014**, 89, 165407.
43. Lin, C. Y.; Lee, M. H.; Lin, M.-F. Coulomb excitations in ABC-stacked trilayer graphene. *Physical Review B* **2018**, 98(4), 041408.
44. Ou, Y. C.; Chiu, Y. H.; Lu, J. M.; Su, W. P.; Lin, M.-F. Electric modulation effect on magneto-optical spectrum of monolayer graphene. *Computer Physics Communications* **2013**, 184, 1821–1826.
45. Lin, C.-L.; Arafune, R.; Kawahara, K.; Kanno, M.; Tsukahara, N.; Minamitani, E.; Kim, Y.; Kawai, M.; Takagi, N. Substrate-induced symmetry breaking in silicene. *Physical Review Letters* **2013**, 110, 076801.
46. Tabert, C. J.; Nicol, E. J. Magneto-optical conductivity of silicene and other buckled honeycomb lattices. *Physical Review B* **2013**, 88, 085434.
47. Davila, M. E.; Xian, L.; Cahangirov, S.; Rubio, A.; Le Lay, G. Germanene: A novel two-dimensional germanium allotrope akin to graphene and silicene. *New Journal of Physics* **2014**, 16, 095002.
48. Devaraju, M. K.; Honma, I. Hydrothermal and solvothermal process towards development of $LiMPO_4$ (M = Fe, Mn) nanomaterials for lithium-ion batteries. *Advanced Energy Materials* **2012**, 2, 284–297.

49. Huang, H.; Yin, S.-C.; Nazar, L. F. Approaching theoretical capacity of LiFePO$_4$ at room temperature at high rates. *Electrochemical and Solid-State Letters* **2001**, 4, A170–A172.

50. Devaraju, M. K.; Truong, Q. D.; Tomai, T.; Hyodo, H.; Honma, I. Antisite defects in LiCoPO$_4$ nanocrystals synthesized via a supercritical fluid process. *RSC Advances* **2014**, 4, 52410–52414.

51. Devaraju, M. K.; Truong, Q. D.; Hyodo, H.; Tomai, T.; Honma, I. Supercritical fluid methods for synthesis of LiCoPO$_4$ nanoparticles and their application to lithium ion battery. *Inorganics* **2014**, 2, 233–247.

52. Truong, Q. D.; Devaraju, M. K.; Tomai, T.; Honma, I. Controlling the shape of LiCoPO$_4$ nanocrystals by supercritical fluid process for enhanced energy storage properties. *Scientific Reports* **2014**, 4, 3975.

53. Devaraju, K.; Truong, Q. D.; Hyodo, H.; Sasaki, Y.; Honma, I. Synthesis, characterization and observation of antisite defects in LiNiPO$_4$ nanomaterials. *Scienetific Reports* **2015**, 5(1), 11041.

54. Manickam, M.; Pritam, S.; Dominique, A.; Danielle, E. M. Synthesis of olivine LiNiPO$_4$ for aqueous rechargeable battery. *Electrochimica Acta* **2011**, 56, 4356–4360.

55. Tang, T.; Chen, P.; Luo, W.; Luo, D.; Wang, Y. Crystalline and electronic structures of lithium silicates: A density functional theory study. *Journal of Nuclear Materials* **2012**, 420, 31–38.

56. Ma, S.-G.; Shen, Y.-H.; Kong, X.-G.; Gao, T.; Chen, X.-J.; Xiao, C.-J.; Lu, T.-C. A new interatomic pair potential for the modeling of crystalline Li$_2$SiO$_3$. *Materials and Design* **2017**, 118, 218–225.

57. Liu, Z.; Deng, H.; Zhang, S.; Hu, W.; Gao, F. Theoretical prediction of LiScO$_2$ nanosheets as a cathode material for Li-O$_2$ batteries. *Physical Chemistry Chemical Physics* **2018**, 20, 22351–22358.

58. Lu, X.; Gu, L.; Hu, Y. S. New insight into the atomic-scale bulk and surface structure evolution of Li$_4$Ti$_5$O$_{12}$ anode. *Journal of the American Chemical Society* **2015**, 137, 1581–1586.

59. Yi, T. F.; Yang, S. Y.; Xie, Y. Recent advances of Li$_4$Ti$_5$O$_{12}$ as a promising next generation anode material for high power lithium-ion batteries. *Journal of Materials Chemistry A* **2015**, 3, 5750–5777.

60. Nair, V. S.; Sreejith, S.; Borah, P.; Hartung, S.; Bucher, N.; Zhao, Y.; Madhavi, S. Crystalline Li$_3$V$_6$O$_{16}$ rods as high-capacity anode materials for aqueous rechargeable lithium batteries (ARLB). *RSC Advances* **2014**, 4, 28601–28605.

61. Goodenough, J. B.; Kim, Y. Challenges for rechargeable Li batteries. *Chemistry of Materials* **2010**, 22, 587.

62. John B.; Park, G.-S. The Li-ion rechargeable battery: A perspective. *Journal of the American Chemical Society* **2013**, 135(4), 1167–1176.

63. Wu, J.; Fenech, M.; Webster, R. F.; Richard, D.; Sharma, N. Electron microscopy and its role in advanced lithium-ion battery research. *Sustainable Energy Fuels* **2019**, 3, 1623.

64. Wang, D.-Y.; Wei, C.-Y.; Lin, M.-C.; Pan, C.-J.; Chou, H.-L.; Chen, H.-A.; Gong, M.; Wu, Y.-P.; Yuan, C.; Angell, M.; Hsieh, Y.-J.; Chen, Y.-H.; Wen, C.-Y.; Chen, C.-W.; Hwang, B.-J.; Chen, C.-C.; Dai, H. Advanced rechargeable aluminium ion battery with a high-quality natural graphite cathode. *Nature Communications* **2017**, 8, 14283.

65. Leisegang, T.; Meutzner, F.; Zschornak, M.; Munchgesang, W.; Schmid, R.; Nestler, T.; Meyer, D. C. The aluminum-ion battery: A sustainable and seminal concept? *Frontiers in Chemistry* **2019**, 7, 268.

66. Rani, J. V.; Kanakaiah, V.; Dadmal, T.; Rao, M. S.; Bhavanarushi, S. Fluorinated natural graphite cathode for rechargeable ionic liquid-based aluminum-ion battery. *Journal of the Electrochemical Society* **2013**, 160, A1781–A1784.

67. Ramaprabhu, S.; Samantaray, S. S. A room temperature multivalent rechargeable Iron ion battery with ether-based electrolyte: New type of post-lithium ion battery. *Chemical Communications* **2019**, 55, 10416–10419.

68. Kuganathan, N.; Iyngaran, P.; Chroneos, A. Lithium diffusion in Li_5FePO_4. *Scientific Report* **2018**, 8, 5832.

69. Okumura, T.; Shikano, M.; Kobayashi, H. Effect of bulk and surface structural changes in Li_5FeO_4 positive electrodes during first charging on subsequent lithium-ion battery performance. *Journal of Materials Chemistry A* **2014**, 2, 11847–11856.

70. Zhu, J.; Zu, W.; Yang, G.; Song, Q. A novel electrochemical supercapacitor based on $Li_4Ti_5O_{12}$ and $LiNi/Co/Mn/O_2$. *Materials Letters* **2014**, 115, 237–240.

71. Thackeray, M. M.; Kang, S. H.; Johnson, C. S.; Vaughey, J. T.; Benedek, R.; Hackney, S. A. Li_2MnO_3-stabilized $LiMO_2$ (M = Mn, Ni, Co) electrodes for lithium-ion batteries. *Journal of Materials Chemistry A* **2007**, 17, 3112–3125.

72. Samin, A.; Kurth, M.; Cao, L. Ab initio study of radiation effects on the $Li_4Ti_5O_{12}$ electrode used in lithium-ion batteries. *AIP Advances* **2015**, 5, 047110.

73. Laubach, S.; Laubach, S.; Schmidt, P. C.; Ensling, D.; Schmid, S.; Jaegermann, W.; Thißen, A.; Nikolowski, K.; Ehrenberg, H. Changes in the crystal and electronic structure of $LiCoO_2$ and $LiNiO_2$ upon Li intercalation and deintercalation. *Physical Chemistry Chemical Physics* **2009**, 11(17), 3278.

74. Ensling, D.; Cherkashinin, G.; Schmid, S.; Bhuvaneswari, S.; Thissen, A.; Jaegermann, W. Nonrigid band behavior of the electronic structure of $LiCoO_2$ thin film during electrochemical Li deintercalation. *Chemistry of Materials* **2014**, 26(13), 3948–3956.

75. Kim, T.; Song, W.; Son, D.-Y.; Ono, L. K.; Qi, Y. Lithium-ion batteries: Outlook on present, future, and hybridized technologies. *Journal of Materials Chemistry A* **2019**, 7, 2942.

76. Lin, M.-C.; Gong, M.; Lu, B.; Wu, Y.; Wang, D.-Y.; Guan, M.; Dai, H. An ultrafast rechargeable aluminium-ion battery. *Nature* **2015**, 520(7547), 324–328.

77. Das, S. K.; Mahapatra, S.; Lahan, H. Aluminium-ion batteries: Developments and challenges. *Journal of Materials Chemistry A* **2017**, 5(14), 6347–6367.

78. Fallahzadeh, R.; Farhadian, N. Molecular dynamics simulation of lithium ion diffusion in $LiCoO_2$ cathode material. *Solid State Ionics* **2015**, 280, 10–17.

79. Wu, X.; Song, K.; Zhang, X.; Hu, N.; Li, L.; Li, W.; Zhang, H. Safety issues in lithium ion batteries: Materials and cell design. *Frontiers in Energy Research* **2019**, 7, 65.

80. Balke, N.; Jesse, S.; Morozovska, A. N.; Eliseev, E.; Chung, D. W.; Kim, Y.; Kalinin, S. V. Nanoscale mapping of ion diffusion in a lithium-ion battery cathode. *Nature Nanotechnology* **2010**, 5(10), 749–754.

81. Macedo, L. J. A.; Lima, F. C. D. A.; Amorim, R. G.; Freitas, R. O.; Yadav, A.; Iost, R. M.; Crespilho, F. N. Interplay of non-uniform charge distribution on the electrochemical modification of graphene. *Nanoscale* **2018**, 10(31), 15048–15057.

82. Gauthier, M.; Carney, T. J.; Grimaud, A.; Giordano, L.; Pour, N.; Chang, H.-H.; Shao-Horn, Y. Electrode-electrolyte interface in Li-ion batteries: Current understanding and new insights. *The Journal of Physical Chemistry Letters* **2015**, 6(22), 4653–4672.

83. Sun, P.; Ma, R.; Ma, W.; Wu, J.; Wang, K.; Sasaki, T.; Zhu, H. Highly selective charge-guided ion transport through a hybrid membrane consisting of anionic graphene oxide and cationic hydroxide nanosheet superlattice units. *NPG Asia Materials* **2016**, 8(4), e259–e259.

84. Ji, H.; Urban, A.; Kitchaev, D. A.; Kwon, D.-H.; Artrith, N.; Ophus, C.; Ceder, G. Hidden structural and chemical order controls lithium transport in cation-disordered oxides for rechargeable batteries. *Nature Communications* **2019**, 10(1), 592.

85. Morimoto, T.; Nagai, M.; Minowa, Y.; Ashida, M.; Yokotani, Y.; Okuyama, Y.; Kani, Y. Microscopic ion migration in solid electrolytes revealed by terahertz time-domain spectroscopy. *Nature Communications* **2019**, 10(1), 2662.

86. Fitzhugh, W.; Ye, L.; Li, X. The effects of mechanical constriction on the operation of sulfide based solid-state batteries. *Journal of Materials Chemistry A* **2019**, 7, 23604–23627.
87. Zhu, X.; Ong, C. S.; Xu, X.; Hu, B.; Shang, J.; Yang, H.; Li, R.-W. Direct observation of lithium-ion transport under an electrical field in Li_xCoO_2 nanograins. *Scientific Reports* **2013**, 3(1), 1084.
88. Zhang, L.; Gong, T.; Wang, H.; Guo, Z.; Zhang, H. Memristive devices based on emerging two-dimensional materials beyond graphene. *Nanoscale* **2019**, 11(26), 12413–12435.
89. Liu, J.; Song, K.; Zhu, C.; Chen, C.-C.; van Aken, P. A.; Maier, J.; Yu, Y. Ge/C nanowires as high-capacity and long-life anode materials for Li-ion batteries. *ACS Nano* **2014**, 8(7), 7051–7059.
90. Iijima, S. Synthesis of carbon nanotubes. *Nature* **1991**, 354, 56–58.
91. Zhang, M.; Li, J. Carbon nanotube in different shapes. *Materials Today* **2009**, 12(6), 12–18.
92. Liu, C.; Cheng, H.-M. Carbon nanotubes: Controlled growth and application. *Materials Today* **2013**, 16(1–2), 19–28.
93. Marganska, M.; Schmid, D. R.; Dirnaichner, A.; Stiller, P. L.; Strunk, C.; Grifoni, M.; Huttel, A. K. Shaping electron wave functions in a carbon nanotube with a parallel magnetic field. *Physical Review Letters* **2019**, 122(8), 086802.
94. Odom, T. W.; Huang, J.-L.; Kim, P.; Ouyang, M.; Lieber, C. M. Scanning tunneling microscopy and spectroscopy studies of single wall carbon nanotubes. *Journal of Materials Research* **1998**, 13(9), 2380–2388.
95. Meunier, V.; Lambin, P. Scanning tunneling microscopy and spectroscopy of topological defects in carbon nanotubes. *Carbon* **2000**, 38(11–12), 1729–1733.
96. Giusca, C. E.; Tison, Y.; Silva, S. R. P. Inter-layer interaction in double-walled carbon nanotubes evidenced by scanning tunneling microscopy and spectroscopy. *Nano* **2008**, 3(2), 65–73.
97. Boumia, L.; Zidour, M.; Benzair, A.; Tounsi, A. A Timoshenko beam model for vibration analysis of chiral single-walled carbon nanotubes. *Physica E: Low-Dimensional Systems and Nanostructures* **2014**, 59, 186–191.
98. MacFarlane, W.; Chakhalian, J.; Kie, R.; Dunsiger, S.; Miller, R.; Sonier, J.; Fischer, J. A SR study of single-walled carbon nanotubes. *Physica B: Condensed Matter* **2000**, 289–290, 589–593.
99. Liu, B.; Wu, F.; Gui, H.; Zheng, M.; Zhou, C. Chirality-controlled synthesis and applications of single-wall carbon nanotubes. *ACS Nano* **2017**, 11(1), 31–53.
100. Wu, B.; Geng, D.; Liu, Y. Evaluation of metallic and semiconducting single-walled carbon nanotube characteristics. *Nanoscale* **2011**, 3(5), 2074.
101. Pillai, S. K.; Ray, S. S.; Moodley, M. Purification of multi-walled carbon nanotubes. *Journal of Nanoscience and Nanotechnology* **2008**, 8(12), 6187–6207.
102. Eatemadi, A.; Daraee, H.; Karimkhanloo, H.; Kouhi, M.; Zarghami, N.; Akbarzadeh, A.; Abasi, M.; Hanifehpour, Y.; Joo, S. Carbon nanotubes: Properties, synthesis, purification, and medical applications. *Nanoscale Research Letters* **2014**, 9(1), 393.
103. Pillewan, V. J.; Raut, D. N.; Patil, K. N.; Shinde, D. K. Carbon to carbon nanotubes synthesis process: An experimental and numerical study. *Materials Today: Proceedings* **2018**, 5(2), 6444–6452.
104. Zhang, Z.; Mu, S.; Zhang, B.; Tao, L.; Huang, S.; Huang, Y.; Zhao, Y. A novel synthesis of carbon nanotubes directly from an indecomposable solid carbon source for electrochemical applications. *Journal of Materials Chemistry A* **2016**, 4(6), 2137–2146.
105. Kanemitsu, Y. Excitons in semiconducting carbon nanotubes: Diameter-dependent photoluminescence spectra. *Physical Chemistry Chemical Physics* **2011**, 13(33), 14879.
106. Velizhanin, K. A. Exciton relaxation in carbon nanotubes via electronic-to-vibrational energy transfer. *The Journal of Chemical Physics* **2019**, 151(14), 144703.

107. Chacon-Torres, J. C.; Dzsaber, S.; Vega-Diaz, S. M.; Akbarzadeh, J.; Peterlik, H.; Kotakoski, J.; Reich, S. Potassium intercalated multiwalled carbon nanotubes. *Carbon* **2016**, 105, 90–95.
108. Chakrapani, N.; Zhang, Y. M.; Nayak, S. K.; Moore, J. A.; Carroll, D. L.; Choi, Y. Y.; Ajayan, P. M. Chemisorption of acetone on carbon nanotubes. *The Journal of Physical Chemistry B* **2003**, 107(35), 9308–9311.
109. Sankaran, M.; Viswanathan, B. Hydrogen storage in boron substituted carbon nanotubes. *Carbon* **2007**, 45(8), 1628–1635.
110. Bai, J.; Zeng, X. C.; Tanaka, H.; Zeng, J. Y. Metallic single-walled silicon nanotubes. *Proceedings of the National Academy of Sciences* **2004**, 101(9), 2664–2668.
111. Li, X.; Meng, G.; Xu, Q.; Kong, M.; Zhu, X.; Chu, Z.; Li, A.-P. Controlled synthesis of germanium nanowires and nanotubes with variable morphologies and sizes. *Nano Letters* **2011**, 11(4), 1704–1709.
112. Khalilov, U.; Bogaerts, A.; Neyts, E. C. Atomic scale simulation of carbon nanotube nucleation from hydrocarbon precursors. *Nature Communications* **2015**, 6(1), 10306.
113. Carreon, M. L.; Thapa, A. K.; Jasinski, J. B.; Sunkara, M. K. The capacity and durability of amorphous silicon nanotube thin film anode for lithium ion battery applications. *ECS Electrochemistry Letters* **2015**, 4(10), A124–A128.
114. Casiello, M.; Picca, R.; Fusco, C.; D'Accolti, L.; Leonardi, A.; Lo Faro, M.; Nacci, A. Catalytic activity of silicon nanowires decorated with gold and copper nanoparticles deposited by pulsed laser ablation. *Nanomaterials* **2018**, 8(2), 78.
115. Tserpes, K. I.; Silvestre, N. *Modeling of Carbon Nanotubes, Graphene and Their Composites*, 1st Edition, Springer, New York, **2014**.
116. Molle, A.; Grazianetti, C.; Tao, L.; Taneja, D.; Alam, M. H.; Akinwande, D. Silicene, silicene derivatives, and their device applications. *Chemical Society Reviews* **2018**, 47(16), 6370–6387.
117. Ezawa, M. Quantum hall effects in silicene. *Journal of the Physical Society of Japan* **2012**, 81(6), 064705.
118. Correa, J. D.; da Silva, A. J. R.; Pacheco, M. Tight-binding model for carbon nanotubes from ab initio calculations. *Journal of Physics: Condensed Matter* **2010**, 22(27), 275503.
119. Zhang, R. Q.; Lee, H.-L.; Li, W.-K.; Teo, B. K. Investigation of possible structures of silicon nanotubes via density-functional tight-binding molecular dynamics simulations and ab initio calculations. *The Journal of Physical Chemistry B* **2005**, 109(18), 8605–8612.
120. White, C. T.; Mintmire, J. W. Fundamental properties of single-wall carbon nanotubes. *The Journal of Physical Chemistry B* **2005**, 109, 1, 52–65.
121. Ezawa, M. Dirac theory and topological phases of silicon nanotube. *Europhysics Letters* **2012**, 98(6), 67001.
122. Ahmadi, N.; Shokri, A. A.; Elahi, S. M. Optical transition of zigzag silicon nanotubes under intrinsic curvature effect. *Silicon* **2015**, 8(2), 217–224.
123. Chen, J.; Liu, M.; Sun, J.; Xu, F. Templated magnesiothermic synthesis of silicon nanotube bundles and their electrochemical performances in lithium ion batteries. *RSC Advances* **2014**, 4(77), 40951–40957.
124. Chitsazan, A.; Monajjemi, M.; Aghaei, H.; Sayadian, M. Neutral gases adsorption with hydrogen on silicon nanotubes: A fuel cell investigation. *Oriental Journal of Chemistry* **2017**, 33(3), 1366–1374.
125. Yan, Y.; Miao, J.; Yang, Z.; Xiao, F.-X.; Yang, H. B.; Liu, B.; Yang, Y. Carbon nanotube catalysts: Recent advances in synthesis, characterization and applications. *Chemical Society Reviews* **2015**, 44(10), 3295–3346.
126. Lan, J.; Cheng, D.; Cao, D.; Wang, W. Silicon nanotube as a promising candidate for hydrogen storage: From the first principle calculations to grand canonical Monte Carlo simulations. *The Journal of Physical Chemistry C* **2008**, 112(14), 5598–5604.

127. Khalili, S.; Asghar, A.; Ghoreyshi; Jahanshahi, M. Carbon dioxide captured by multiwalled carbon nanotube and activated charcoal: A comparative study. *Chemical Industry & Chemical Engineering Quarterly* **2013**, 19(1), 153–164.

128. Gao, W.; Kono, J. Science and applications of wafer-scale crystalline carbon nanotube films prepared through controlled vacuum filtration. *Royal Society Open Science* **2019**, 6(3), 181605.

129. Grabowski, K.; Zbyrad, P.; Uhl, T.; Staszewski, W. J.; Packo, P. Multiscale electromechanical modeling of carbon nanotube composites. *Computational Materials Science* **2017**, 135, 169–180.

130. Fathi, D. A review of electronic band structure of graphene and carbon nanotubes using tight binding. *Journal of Nanotechnology* **2011**, 2011, 1–6.

131. Yue, L.; Ma, J.; Zhang, J.; Zhao, J.; Dong, S., Liu, Z.; Chen, L. All solid-state polymer electrolytes for high-performance lithium ion batteries. *Energy Storage Materials* **2016**, 5, 139–164.

132. Kim, J. G.; Son, B.; Mukherjee, S.; Schuppert, N.; Bates, A.; Kwon, O.; Park, S. A review of lithium and non-lithium based solid state batteries. *Journal of Power Sources* **2015**, 282, 299–322.

133. Li, J.; Ma, C.; Chi, M.; Liang, C.; Dudney, N. J. Solid electrolyte: The key for high-voltage lithium batteries. *Advanced Energy Materials* **2015**, 5(4), 1401408.

134. Zheng, J. Q.; Li, Y. F.; Yang, R.; Li, G; Ding, X. K. Lithium ion conductivity in the solid electrolytes $(Li_{0.25}La_{0.25})_{1-x}M_{0.5x}NbO_3$ (M = Sr, Ba, Ca, x = 0.125) with perovskite-type structure. *Ceramics International* **2017**, 43(2), 1716–1721.

135. Kim, H. S.; Oh, Y.; Kang, K. H.; Kim, J. H.; Kim, J.; Yoon, C. S. Characterization of sputter-deposited $LiCoO_2$ thin film grown on NASICON-type electrolyte for application in all-solid-state rechargeable lithium battery. *ACS Applied Materials and Interfaces* **2017**, 9(19), 16063–16070.

136. Han, F.; Zhu, Y.; He, X.; Mo, Y.; Wang, C. Electrochemical stability of $Li_{10}GeP_2S_{12}$ and $Li_7La_3Zr_2O_{12}$ solid electrolytes. *Advanced Energy Materials* **2016**, 6(8), 1501590.

137. Deng, Y.; Eames, C.; Fleutot, B.; David, R.; Chotard, J. N., Suard, E.; Islam, M. S. Enhancing the lithium ion conductivity in lithium superionic conductor (LISICON) solid electrolytes through a mixed polyanion effect. *ACS Applied Materials Interfaces* **2017**, 9(8), 7050–7058.

138. Tron, A., Nosenko, A., Park, Y. D.; Mun, J. Synthesis of the solid electrolyte Li_2O-LiF-P_2O_5 and its application for lithium-ion batteries. *Solid State Ionics* **2017**, 308, 40–45.

139. Liu, D.; Zhu, W.; Feng, Z.; Guerfi, A.; Vijh, A.; Zaghib, K. Recent progress in sulfide-based solid electrolytes for Li-ion batteries. *Materials Science and Engineering: B* **2016**, 213, 169–176.

140. Tarhouchi, I.; Viallet, V.; Vinatier, P.; Ménétrier, M. Electrochemical characterization of $Li_{10}SnP_2S_{12}$: An electrolyte or a negative electrode for solid state Li-ion batteries? *Solid State Ionics* **2016**, 296, 18–25.

141. Le Van-Jodin, L.; Claudel, A.; Secouard, C.; Sabary, F.; Barnes, J. P.; Martin, S. Role of the chemical composition and structure on the electrical properties of a solid-state electrolyte: Case of a highly conductive LiPON. *Electrochimica Acta* **2018**, 259, 742–751.

142. Xu, R. C.; Xia, X. H.; Yao, Z. J.; Wang, X. L.; Gu, C. D.; Tu, J. P. (2016). Preparation of $Li_7P_3S_{11}$ glass-ceramic electrolyte by dissolution-evaporation method for all-solid-state lithium ion batteries. *Electrochimica Acta* **2016**, 219, 235–240.

143. Meesala, Y. Recent advancements in Li-ion conductors for all-solid-state Li-ion batteries. *ACS Energy Letters* **2017**, 2(12), 2734–2751.

144. Larraz, G.; Orera, A.; Sanjuan, M. L. Cubic phases of garnet-type $Li_7La_3Zr_2O_{12}$: The role of hydration. *Journal of Materials Chemistry A* **2013**, 1(37), 11419–11428.

145. Jayaprakash, N.; Das, S. K.; Archer, L. A. The rechargeable aluminum-ion battery. *Chemical Communications* **2011**, 47, 12610–12612.

146. Yang, Z.; Zhang, J.; Kintner-Meyer, M. C. W.; Lu, X., Choi, D; Lemon, J. P.; Liu, J. Electrochemical energy storage for green grid. *Chemical Reviews* **2011**, 111(5), 3577–3613.
147. Hudak, N. S. Chloroaluminate-doped conducting polymers as positive electrodes in rechargeable aluminum batteries. *The Journal of Physical Chemistry C* **2014**, 118(10), 5203–5215.
148. Li, Z.; Xiang, K.; Xing, W.; Carter, W. C.; Chiang, Y. Reversible aluminum-ion intercalation in Prussian blue analogs and demonstration of a high-power aluminum-ion asymmetric capacitor. *Advanced Energy Materials* **2015**, 5, 1401410.
149. Wang, H. L.; Gu, S. C.; Bai, Y.; Chen, S.; Wu, F.; Wu, C. High-voltage and noncorrosive ionic liquid electrolyte used in rechargeable aluminum battery. *ACS Applied Materials & Interfaces* **2016**, 8, 27444–27448.
150. Wang, H. L.; Gu, S. C.; Bai, Y.; Chen, S.; Zhu, N.; Wu, F.; Wu, C. Anion-effects on electrochemical properties of ionic liquid electrolytes for rechargeable aluminum batteries. *Journal of Materials Chemistry A* **2015**, 3, 22677–22686.
151. Sun, H.; Wang, W.; Yu, Z.; Yuan, Y.; Wang, S.; Jiao, S. A new aluminium-ion battery with high voltage, high safety and low cost. *Chemical Communications* **2015**, 51(59), 11892–11895.
152. Jiao, H.; Wang, C.; Tu, J.; Tian, D.; Jiao, S. A rechargeable Al-ion battery: Al/molten $AlCl_3$-urea/graphite. *Chemical Communications* **2017**, 53, 2331–2334.
153. Zhang, Y. U.; Liu S.; Ji, Y., Ma, J.; Yu, H. Emerging nonaqueous aluminum-ion batteries: Challenges, status, and perspectives. *Advanced Materials* **2018**, 30, 17063.
154. Li, Z.; Niu, B.; Liu, J.; Li, J.; Kang, F. Rechargeable aluminum-ion battery based on MoS_2 microsphere cathode. *ACS Applied Materials & Interfaces* **2018**, 10(11), 9451–9459.
155. Chen, J.; Kuriyama, N.; Yuan, H.; Takeshita, H. T.; Sakai, T. Electrochemical hydrogen storage in MoS_2 nanotubes. *Journal of the American Chemical Society* **2001**, 123, 11813–11814.
156. Nestler, T.; Fedotov, S.; Leisegang, T.; Meyer, D. C. Towards Al^{3+} mobility in crystalline solids: Critical review and analysis. *Critical Reviews in Solid State and Materials Sciences* **2019**, 44(4), 298–323.
157. Wang, H.; Gu, S.; Bai, Y.; Chen, S.; Zhu, N.; Wu, C.; Wu, F. Anion-effects on electrochemical properties of ionic liquid electrolytes for rechargeable aluminum batteries. *Journal of Materials Chemistry A* **2015**, 3(45), 22677–22686.
158. Elia, G. A.; Marquardt, K.; Hoeppner, K.; Fantini, S.; Lin, R.; Knipping, E.; Peters, W.; Drillet, J. F.; Passerini, S.; Hahn, R. An overview and future perspectives of aluminum batteries. *Advanced Materials* **2016**, 28(35), 7564–7579.
159. Zafar, Z. A.; Imtiaz, S.; Razaq, R.; Ji, S.; Huang, T.; Zhang, Z. Cathode materials for rechargeable aluminum batteries: Current status and progress. *Journal of Materials Chemistry A* **2017**, 5, 5646–5660.
160. Liu, C.; Neale, Z. G.; Cao, G. Understanding electrochemical potentials of cathode materials in rechargeable batteries. *Materials Today* **2016**, 19(2), 109–123.
161. Gu, S. C.; Wang, H. L.; Wu, C.; Bai, Y.; Li, H.; Wu, F. Confirming reversible Al^{3+} storage mechanism through intercalation of Al^{3+} into V_2O_5 nanowires in a rechargeable aluminum battery. *Energy Storage Materials* **2017**, 6, 9–17.
162. Yue, Y.; Liang, H. Micro- and nano-structured vanadium pentoxide (V_2O_5) for electrodes of lithium-ion batteries. *Advanced Energy Materials* **2017**, 7, 1–32.

16 Problems

Thi Dieu Hien Nguyen,
Ngoc Thanh Thuy Tran, and Ming-Fa Lin
National Cheng Kung University

CONTENT

The featured physical, chemical, and material properties could be solved by the first-principles method (the VASP calculations) and the phenomenological model (the tight-binding model).

1. Exploring the low-lying band structure for a monolayer graphene (Figure 16.1a: [1]) by the tight-binding model under the carbon-$2p_z$ orbital bondings. Determine (a) a primitive unit cell and two lattice vectors, (b) the first Brillouin zone and two reciprocal lattice vectors, (c) a 2×2 Hamiltonian matrix, (d) valence and conduction energy dispersions, and (e) wave-vector-dependent wave functions.
2. The similar calculations, as revealed in problems 1(a)–1(e), applied for a monolayer silicene/germanene with buckling in A and B sublattices (Figure 16.1b; [2–5]). Here, the spin-orbital coupling plays a critical role in the electronic properties through the 4×4 Hamiltonian matrix. (f) The gate-voltage effects, being induced a uniform perpendicular electric field, could be achieved by changing its diagonal matrix elements. (g) A detailed comparison is also made between grapheme and silicene.

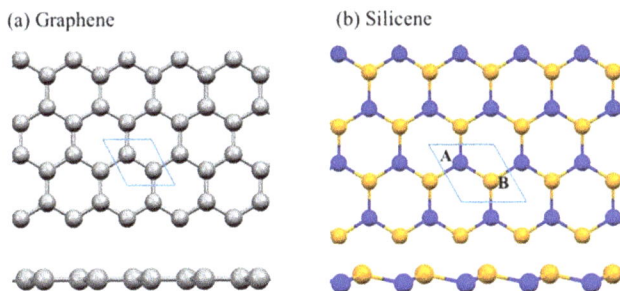

FIGURE 16.1 The geometric structures for monolayer: (a) graphene and silicene with the (b)/(c) top/side view.

3. The above-mentioned phenomenological model is suitable for studying the electronic properties of AA-, AB-, and ABC-stacked graphites [6–13], in which the second system serves as the anode material of lithium-ion-based batteries. Thoroughly investigate (a) their primitive unit cells under the different stacking configurations, (b) the distinct overlaps in valence and conduction bands, (c) the different densities of states across the Fermi level, (d) the existence of Dirac-cone structures, and (e) the important differences between 3D and 2D layered honeycomb lattices.

4. The rich and unique magnetic quantization in monolayer graphene and bulk graphites through the generalized tight-binding model [14]. The vector potential, being created by a uniform perpendicular magnetic field, leads to the periodical Peierls phases, and thus an extra magnetic-field-dependent period characterizes/evaluates/discusses (a) an enlarged magnetic unit cell, (b) the modifications on the intralayer and interlayer hopping integrals of the Hamiltonian matrix elements, (c) the Landau levels/sub-bands in 2D/3D systems, (d) the localized magnetic wave functions, and (e) the dimension- and stacking-diversified phenomena.

5. The generalized tight-binding model is also reliable for the electronic and magneto-electronic properties of the AAA-[14,15], ABA-[14,16,17], ABC-[14,17,18], and AAB-stacked [14,19] trilayer graphene systems (Figure 16.2). Fully explore (a) the low-energy features in terms of energy dispersions, band overlaps, and electron and hole energy spectra, (b) the diverse van Hove singularities of density of states, (c) the unusual Landau levels with the high degeneracy, (d) the normal or irregular magnetic wave functions with the well-behaved or undefined modes, (e) the abnormal anti-crossing behaviors in the field-induced energy spectra, and (f) the combined effects due to the perpendicular magnetic and electric fields.

6. The VASP calculations for studying the lithium-based battery anode materials of the 1D silicene/germanene nanotubes [20–24]. Delicately identify/characterize/calculate (a) the arm-chair, zigzag, and chiral hexagon arrangement on a hollow cylinder with the buckled A and B sublattices, (b) angular-momentum-decoupled band structures under the periodical boundary condition, (c) the standing waves along the azimuthal direction, (d) the critical roles of orbital hybridizations on a cylindrical surface, and (e) the radius- and chiral-angle-dependent band gaps and compare (f) the significant similarities and differences between silicene and germanene nanotubes.

7. Using the first-principles method to fully investigate the quantum-confinement effects in 1D silicene/germanene nanoribbons in the presence or absence of hydrogen passivation [25–27]. Define/evaluate/illustrate (a) the armchair, zigzag, and chiral open edges, (b) the modulated bond lengths near two boundaries, (c) the feature band structures with the transverse translation symmetry, (d) the unique wave functions under the open boundary condition, (e) the strong dependences of energy gaps on the spin-orbital coupling, buckled structure, edge structure, and ribbon widths, (f) the rich chemical bondings from the spatial charge density distributions, (g) the

multi-orbital hybridizations through the atom- and orbital-decomposed van Hove singularities of density of states, and (h) the dimensionality-enriched phenomena.

8. Intercalations/de-intercalations of $Li^+/AlCl_4^-/Al_2Cl_7^-$ into/from the inter-layer spacings of graphitic sheets, serving as the anode or cathode materials in the lithium-/aluminum-ion-based batteries [28–31]. The delicate VASP calculations and analyses on graphite intercalation ion compounds are able to provide (a) the diverse geometric symmetries, (b) the semimetallic or metallic behaviors, (c) the spin-degenerate or spin-split energy bands near the Fermi level, (d) the chemical bonding strengths of carbon-ion bonds from the spatial charge density distributions, (e) the existence of ferromagnetic spin configurations related to the anions, and (f) the minor or major modifications after the cation or anion intercalations.

9. The 3D binary V_2O_5 could be utilized as the cathode material of iron-ion-based batteries [32]. Its rich properties of vanadium pentoxide will be clearly revealed under the first-principles calculations: (a) the crystal symmetry with or without the modulation of bond length, (b) the band gap of semiconductor, (c) the existence of the vanadium-induced spin configuration, (d) the multi-orbital hybridizations in V–O chemical bonds and their drastic changes after certain substitutions of vanadium by iron (or the important differences between binary and ternary compounds).

10. The 3D $LiFeO_2$ and $LiFePO_4$ materials are used as cathodes in lithium-ion-based batteries [33–36]. The VASP numerical simulations are capable of identifying (a) the extremely nonuniform physical and chemical environments (e.g., the great modulation of bond lengths), (b) the strong three-/four-atom dependences for each energy band, (c) the multi-orbital hybridization in Li–O, Fe–O, and P–O bonds from the significant overlaps in the spatial charge density distributions, (d) the atom-dominated energy ranges due to the different strengths of van Hove singularities, (e) the atom- and orbital-decomposed magnetic moments, (f) the spin-split energy band close to the Fermi level, (g) the ferromagnetic metals or semiconductors, (h) the ferromagnetic spin density arrangements around Li, Fe, P, and O atoms, and (i) the enhanced battery functionalities after the dramatic transformation from three into four components.

11. The 3D ternary LiGe/SnO compounds are used as electrolytes in lithium-ion-based batteries [37–39]. The VASP numerical simulations could calculate/examine/illustrate (a) the highly anisotropic chemical environments in the presence of various chemical bonds, (b) the strong correlations of Li–O and Ge–/Sn–O bonds from the atom-dominated valence and conduction bands, (c) the different contributions of oxygen orbitals from the spatial charge densities and van Hove singularities, and (d) the important differences or similarities among LiGeO, LiSnO, and LiSiO (details in Chapter 5) compounds.

12. The unusual essential properties in binary 3D silicon carbide (SiC; [40,41]). The diverse phenomena are created by the different silicon/carbon concentrations and dimensionalities. The VASP calculations are mainly focused on

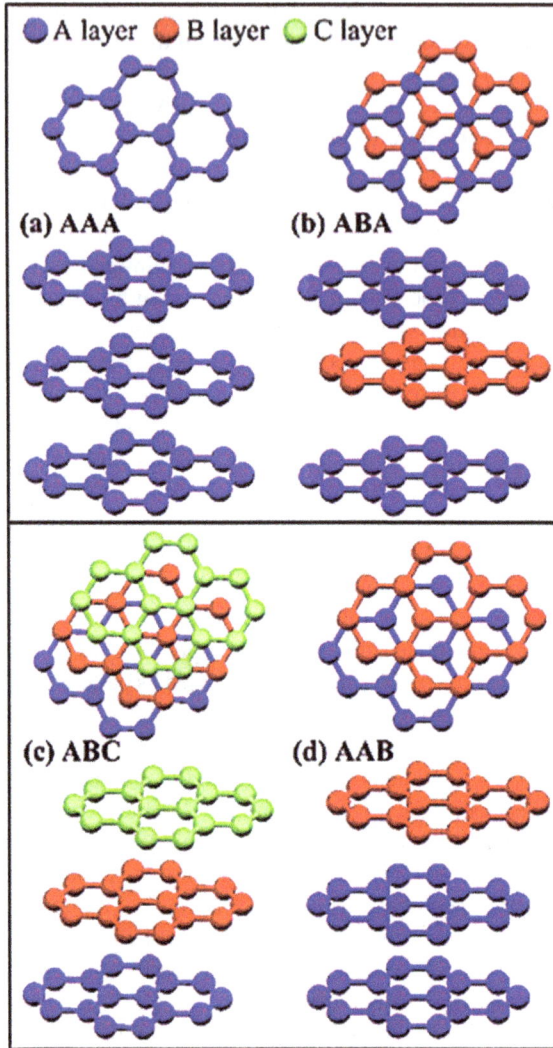

FIGURE 16.2 The geometric structures of the (a) AAA-, (b) ABA-, (c) ABC-, and (d) AAB-stacked trilayer graphene systems.

(a) the modulated chemical bonds, (b) band gaps, (c) featured energy spectra/wave functions, (d) atom- and orbital-generated charge densities, (e) concise multi-orbital hybridizations, (f) distinct special structures of van Hove singularities, and (g) significant differences among 3D, 2D (honeycomb lattices; [42]), and 1D (nanotubes and nanoribbons; [43–46]) compounds.

13. The rich and unique phenomena in 3D silicon/germanium/diamond [47–51]. Concerning mono-element crystals, the delicate VASP simulations are thoroughly made on (a) the symmetric geometries with the sp³ chemical

bondings, (b) the atom- and orbital-diversified energy bands, (c) the unusual charge density distributions, (d) the different van Hove singularities, and (e) the similarities and differences among three kinds of materials, especially for the significant effects being induced by the spin-orbital couplings. Also, a detailed comparison is made for the first-principles method and the tight-binding model in terms of the essential properties [52], such as the hole bands nearest to the Fermi level.

14. The exact identification of multi-orbital hybridizations in 3D SiO_2 [53]. This material frequently serves as a very stable substrate; therefore, its intrinsic properties play a critical role in the crystal growth. The first-principles simulations will examine/solve/identify the optimal crystal structure, (b) the distinct Si–, O– or (Si, O)-dominated energy ranges, the highly anisotropic charge density distribution associated with the strong chemical bondings, and (d) the concise orbital hybridizations from the atom- and orbital-projected density of states. Very interestingly, (e) geometric structures and (f) electronic properties are significantly diversified by the two- and three-component compounds, respectively, corresponding to SiO_2 and $LiSiO_2$ [54–56].

15. The VASP calculations are very suitable for the rich properties of 3D binary iron/cobalt/nickel oxide compounds [57]: (a) the distinct bond lengths and ground state energies, (b) the atom- and orbital-created magnetic moments, (c) the atom- and spin-dependent energy bands, the different spin densities around transition metal and oxygen atoms, (e) the multi-orbital overlaps in charge density distributions, and (f) the atom-, orbital-, and spin-dominated energy ranges. Such characteristics are dramatically changed under the transformation from binary materials into ternary ones (lithium transition metal oxide compounds; problem 10).

REFERENCES

1. Torbatian, Z.; Asgari, R. Plasmonic physics of 2D crystalline materials. *Applied Sciences* **2018**, 8(2), 238.
2. Roome, N. J.; Carey, J. D. Beyond graphene: Stable elemental monolayers of silicene and germanene. *ACS Applied Materials & Interfaces* **2014**, 6(10), 7743–7750.
3. Ye, X.; Shao, Z.; Zhao, H.; Yang, L.; Wang, C. Intrinsic carrier mobility of germanene is larger than graphene's: First-principle calculations. *RSC Advances* **2014**, 4, 21216–21220.
4. Dimoulas, A. Silicene and germanene: Silicon and germanium in the "flatland". *Microelectronic Engineering* **2015**, 131, 68–78.
5. Ezawa, M. Monolayer topological insulators: Silicene, germanene, and stanene. *Journal of the Physical Society of Japan* **2015**, 84(12), 121003.
6. Chuang, Y. C.; Wu, J. Y.; Lin, M. F. Analytical calculations on low-frequency excitations in AA-stacked bilayer graphene. *Journal of Physical Society Japan* **2012**, 81, 124713.
7. Lin, M. F.; Chuang, Y. C.; Wu, J. Y. Electrically tunable plasma excitations in AA-stacked multilayer graphene. *Physics Review B* **2012**, 86, 125434.
8. Chuang, Y. C.; Wu, J. Y.; Lin, M. F. Electric-field-induced plasmon in AA-stacked bilayer graphene. *Annals of Physics* **2013**, 339, 298.

9. Ho, J. H.; Lu, C. L.; Hwang, C. C.; Chang, C. P.; Lin, M. F. Coulomb excitations in AA- and AB-stacked bilayer graphites. *Physics Review B* **2006**, 74, 085406.

10. Chuang, Y. C.; Wu, J. Y.; Lin, M. F. Electric field dependence of excitation spectra in AB-stacked bilayer graphene. *Scientific Reports* **2013**, 3, 1368.

11. Sugawara, K.; Yamamura, N.; Matsuda, K.; Norimatsu, W.; Kusunoki, M.; Sato, T.; Takahashi, T. Selective fabrication of free-standing ABA and ABC trilayer graphene with/without Dirac-cone energy bands. *NPG Asia Materials* **2018**, 10(2), e466.

12. Shan, Y.; Li, Y.; Huang, D.; Tong, Q.; Yao, W.; Liu, W.-T.; Wu, S. Stacking symmetry governed second harmonic generation in graphene trilayers. *Science Advances* **2018**, 4(6), eaat0074.

13. Yelgel, C. Electronic structure of ABC-stacked multilayer graphene and trigonal warping: A first principles calculation. *Journal of Physics* **2016**, 707, 012022.

14. Lin, C.-Y.; Wu, J.-Y.; Ou, Y.-J.; Chiu, Y.-H.; Lin, M.-F. Magneto-electronic properties of multilayer graphenes. *Physical Chemistry Chemical Physics* **2015**, 17(39), 26008–26035.

15. Redouani, I.; Jellal, A.; Bahaoui, A.; Bahlouli, H. Multibands tunneling in AAA-stacked trilayer graphene. *Superlattices and Microstructures* **2018**, 116, 44–53.

16. Lin, C.-Y.; Wu, J.-Y.; Chiu, Y.-H.; Chang, C.-P.; Lin, M.-F. Stacking-dependent magnetoelectronic properties in multilayer graphene. *Physical Review B* **2014**, 90, 205434–205444.

17. Van Duppen, B.; Sena, S. H. R.; Peeters, F. M. Multiband tunneling in trilayer graphene. *Physical Review B* **2013**, 87(19), 195439(10).

18. Lin, Y.-P.; Lin, C.-Y.; Ho, Y.-H.; Do, T.-N.; Lin, M.-F. Magneto-optical properties of ABC-stacked trilayer graphene. *Physical Chemistry Chemical Physics* **2015**, 17(24), 15921–15927.

19. Do, T.-N.; Shih, P.-H.; Chang, C.-P.; Lin, C.-Y.; Lin, M.-F. Rich magneto-absorption spectra of AAB-stacked trilayer graphene. *Physical Chemistry Chemical Physics* **2016**, 18(26), 17597–17605.

20. Ezawa, M. Dirac theory and topological phases of silicon nanotube. *Europhysics Letters* **2012**, 98(6), 67001.

21. Lima, M. P. Double-walled silicon nanotubes: An ab initio investigation. *Nanotechnology* **2018**, 29(7), 075703.

22. Zhang, R. Q.; Lee, S. T.; Law, C.-K., Li, W.-K. and Teo, B. K. Silicon nanotubes: Why not? *Chemical Physics Letters* 2002, 364, 251–258.

23. Srimathi, U.; Nagarajan, V.; Chandiramouli, R. Adsorption studies of volatile organic compounds on germanene nanotube emitted from banana fruit for quality assessment-A density functional application. *Journal of Molecular Graphics and Modelling* **2018**, 82, 129–136.

24. Snehha, P.; Nagarajan, V.; Chandiramouli, R. Germanene nanotube electroresistive molecular device for detection of NO_2 and SO_2 gas molecules: A first-principles investigation. *Journal of Computational Electronics* **2018**, 18(40), 308–318.

25. Monshi, M. M.; Aghaei, S. M.; Calizo, I. Edge functionalized germanene nanoribbons: Impact on electronic and magnetic properties. *RSC Advances* **2017**, 7(31), 18900–18908.

26. Yagmurcukardes, M.; Peeters, F. M.; Senger, R. T.; Sahin, H. Nanoribbons: From fundamentals to state-of-the-art applications. *Applied Physics Reviews* **2016**, 3(4), 041302.

27. Mehdi Aghaei, S.; Calizo, I. Band gap tuning of armchair silicene nanoribbons using periodic hexagonal holes. *Journal of Applied Physics* **2015**, 118(10), 104304.

28. Abe, T.; Sagane, F.; Ohtsuka, M.; Iriyama, Y.; Ogumi, Z. Lithium-ion transfer at the interface between lithium-ion conductive ceramic electrolyte and liquid electrolyte – A key to enhancing the rate capability of lithium-ion batteries. *Journal of the Electrochemical Society* **2005**, 152(11), A2151.

29. Maruyama, S.; Fukutsuka, T.; Miyazaki, K.; Abe, Y.; Yoshizawa, N.; Abe, T. Lithium-ion intercalation and deintercalation behaviors of graphitized carbon nanospheres. *Journal of Materials Chemistry A* **2018**, 6(3), 1128–1137.

30. Das, S. K.; Mahapatra, S.; Lahan, H. Aluminium-ion batteries: Developments and challenges. *Journal of Materials Chemistry A* **2017**, 5(14), 6347–6367.

31. Angell, M.; Pan, C.-J.; Rong, Y.; Yuan, C.; Lin, M.-C.; Hwang, B.-J.; Dai, H. High Coulombic efficiency aluminum-ion battery using an $AlCl_3$-urea ionic liquid analog electrolyte. *Proceedings of the National Academy of Sciences* **2017**, 114(5), 834–839.

32. Ramaprabhu, S.; Samantaray, S. S. A room temperature multivalent rechargeable iron ion battery with ether-based electrolyte: New type of post-lithium ion battery. *Chemical Communications* **2019**, 55, 10416–10419.

33. Li, J.; Luo, J.; Wang, L.; He, X. Recent advances in the $LiFeO_2$-based materials for Li-ion batteries. *International Journal of Electrochemical Science* **2011**, 6(5), 1550–1561.

34. Li, K.; Chen, H.; Shua, F.; Chen, K.; Xue, D. Low temperature synthesis of Fe_2O_3 and $LiFeO_2$ as cathode materials for lithium-ion batteries. *Electrochimica Acta* **2014**, 136, 10–18.

35. Yuan, L.-X.; Wang, Z.-H.; Zhang, W.-X.; Hu, X.-L.; Chen, J.-T.; Huang, Y.-H.; Goodenough, J. B. Development and challenges of $LiFePO_4$ cathode material for lithium-ion batteries. *Energy & Environmental Science* **2011**, 4(2), 269–284.

36. Guo, K. W. Urgency of $LiFePO_4$ as cathode material for Li-ion batteries. *Advances in Materials Research* **2015**, 4(2), 63–76.

37. Xiao, R.; Li, H.; Chen, L. Candidate structures for inorganic lithium solid-state electrolytes identified by high-throughput bond-valence calculations. *Journal of Materiomics* **2015**, 1(4), 325–332.

38. Zhu, Y.; He, X.-F. Mo, Y.-F. Origin of outstanding stability in the lithium solid electrolyte materials: Insights from thermodynamic analyses based on first principles calculations. *ACS Applied Materials & Interfaces* **2018**, 7, 23685–23693.

39. Park, J. S.; Park, Y. J. Surface modification of a $Li[Ni_0:8Co_0:15Al_0:05]O_2$ cathode using Li_2SiO_3 solid electrolyte. *Journal of Electrochemical Science and Technology* **2017**, 8(2), 101–106.

40. Kim, K.; Ju, H.; Kim, J. Vertical particle alignment of boron nitride and silicon carbide binary filler system for thermal conductivity enhancement. *Composites Science and Technology* **2016**, 123, 99–105.

41. Fan, Q.; Chai, C., Wei, Q.; Yang, Y. The mechanical and electronic properties of carbon-rich silicon carbide. *Materials* **2016**, 9(5), 333.

42. Milowska, K.; Magdalena Birowska, M.; Majewski, J. A. Mechanical and electrical properties of carbon nanotubes and graphene layers functionalized with amines. *Diamond and Related Materials* **2012**, 23, 167–171.

43. Wu, X.-J.; Pei, Y.; Zeng, X. Z. B_2C graphene, nanotubes, and nanoribbons. *Nano Letters* **2009**, 9(4), 1577–1582.

44. Soldano, C.; Talapatra, S.; Kar, S. Carbon nanotubes and graphene nanoribbons: Potentials for nanoscale electrical interconnects. *Electronics* **2013**, 2(4), 280–314.

45. Zheng, W.-T.; Sun, C. Q. Underneath the fascinations of carbon nanotubes and graphene nanoribbons. *Energy Environmental Science* **2011**, 4, 627–655.

46. Hartmann, R. R.; Saroka, V. A.; Portnoi, M. E. Interband transitions in narrow-gap carbon nanotubes and graphene nanoribbons. *Journal of Applied Physics* **2019**, 125(15), 151607.

47. Guo, Y.; Wang, Q.; Kawazoe, Y.; Jena, P. A new silicon phase with direct band gap and novel optoelectronic properties. *Scientific Reports* **2015**, 5(1), 14342.

48. Malone, B. D.; Sau, J. D.; Cohen, M. L. Ab initio survey of the electronic structure of tetrahedrally bonded phases of silicon. *Physical Review B* **2008**, 78, 035210.

49. Deinzer, G.; Birner, G.; Strauch, D. Ab initio calculation of the line width of various phonon modes in germanium and silicon. *Physical Review B* **2003**, 67, 144304.

50. Kiefer, F.; Karttunen, A. J.; Doblinger, M.; Fassler, T. F. Bulk synthesis and structure of a microcrystalline allotrope of germanium (m-allo-Ge). *Chemistry of Materials* **2011**, 23, 4578–86.

51. Shenderova, O. A.; Shames, A. I.; Nunn, N. A.; Torelli, M. D.; Vlasov, I.; Zaitsev, A. Review article: Synthesis, properties, and applications of fluorescent diamond particles. *Journal of Vacuum Science & Technology B* **2019**, 37(3), 030802.

52. Zolyomi, V.; Wallbank, J. R.; Fal'ko, V. I. Silicane and germanane: Tight-binding and first-principles studies. *2D Materials* **2014**, 1(1), 011005.

53. Sciuto, E. L.; Bongiorno, C.; Scandurra, A.; Petralia, S.; Cosentino, T.; Conoci, S.; Libertino, S. Functionalization of bulk SiO_2 surface with biomolecules for sensing applications: Structural and functional characterizations. *Chemosensors* **2018**, 6(4), 59.

54. Martin-Samos, L.; Bussi, G.; Ruini, A.; Molinari, E.; Caldas, M. J. SiO_2 in density functional theory and beyond. *Physica Status Solidi* **2010**, 248(5), 1061–1066.

55. Zhang, Y.; Li, Y.; Wang, Z.; Zhao, K. Lithiation of SiO_2 in Li-ion batteries: In situ transmission electron microscopy experiments and theoretical studies. *Nano Letters* **2014**, 14(12), 7161–7170.

56. Wang, J.-Q.; Kang, H.-L.; Zhang, Y. Simulation calculation and analysis of electronic structure and electrical properties of metal-doped SnO_2. *Advances in Materials* **2018**, 6, 1–10.

57. Zhang, J.; Liu, F.; Cheng, J. P.; Zhang, X. B. Binary nickel-cobalt oxides electrode materials for high-performance supercapacitors: Influence of its composition and porous nature. *ACS Applied Materials & Interfaces* **2015**, 7(32), 17630–17640.

Index

Note: **Bold** page numbers refer to tables and *italic* page numbers refer to figures.

.

For Product Safety Concerns and Information please contact our EU
representative GPSR@taylorandfrancis.com
Taylor & Francis Verlag GmbH, Kaufingerstraße 24, 80331 München, Germany

www.ingramcontent.com/pod-product-compliance
Lightning Source LLC
Chambersburg PA
CBHW060336220326
41598CB00023B/2723

* 9 7 8 0 3 6 7 6 8 6 2 6 0 *